Analytic Methods of Orbit Prediction and Control

Analytic Methods of Orbit Prediction and Control: Low-Thrust and Impulsive Propulsion Applications

Jean Albert Kéchichian

Volume 265
Progress in Astronautics and Aeronautics

Timothy C. Lieuwen, Editor-in-Chief
Georgia Institute of Technology
Atlanta, Georgia

AIAA
American Institute of Aeronautics and Astronautics, Inc.

Cover image: The Cygnus space freighter's cymbal-shaped UltraFlex solar array. Courtesy NASA.

ISBN 978-1-62410-697-2
Ebook ISBN 978-1-62410-698-9

PROGRESS IN ASTRONAUTICS AND AERONAUTICS

EDITOR-IN-CHIEF

Timothy C. Lieuwen
Georgia Institute of Technology

EDITORIAL BOARD

Paul M. Bevilaqua
Lockheed Martin (Ret.)

Steven A. Brandt
U.S. Air Force Academy (Ret.)

José Camberos
Air Force Institute of Technology

Richard Curran
Delft University of Technology

Carolin Frueh
Purdue University

James Hileman
Federal Aviation Administration

Micah Howard
Sandia National Laboratories

Christopher H. M. Jenkins
Montana State University

Thomas Kurfess
Oak Ridge National Laboratory

Mark J. Lewis
NDIA Emerging Technologies Institute

Nateri Madavan
NASA (ARMD)

Dimitri Mavris
Georgia Institute of Technology

Kristi Morgansen
University of Washington

Lana Osusky
GE Research

Sukesh Roy
Spectral Energies, LLC

Julie Shah
Massachusetts Institute of Technology

Kelly Stephani
University of Illinois at Urbana-Champaign

Karen Thole
The Pennsylvania State University

Oleg Yakimenko
Naval Postgraduate School

Namiko Yamamoto
The Pennsylvania State University

To the memory of Sir Alexander Fleming of Scotland for his discovery and isolation of penicillin that saved the lives of countless millions including mine as a child.

TABLE OF CONTENTS

PREFACE

In this book, analytic methods of orbit prediction and control are investigated for important low-thrust and impulsive orbit change and stationkeeping applications, and the results are compared with actual exact numerically integrated counterparts to evaluate their accuracy and applicability. With one eye on future autonomous on-orbit navigation applications in both near-circular and general elliptical orbits, with minimal computational effort and high accuracy, the orbit prediction and control segments of such autonomous navigation systems are explored by also taking into account the effect of shadowing where no thrust is allowed, such as for electric propulsion applications. This book shows that many of these orbit prediction and stationkeeping problems can be solved entirely in analytic form with very high accuracy. It also allows practitioners to solve similar problems by tailoring them to their own applications and to better understand and evaluate the pertinent variables and parameters that drive and influence the resulting description and evolution of the orbits.

Starting with the problem of low-thrust tangential thrusting along small- to moderate-eccentricity orbits, a given orbit is updated after each revolution if shadow is present, to account for the changing shadow geometry, and the secular variation of the node due to the second zonal harmonic of Earth's potential J_2 is updated analytically. Analytic expressions for the variations of the pertinent elements due to the combined effects of J_2 and the thrust acceleration are also obtained for the near-circular case in low Earth orbit (LEO) using continuous, constant, low-thrust acceleration along the tangential direction. Furthermore, for this problem, the analytic integration of the orbit equations is shown to be accurate for several tens of revolutions in LEO and about 10 revolutions in geostationary Earth orbit (GEO). The analytic integration is further extended to include the effect of Earth's oblateness on the expanding orbit. This analytic long-term orbit prediction capability will minimize the computational loads of an onboard computer for autonomous orbit-transfer applications and allow, among other things, the consideration of long multi-orbit data arcs for analytic orbit determination updates, thereby significantly decreasing the frequency of these updates.

The problem of zero-eccentricity-constrained orbit raising in a circular orbit in the presence of shadowing is analyzed next, using both numerical and analytical methods. Given the shadow arc length, piecewise-constant pitch angles are selected, and the location along the orbit where the pitch angle switches is optimized to effect the largest change in semimajor axis per revolution. These suboptimal analytic solutions are almost as effective as the optimal solution, especially in higher orbit. These strategies remove the need for continuously reorienting the spacecraft attitude in pitch, providing robust real-time onboard guidance capability for electric orbit transfer vehicle (EOTV) orbit-raising applications. Exact numerical simulations for typical six-month solar-electric transfers from LEO to GEO show that, in the worst case of shadowing geometry, the intermediate orbits could reach eccentricities on the order of 0.2 at most. In view of the small eccentricity buildup, it is convenient to force eccentricity to remain at zero by developing analytic methods that result in simple but suboptimal steering laws that are easily implementable in mission analysis software and by fully accounting for the varying shadow geometry during the transfer. Constant relative pitch angles are easily implemented by the onboard attitude control system, which, in this case, must hold the pitch attitude constant relative to the local horizon. Our approach is based on the theory of function minimization and is purely analytic, requiring minimal computational effort.

The problem of low-thrust inclination control in near-circular orbit ($0 \leq e \leq 10^{-2}$) and in the presence of Earth shadow is analyzed next, using simple steering laws consisting of piecewise-constant yaw angle selection. The linearized form of the variation-of-parameters equations is used, providing analytic expressions for the components of the inclination change vector due to out-of-plane thrusting. Two-switch transfer algorithms are analyzed, and their performances are compared, providing a robust suboptimal transfer mode amenable to onboard autonomous guidance applications for future EOTVs. These algorithms are simple to implement and achieve maximum rotation of the orbit plane about the desired line of nodes of the current and final orbits using constant yaw angles, with the thrust turned off during shadowing. The shadow geometry is easily calculated once the Sun look angle β_s is evaluated from the knowledge of the solar right ascension and declination, as well as the spacecraft orbit equatorial inclination and ascending node. The node is allowed to regress because of J_2, and the amount by which β_s changes is calculated from the analytic expression defining the regression rate. This rate is dependent on the orbital equatorial inclination as well as the orbital semimajor axis.

The dynamics and control of the motion of the orbit planes of a constellation of five satellites in inclined geosynchronous orbits are analyzed next, such that orbit maintenance strategies that confine each individual orbit inclination within a predefined tolerance deadband are designed, using fast analytic orbit prediction approximations. Given a constellation lifetime, it is desired to maintain each satellite orbit inclination within a small tolerance band centered at its nominal inclination and determine the total accumulated impulsive velocity change (ΔV) required by each satellite during its lifetime. The control strategies developed for the geostationary satellites are no longer applicable here, because the nominal inclination is not near zero but at 12° for the application at hand. Furthermore, the ideal even spacing in nodes at the initial time is distorted if left uncontrolled, because each satellite orbit starts from a different node, experiencing a different nodal rate, which, in time, upsets the evenly spaced nodal distribution. Using a combination of graphical, analytical, and numerical techniques, orbit maintenance strategies are devised to control the inclination, or both the inclination and the node, of each satellite such that the tolerance band in inclination is never violated and, in the latter case, the evenly spaced nodal configuration is maintained over the constellation lifetime. These strategies for controlling the orbit plane perform intermittent impulsive maneuvers that either adjust only the inclination to confine each satellite orbit within a user-defined inclination tolerance deadband centered at the common nominal inclination at all times or adjust both the inclination and node to also keep the initially evenly spaced nodal configuration from being distorted. The method is applicable to constellations using an arbitrarily large number of satellites, with arbitrary identical equatorial inclination.

The problem of north–south stationkeeping of geostationary spacecraft using electric thrusters is analyzed next. Pure yawing with short-duration low-thrust arcs applied infrequently is assumed, and the dynamics are cast in continuous form to obtain an analytic steering law in inclination–node (i, Ω) space that brings the spacecraft back to the ideal initial orbit orientation for the initiation of an optimal free-drift period that satisfies the inclination constraint for the longest possible duration. This problem is posed as a minimum-time navigation problem between two (i, Ω) pairs and is similar to the Zermelo problem of navigating a ship in strong variable currents. Considerable amounts of fuel can be saved if the low-specific-impulse chemical rockets are replaced by high-specific-impulse electric engines to execute the same stationkeeping maneuvers, thereby extending the operational life of these satellites. However, these maneuvers cannot be carried out in an essentially instantaneous manner, but must be implemented in small incremental steps spanning several weeks or more, depending on level of the thrust acceleration and frequency of the incremental maneuvering. The inherent long durations of these low-thrust maneuvers are factored into the design of the maneuver strategy, inasmuch as the strategy must account for the natural drift that occurs before the completion of the maneuver sequence.

Next, optimal low-thrust pitch profiles that maximize the change in the semimajor axis while constraining the eccentricity change to zero over a revolution, in the presence of shadowing where no thrust is applied, are generated by using numerical quadrature methods. The procedure is extended to the dual problem of maximizing the change in eccentricity while constraining the semimajor axis change to zero for the near-circular case with shadowing. The method is further applied to the more general elliptical case using continuous thrust to circularize a highly elliptical synchronous orbit that remains synchronous during the circularization maneuver. This particular example is currently used in the final steps of transferring certain communications satellites to their geostationary orbits after their proper placement in an intermediate elliptical synchronous orbit using chemical propulsion. The solution is obtained through direct application of the theory of the maxima and minima and through a numerical search for the value of the appropriate constant Lagrange multiplier such that a certain integral is driven to zero. Simple analytic expressions for the constant out-of-plane thrust angle as well as the total required velocity change to circularize and rotate a near-circular orbit simultaneously are also derived for preliminary analyses.

In Chapter 7, the optimal variation in thrust pitch angle that results in the maximum change in the eccentricity of a general elliptical orbit using continuous, constant, low-thrust acceleration while keeping the orbit energy unchanged after a full thrust cycle is determined through the direct use of the theory of maxima and through numerical quadrature and search techniques. The analysis takes into account the presence of a shadow arc arbitrarily positioned along the elliptical orbit, where thrust is cut off. Unlike the well-known nonoptimal scheme that uses a thrust orientation perpendicular to the line of apsides at all times, the present optimal scheme allows for the maximum change in eccentricity for a more efficient orbit circularization. Approximate but highly accurate analytic expressions for the changes in the eccentricity and semimajor axis of a general elliptical orbit, perturbed by a constant low-thrust acceleration applied along the fixed inertial direction normal to the orbit major axis, are also derived for general use and rapid calculations. The two-body thrust-perturbed orbit is, thus, constrained to remain circular during the transfer, and the optimal control law for the thrust direction, derived for the fast-time-scale problem of maximizing the inclination change for a given change in the semimajor axis, is used in an averaging procedure to solve the overall slow-time-scale transfer problem. The present analysis tackles the fast-time-scale planar problem in the more general elliptical case, by deriving the optimal thrust pitch profile that maximizes the change in eccentricity while holding the semimajor axis constant and by also extending the analysis to the dual problem of maximizing the orbit semimajor axis while keeping the eccentricity unchanged after one cycle of intermittent thrust. The change in eccentricity over a single orbit can then be used in conjunction with an inclination change to produce average rates of change in the eccentricity and inclination, to solve the overall slow-time-scale minimum-time transfer problem of circularizing and rotating an initial elliptical orbit without changing its orbital energy. The presence of a shadow arc disrupts the constancy of the semimajor axis when the inertially fixed firing direction mode is used. The optimal mode thus derived extends the results obtained in the circular case, and it ensures that the tailored thrust pitch profile not only optimizes the change in the eccentricity but also satisfies the semimajor axis constraint, regardless of the size and orbital location of the shadow arc.

Chapter 8 provides the analytic solution of the fraction of debris from an isotropic explosion in general elliptical orbit that would fly in orbits whose perigees are below a certain given altitude. Given an explosion velocity, a spacecraft idealized as a sphere breaks up into a myriad of small fragments that either remain in Earth orbit or reenter the atmosphere. The percentages of the debris fragments emanating from either side of the separating curve on the idealized sphere are thus evaluated rapidly through a few calls to a quartic solver routine. This analytic method provides percentage counts that are identical in accuracy to those obtained by a purely numerical method, and it can readily be extended to analyze nonisotropic explosions as well.

An entire chapter (Chapter 9) is devoted to Edelbaum's classic problem of minimum-time low-thrust transfer between inclined circular orbits, within the context of the additional perturbation due to Earth's oblateness. The original analytic theory of Edelbaum is extended by also considering the right ascension of the ascending node variable, which is needed to account for the precession of the instantaneous orbit during the transfer, due to the perturbation in the second zonal harmonic of Earth's potential J_2. The J_2 perturbation is then taken into account, both in the "exact" or unaveraged sense with optimal thrust vectoring and within the context of an analytic averaging scheme that uses the piecewise-constant yaw strategy. The addition of J_2 will precess the orbit plane and starts to vary the orientation of the line of nodes, which effectively remained fixed in the Edelbaum thrust-only problem. Therefore, the instantaneous line of nodes is continuously updated to continue the process of analytic averaging in a reliable manner. Numerical comparisons for an example of large multiday noncoplanar transfer show that, for the real-world case of perturbations in thrust and J_2 case, the analytic averaging is about 13% less economical to carry out in terms of the change in velocity ΔV, but it is much more easily implemented in actual operations, because the thrust vector orientation maintains a constant direction with respect to rotating axes, unlike the exact case in which this orientation must be continuously steered in time.

The last three chapters are focused on the derivations that lead to many of the differential equations that are used throughout the book, to describe the variation of the orbit elements in polar coordinates, to provide useful analytic forms that account for the effects of air drag and zonal harmonic perturbations in near-circular orbit using relative motion coordinates, and to derive the general form of the variation of the classical orbit elements using Breakwell's analysis as shown in his lecture notes from Stanford University. Analysis of the effects of solar gravitational perturbation on a spacecraft in near-circular orbit about a planet also draws from Breakwell's lecture notes and applies the equations to the example of the Venus Orbiting Imaging Radar (VOIR) Mission case study. Finally, the last chapter, based once again on Breakwell's notes, describes his vacant focus theory and the generation of the changes in the orbit elements due to a small impulse at a given location on a general elliptical orbit, as well as other useful expressions that describe the relative motion of two close satellites in near-circular orbit. Other useful pieces of analysis are also provided in analytic form for both educational and practical purposes and for first-order assessments and insight useful in mission analysis.

The research described in this book was carried out initially at NASA Jet Propulsion Laboratory and later mostly at The Aerospace Corporation in California under contract with the U.S. Air Force Space and Missile Systems Center. Mary Villanueva, previously of The Aerospace Corporation and now with Raytheon Space and Airborne Systems in El Segundo, CA, carried out the substantial and complex typesetting of this book, and Jason Perez, also previously of The Aerospace Corporation, generated the transcription of the figures to electronic Adobe Illustrator files. My special thanks to Katrina Buckley, AIAA's Managing Editor, for reformatting and editing the entire original manuscript in a masterclass effort to comply with AIAA's exacting and very high publication standards. Their contributions in making this book a reality are gratefully acknowledged.

La Cañada Flintridge, CA

Jean Albert Kéchichian

CHAPTER 1

Orbit Raising with Low-Thrust Tangential Acceleration in the Presence of Earth Shadow

NOMENCLATURE

a	= semimajor axis, km
c_α, s_α	= cos α, sin α
E	= eccentric anomaly
e	= eccentricity
f_r, f_θ, f_h	= components of the perturbation acceleration vector along the radial, perpendicular, and out-of-plane directions
f_t, f_n	= components of the acceleration vector along the tangential and normal directions
h	= orbit angular momentum
i	= inclination
J_2	= second zonal harmonic of Earth's potential, 1.082×10^{-3}
M	= mean anomaly
n	= spacecraft mean motion, $(\mu/a^3)^{1/2}$, rad/s
p	= orbit parameter, km
R	= equatorial radius of Earth, 6378.14 km
r	= radial distance
t	= time, s
V	= velocity, km/s
α	= mean longitude, $\omega + M$
θ	= angular position measured from the ascending node
θ^*	= true anomaly
μ	= gravitational constant of Earth, 398601.3 km^3/s^2
χ	= angle between the radial direction and the velocity vector
Ω	= longitude of the ascending node
ω	= argument of perigee

1.1 INTRODUCTION

The problem of low-thrust tangential thrusting along small- to moderate-eccentricity orbits in the presence of Earth shadow is analyzed in this chapter. Given the orbital elements and the shadow geometry at the start of each revolution, the changes in the in-plane orbit elements after one revolution of intermittent thrusting are evaluated analytically for a given level of constant acceleration. These perturbation equations are valid for small to moderate eccentricities $(0 \leq e \leq 0.2)$, except for the argument of perigee, which is valid for any eccentricity larger than 0.01 because of the well-known singularity at $e = 0$ associated with the use of the classical elements. When e is less than 0.01, a nonsingular set of equations is used instead so that the orbit is continuously updated with negligible computational effort. These analytic guidance equations valid for low-thrust accelerations on the order of 10^{-4} g and less, are developed for implementation in efficient transfer simulation programs for systems design optimization and preliminary mission analysis work. Furthermore, for the problem of continuous, constant, low-thrust tangential acceleration, the analytic integration of the orbit equations is shown to be accurate for several tens of revolutions in LEO and about 10 revolutions in geosynchronous Earth orbit. The analytic integration is further extended to include the effect of Earth's oblateness on the expanding orbit. This analytic long-term orbit prediction capability will minimize the computational loads of an onboard computer for autonomous orbit-transfer applications and allow, among other things, the consideration of long multi-orbit data arcs for analytic orbit determination updates, thereby decreasing considerably the frequency of these updates. The optimal low-thrust transfer between LEO and a higher orbit not lying in the same plane is such that the orbit raising and inclination change are carried out simultaneously during the transfer with an optimal and varying combination of in-plane and out-of-plane accelerations. Furthermore, when shadowing is present during the initial phase of the transfer, because of the subsequent intermittent thrusting, the transfer orbit builds up eccentricity with apogee along the shadow arc. Numerical simulations show that, for the most severe shadow geometry, the eccentricity can build up from zero to a maximum of about 0.2, which renders problematic the use of most of the analytic and semianalytic optimal transfer guidance algorithms, such as the Edelbaum [1] and Wiesel–Alfano [2] algorithms, which are valid for near-circular orbits and no shadowing. These algorithms are simple to use and calculate, as opposed to the exact optimal transfer guidance algorithm of the SECKSPOT software [3], which requires extensive computational capability. A theory of orbit raising in the presence of Earth shadow was developed [4] with the additional constraint that the eccentricity be held at zero during the transfer. This will tend to lengthen the total transfer time slightly, resulting in a proportional increase in the propellant expenditure. However, these constrained guidance laws are calculated analytically, and once the orbit is in full sunlight, the Edelbaum and Wiesel–Alfano steering laws can once again be used to carry out the remaining part of the transfer. Furthermore, in addition to requiring longer transfer times, these eccentricity-constrained strategies [4] also result in increased dwell times in the Van Allen radiation belts, which further degrades the solar panels because of an increase in the total accumulated fluence. Thus, it is more appropriate to carry out the first leg of the transfer by applying the thrust along the velocity vector without any inclination change, so that the transfer is initially coplanar and the dwell time in the radiation belts is minimized. This strategy will result in the increase in eccentricity mentioned previously, as long as shadowing is present. Although semianalytic methods were developed for the discontinuous-thrust transfer problem [5, 6], efficient codes to support systems optimization and preliminary mission analyses that require a large number of iterations necessitate fast analytic guidance algorithms that also model all other important effects such as degradation of the solar power output due to the Van Allen radiation belts. An effort to create such a computer program was undertaken using the assumption of a circular orbit and a constant yaw profile during the transfer [7]. The strategy of coplanar tangential thrusting in the initial phases of the transfer described here can easily be implemented in the aforementioned code regardless of whether shadowing is present. Once the orbit is safely outside the radiation belts, other analytic guidance laws can be applied to slowly raise, circularize, and rotate the orbit to reach the destination geostationary condition. This chapter concentrates on the initial coplanar phase of the transfer and develops analytic expressions for the changes in the in-plane orbit elements after each revolution of intermittent tangential thrusting with constant acceleration. The orbit is thus updated from revolution to revolution until perigee is safely outside the belts, after which the orbit can be circularized following the same techniques as used previously by the author [4], and the transfer can be completed with simultaneous orbit raising and plane change. The analysis in this chapter first appeared as a conference paper [8] and later as a journal article [9].

1.2 GENERAL ANALYSIS

The first-order differential equations for the classical in-plane orbital elements, namely, the semimajor axis, eccentricity, and argument of perigee are given in terms of the components of the acceleration vector along the instantaneous radial, orthogonal, and out-of-plane directions [10] as

$$\frac{\mathrm{d}a}{\mathrm{d}t} = \frac{2na^2}{\mu}\left[\frac{aes_{\theta^*}}{(1-e^2)^{1/2}}f_r + \frac{a^2(1-e^2)^{1/2}}{r}f_\theta\right] \tag{1.1}$$

$$\frac{\mathrm{d}e}{\mathrm{d}t} = \frac{na^2}{\mu}(1-e^2)^{1/2}[s_{\theta^*}f_r + (c_{\theta^*}+c_E)f_\theta] \tag{1.2}$$

$$\frac{d\omega}{dt} = \frac{na^2}{\mu e}(1-e^2)^{1/2}\left[-c_{\theta^*}f_r + s_{\theta^*}\left(1+\frac{r}{p}\right)f_\theta\right] - c_i\dot{\Omega} \tag{1.3}$$

$$\frac{d\Omega}{dt} = \frac{a^{1/2}(1-e^2)^{1/2}}{\mu^{1/2}}\frac{(1+ec_{\theta^*})^{-1}s_\theta f_h}{s_i} \tag{1.4}$$

In the preceding equations, $\theta = \omega + \theta^*$, $r = a(1-e^2)/(1+ec_{\theta^*})$, and $p = a(1-e^2) = h^2/\mu$. These equations, which are widely used in the literature, have been specialized and converted to a form that is more convenient for the analysis of drag and thrust perturbation effects. If we assume that the thrust acceleration is essentially along the velocity vector, it is preferable to transform the equations of motion in Eqs. (1.1–1.3) so that they are given in terms of the tangential and normal acceleration components f_t and f_n, respectively. The component f_t is along the orbit tangent or velocity vector direction, whereas f_n is perpendicular to the tangent in the orbit plane and pointing away from the origin. The transformation from f_r and f_θ to f_t and f_n is given by [10]

$$f_r = c_\chi f_t + s_\chi f_n \tag{1.5}$$

$$f_\theta = s_\chi f_t - c_\chi f_n \tag{1.6}$$

These equations can also be written as

$$f_r = \frac{e\mu s_{\theta^*}}{hV}f_t + \frac{h}{rV}f_n \tag{1.7}$$

$$f_\theta = \frac{h}{rV}f_t - \frac{e\mu s_{\theta^*}}{hV}f_n \tag{1.8}$$

with the velocity V given by

$$V^2 = \frac{\frac{\mu}{a}(1+e^2+2ec_{\theta^*})}{(1-e^2)} \tag{1.9}$$

This expression for the velocity in terms of the orbital elements is easily obtained from the energy equation $V^2/2 - \mu/r = -\mu/(2a)$, in which r is replaced by $a(1-e^2)(1+ec_{\theta^*})^{-1}$. Substituting Eqs. (1.7) and (1.8) into Eq. (1.1) and using the expression $h = \mu^{1/2}a^{1/2}(1-e^2)^{1/2}$, one can obtain the following simple form for da/dt

$$\frac{da}{dt} = \frac{2na^3}{\mu(1-e^2)^{1/2}}(1+e^2+2ec_{\theta^*})^{1/2}f_t \tag{1.10}$$

which is valid for any elliptical orbit. This can also be cast into the following form involving V:

$$\frac{\mathrm{d}a}{\mathrm{d}t} = \frac{2a^2}{\mu} V f_t \tag{1.11}$$

The normal component has effectively canceled out, as it should because it cannot contribute to any changes in the orbital energy. In a similar manner, Eq. (1.2) can also be transformed into a form involving f_t and f_n by first observing that, from the orbit equation $r = a(1 - ec_E)$, it follows that $c_E = (1 - r/a)/e = (e + c_{\theta^*})/(1 + ec_{\theta^*})$. After some manipulation, we obtain

$$\frac{\mathrm{d}e}{\mathrm{d}t} = \frac{a^{1/2}(1-e^2)^{1/2}}{\mu^{1/2}} \left[\frac{2(e + c_{\theta^*})}{(1 + e^2 + 2ec_{\theta^*})^{1/2}} f_t + \frac{s_{\theta^*}(1 - e^2)}{(1 + ec_{\theta^*})(1 + e^2 + 2ec_{\theta^*})^{1/2}} f_n \right] \tag{1.12}$$

Finally, Eq. (1.3) for $\mathrm{d}\omega/\mathrm{d}t$ is transformed into

$$\frac{\mathrm{d}\omega}{\mathrm{d}t} = \frac{na^2(1-e^2)^{1/2}}{\mu e} \left[-c_{\theta^*} f_r + \frac{(2 + ec_{\theta^*})s_{\theta^*}}{(1 + ec_{\theta^*})} f_\theta \right] \tag{1.13}$$

where $f_h = 0$ has been used to eliminate the $\dot{\Omega}$ contribution. Incidentally, expressions (1.7) and (1.8) can also be written as

$$f_r = \frac{es_{\theta^*}}{(1 + e^2 + 2ec_{\theta^*})^{1/2}} f_t + \frac{(1 + ec_{\theta^*})}{(1 + e^2 + 2ec_{\theta^*})^{1/2}} f_n \tag{1.14}$$

$$f_\theta = \frac{(1 + ec_{\theta^*})}{(1 + e^2 + 2ec_{\theta^*})^{1/2}} f_t - \frac{es_{\theta^*}}{(1 + e^2 + 2ec_{\theta^*})^{1/2}} f_n \tag{1.15}$$

Use of f_r and f_θ in Eq. (1.13) results, after some manipulation, in

$$\frac{\mathrm{d}\omega}{\mathrm{d}t} = \frac{na^2(1-e^2)^{1/2}}{\mu e} \left[\frac{2s_{\theta^*}}{(1 + e^2 + 2ec_{\theta^*})^{1/2}} f_t - \frac{2e + c_{\theta^*}(1 + e^2)}{(1 + ec_{\theta^*})(1 + e^2 + 2ec_{\theta^*})^{1/2}} f_n \right] \tag{1.16}$$

Finally, the perturbation of the true anomaly θ^* can be obtained from

$$\frac{\mathrm{d}\theta^*}{\mathrm{d}t} = \frac{\mathrm{d}\theta}{\mathrm{d}t} - \frac{\mathrm{d}\omega}{\mathrm{d}t} \tag{1.17}$$

where

$$\frac{d\theta}{dt} = \frac{h}{r^2} - \dot{\Omega} c_i \tag{1.18}$$

Substitution of Eqs. (1.3) and (1.18) into Eq. (1.17) yields

$$\frac{d\theta^*}{dt} = \frac{h}{r^2} - \frac{r}{he}[-c_{\theta^*}(1 + ec_{\theta^*})f_r + s_{\theta^*}(2 + ec_{\theta^*})\, f_\theta] \tag{1.19}$$

which reduces to the following form involving f_t and f_n:

$$\begin{aligned}\frac{d\theta^*}{dt} &= \frac{\mu^{1/2}a^{-3/2}}{(1-e^2)^{3/2}}(1 + ec_{\theta^*})^2 - \frac{a^{1/2}(1-e^2)^{1/2}}{e\mu^{1/2}(1 + ec_{\theta^*})} \\ &\quad \times \left[\frac{2s_{\theta^*}(1 + ec_{\theta^*})}{(1 + e^2 + 2ec_{\theta^*})^{1/2}} f_t - \frac{(2e + e^2 c_{\theta^*} + c_{\theta^*})}{(1 + e^2 + 2ec_{\theta^*})^{1/2}} f_n\right]\end{aligned} \tag{1.20}$$

This equation is equivalent to $d\theta^*/dt = h/r^2 - \dot{\omega}$, because $\dot{\Omega} = 0$ and only in-plane accelerations are applied. Equation (1.20) can be replaced by a perturbation equation for the mean anomaly M because, after transformation of

$$\frac{dM}{dt} = n + \frac{(1-e^2)}{nae}\left[\left(c_{\theta^*} - \frac{2e}{1 + ec_{\theta^*}}\right)f_r - \left(\frac{2 + ec_{\theta^*}}{1 + ec_{\theta^*}}\right)s_{\theta^*} f_\theta\right] \tag{1.21}$$

it follows that

$$\frac{dM}{dt} = n - \frac{(1-e^2)a^{1/2}}{e\mu^{1/2}(1 + e^2 + 2ec_{\theta^*})^{1/2}}\left[\frac{2(1 + ec_{\theta^*} + e^2)}{(1 + ec_{\theta^*})} s_{\theta^*} f_t - \frac{(1-e^2)}{(1 + ec_{\theta^*})} c_{\theta^*} f_n\right] \tag{1.22}$$

For each time t during the numerical integration, $M = n(t - t_0) + M_0$, where M_0 is the mean anomaly at time t_0. Then, from Kepler's equation $M = E - es_E$, the eccentric anomaly E is computed, from which the true anomaly θ^* is evaluated using the equation

$$\tan\left(\frac{\theta^*}{2}\right) = \left(\frac{1+e}{1-e}\right)^{1/2} \tan\left(\frac{E}{2}\right) \tag{1.23}$$

If we consider only tangential thrusting, then $f_n = 0$, and the relevant equations of motion reduce to the set

$$\frac{da}{dt} = \frac{2na^3}{\mu(1-e^2)^{1/2}}(1 + e^2 + 2ec_{\theta^*})^{1/2} f_t \tag{1.24}$$

$$\frac{de}{dt} = \frac{2a^{1/2}(1-e^2)^{1/2}}{\mu^{1/2}} \frac{(e + c_{\theta^*})}{(1 + e^2 + 2ec_{\theta^*})^{1/2}} f_t \tag{1.25}$$

$$\frac{d\omega}{dt} = \frac{2na^2(1-e^2)^{1/2}}{\mu e} \frac{s_{\theta^*}}{(1 + e^2 + 2ec_{\theta^*})^{1/2}} f_t \tag{1.26}$$

$$\frac{d\theta^*}{dt} = \frac{n}{(1-e^2)^{3/2}}(1 + ec_{\theta^*})^2 - \frac{2a^{1/2}(1-e^2)^{1/2}}{e\mu^{1/2}} \frac{s_{\theta^*}}{(1 + e^2 + 2ec_{\theta^*})^{1/2}} f_t \tag{1.27}$$

Both $d\omega/dt$ and $d\theta^*/dt$ have singularities at $e = 0$ because they involve division by e. These equations then break down for near-circular orbits or $e \leq 10^{-3}$, but they are otherwise valid for any eccentricity. Given a constant tangential acceleration f_t, these equations can be integrated numerically to provide the changes experienced by the in-plane classical elements as functions of time. They are a set of fully coupled, nonlinear, first-order differential equations in the osculating elements a, e, ω, and θ^*. If we neglect the osculation of θ^* due to the acceleration f_t, then we can change the independent variable from t to θ^* because then, from Eq. (1.27), we can use the relation

$$dt = \frac{(1-e^2)^{3/2}}{n(1 + ec_{\theta^*})^2} d\theta^* \tag{1.28}$$

Substituting Eq. (1.28) into Eqs. (1.24–1.26), we obtain

$$da = \frac{2(1-e^2)(1 + e^2 + 2ec_{\theta^*})^{1/2}}{n^2(1 + ec_{\theta^*})^2} f_t d\theta^* \tag{1.29}$$

$$de = \frac{2(1-e^2)^2(e + c_{\theta^*})}{an^2(1 + ec_{\theta^*})^2(1 + e^2 + 2ec_{\theta^*})^{1/2}} f_t d\theta^* \tag{1.30}$$

$$d\omega = \frac{2a^2(1-e^2)^2 s_{\theta^*}}{\mu e(1 + ec_{\theta^*})^2(1 + e^2 + 2ec_{\theta^*})^{1/2}} f_t d\theta^* \tag{1.31}$$

If a, e, and f_t on the right-hand sides of these expressions are held constant, these equations can now be integrated analytically. The mean motion n is also held constant because it is a function of a.

1.3 ANALYTIC INTEGRATION WITH INTERMITTENT THRUSTING

Because the tangential acceleration f_t is on the order of 10^{-4} g and less, it is sufficient to evaluate the first-order effects on a, e, and ω by holding a, e, and f_t constant during the integration over one revolution. Then, we can update the elements and integrate analytically over the next revolution and so on until the required transfer is achieved. No closed-form solution exists for $\mathrm{d}a$ and $\mathrm{d}e$, and therefore, we must expand them in powers of the small quantity the eccentricity before carrying out the integration. Let us observe that

$$(1+e^2+2ec_{\theta^*})^{1/2} = (1+e^2)^{1/2}(1+a_1c_{\theta^*})^{1/2} \tag{1.32}$$

with $a_1 = 2e/(1+e^2)$. Because a_1 is on the order of e, let us expand the following terms up to $\mathcal{O}(e^4)$

$$(1+a_1c_{\theta^*})^{1/2} \simeq 1+\frac{1}{2}a_1c_{\theta^*}-\frac{1}{8}a_1^2c_{\theta^*}^2+\frac{1}{16}a_1^3c_{\theta^*}^3-\frac{5}{128}a_1^4c_{\theta^*}^4+\mathrm{HOT}$$

where HOT indicates higher-order terms in the binomial expansion. Similarly

$$(1+ec_{\theta^*})^{-2} \simeq 1-2ec_{\theta^*}+3e^2c_{\theta^*}^2-4e^3c_{\theta^*}^3+5e^4c_{\theta^*}^4+\mathrm{HOT}$$

Let

$$K = \frac{2(1-e^2)}{n^2}f_t \tag{1.33}$$

Then

$$\begin{aligned}\Delta a &= K\int_0^{2\pi}\frac{(1+e^2+2ec_{\theta^*})^{1/2}}{(1+ec_{\theta^*})^2}\mathrm{d}\theta^* \\ &= K\int_0^{2\pi}(1+b_1c_{\theta^*}+b_2c_{\theta^*}^2+b_3c_{\theta^*}^3+b_4c_{\theta^*}^4)\mathrm{d}\theta^* = 2\pi K\left(1+\frac{b_2}{2}+\frac{3}{8}b_4\right)\end{aligned} \tag{1.34}$$

where

$$b_1 = \frac{1}{2}a_1 - 2e \tag{1.35}$$

$$b_2 = 3e^2 - a_1e - \frac{1}{8}a_1^2 \tag{1.36}$$

$$b_3 = \frac{3}{2}a_1e^2 - 4e^3 + \frac{1}{4}a_1^2e + \frac{1}{16}a_1^3 \tag{1.37}$$

$$b_4 = 5e^4 - 2a_1e^3 - \frac{3}{8}a_1^2e^2 - \frac{1}{8}a_1^3e - \frac{5}{128}a_1^4 \tag{1.38}$$

We can expand b_2 and b_4 in powers of e to obtain $b_2 \simeq e^2/2 + 3e^4$ and $b_4 \simeq -(17/8)e^4$, yielding the following final form for Δa after one revolution of continuous thrust:

$$\Delta a = 2\pi K\left(1+\frac{e^2}{4}+\frac{45}{64}e^4\right) \tag{1.39}$$

This equation is accurate for small to moderate eccentricities up to 0.2 with an error of less than 1%. If we wish to integrate between 0 and θ_1^* and then between

θ_2^* and 2π, where θ_1^* and θ_2^* represent the true anomalies of the shadow entry and exit points, respectively, then from

$$\Delta a = K \int_0^{\theta_1^*} \frac{(1+e^2+2ec_{\theta^*})^{1/2}}{(1+ec_{\theta^*})^2} \mathrm{d}\theta^* + K \int_{\theta_2^*}^{2\pi} \frac{(1+e^2+2ec_{\theta^*})^{1/2}}{(1+ec_{\theta^*})^2} \mathrm{d}\theta^*$$

we obtain the following expression for Δa valid for intermittent thrusting due to the presence of Earth shadow along the $\theta_2^* - \theta_1^*$ arc:

$$\begin{aligned}\Delta a = {} & 2\pi K\left(1 + \frac{b_2}{2} + \frac{3}{8}b_4\right) + K\Big\{(\theta_1^* - \theta_2^*) + b_1(s_{\theta_1^*} - s_{\theta_2^*}) \\ & + b_2\left[\frac{(\theta_1^* - \theta_2^*)}{2} + \frac{1}{4}(s_{2\theta_1^*} - s_{2\theta_2^*})\right] \\ & + \frac{b_3}{3}\left[s_{\theta_1^*}(c_{\theta_1^*}^2 + 2) - s_{\theta_2^*}(c_{\theta_2^*}^2 + 2)\right] \\ & + b_4\left[\frac{3}{8}(\theta_1^* - \theta_2^*) + \frac{1}{4}(s_{2\theta_1^*} - s_{2\theta_2^*}) + \frac{1}{32}(s_{4\theta_1^*} - s_{4\theta_2^*})\right]\Big\}\end{aligned} \tag{1.40}$$

For the integration of the differential equation for eccentricity in terms of time de/dt, we need the additional expansions

$$(1 + a_1 c_{\theta^*})^{-1/2} \simeq 1 - \frac{1}{2}a_1 c_{\theta^*} + \frac{3}{8}a_1^2 c_{\theta^*}^2 - \frac{5}{16}a_1^3 c_{\theta^*}^3 + \frac{35}{128}a_1^4 c_{\theta^*}^4$$

$$(1 + a_1 c_{\theta^*})^{-1/2}(1 + ec_{\theta^*})^{-2} \simeq 1 + c_1 c_{\theta^*} + c_2 c_{\theta^*}^2 + c_3 c_{\theta^*}^3 + c_4 c_{\theta^*}^4$$

where

$$c_1 = -2e - \frac{1}{2}a_1 \tag{1.41}$$

$$c_2 = 3e^2 + a_1 e + \frac{3}{8}a_1^2 \tag{1.42}$$

$$c_3 = -4e^3 - \frac{3}{2}a_1 e^2 - \frac{3}{4}a_1^2 e - \frac{5}{16}a_1^3 \tag{1.43}$$

$$c_4 = 5e^4 + 2a_1 e^3 + \frac{9}{8}a_1^2 e^2 + \frac{5}{8}a_1^3 e \tag{1.44}$$

Letting

$$K' = \frac{2(1-e^2)^2}{an^2(1+e^2)^{1/2}} f_t \tag{1.45}$$

and integrating Eq. (1.30) between 0 and 2π yields

$$\Delta e = K' \int_0^{2\pi} (e + c_{\theta^*})(1 + c_1 c_{\theta^*} + c_2 c_{\theta^*}^2 + c_3 c_{\theta^*}^3 + c_4 c_{\theta^*}^4)\mathrm{d}\theta^*$$

which, after some manipulation, reduces to

$$\Delta e = 2\pi K'\left[e\left(1+\frac{c_2}{2}+\frac{3}{8}c_4\right)+\left(\frac{c_1}{2}+\frac{3}{8}c_3\right)\right]$$

$$\Delta e = -\pi K' e\left(1+\frac{15}{8}e^2\right) \tag{1.46}$$

If $e = 0$, then Δe will vanish at the end of each revolution with continuous tangential thrusting. If $e \neq 0$, then from Eq. (1.46), Δe will always be negative, indicating that the orbit will be less eccentric after each revolution with continuous tangential thrusting. If, on the other hand, we consider intermittent thrusting from 0 to θ_1^* and from θ_2^* to 2π, then

$$\begin{aligned}\Delta e = K'\Bigg\{&e(\theta_1^*-\theta_2^*)+d_1(s_{\theta_1^*}-s_{\theta_2^*})+d_2\left[\frac{(\theta_1^*-\theta_2^*)}{2}+\frac{1}{4}(s_{\theta_1^*}-s_{\theta_2^*})\right]\\&+d_3\left[2(s_{\theta_1^*}-s_{\theta_2^*})+s_{\theta_1^*}c^2_{\theta_1^*}-s_{\theta_2^*}c^2_{\theta_2^*}\right]\\&+d_4\left[\frac{3}{8}(\theta_1^*-\theta_2^*)+\frac{1}{4}(s_{2\theta_1^*}-s_{2\theta_2^*})+\frac{1}{32}(s_{4\theta_1^*}-s_{4\theta_2^*})\right]\\&+d_5\left[s_{\theta_1^*}c^4_{\theta_1^*}-s_{\theta_2^*}c^4_{\theta_2^*}+\frac{8}{3}(s_{\theta_1^*}-s_{\theta_2^*})+\frac{4}{3}(s_{\theta_1^*}c^2_{\theta_1^*}-s_{\theta_2^*}c^2_{\theta_2^*})\right]\Bigg\}\\&+\pi K'\left(2e+d_2+\frac{3}{4}d_4\right)\end{aligned} \tag{1.47}$$

where

$$d_1 = 1 + ec_1 \tag{1.48}$$

$$d_2 = c_1 + ec_2 \tag{1.49}$$

$$d_3 = \frac{1}{3}(c_2 + ec_3) \tag{1.50}$$

$$d_4 = c_3 + ec_4 \tag{1.51}$$

$$d_5 = c_4/5 \tag{1.52}$$

When shadowing is present, Δe can be either positive or negative, depending on the relative geometry of the orbit and the shadow arc. Finally, the integration of $d\omega/dt$ in Eq. (1.26) can be carried out in closed form without the need for expansions in powers of the eccentricity such that the resulting formula will be valid for any eccentricity $e \geq 10^{-2}$ and will not be restricted to the range $10^{-2} \leq e \leq 0.2$, as for Δa and Δe. Therefore

$$\Delta\omega = \frac{2a^2(1-e^2)^2}{\mu e}f_t\int_0^{2\pi}\frac{s_{\theta^*}}{(1+ec_{\theta^*})^2(1+e^2+2ec_{\theta^*})^{1/2}}d\theta^* \tag{1.53}$$

The integral in Eq. (1.53) can be written as $k' \int 1/[(1+ex)^2(1+a_1x)^{1/2}]\mathrm{d}x$, with a_1 defined as before, $k' = -1/(1+e^2)^{1/2}$, and $x = c_{\theta^*}$. However

$$\int \frac{\mathrm{d}x}{(1+ex)^2(1+a_1x)^{1/2}} = \frac{-1}{k}\left[\frac{(1+a_1x)^{1/2}}{(1+ex)} + \frac{a_1}{2}\int \frac{\mathrm{d}x}{(1+ex)(1+a_1x)^{1/2}}\right]$$

such that, after a few additional steps, $\Delta\omega$ can be written as

$$\Delta\omega = \frac{-2a^2(1-e^2)(1+e^2)^{1/2}}{\mu e^2} f_t \times \left[\frac{(1+a_1c_{\theta^*})^{1/2}}{(1+ec_{\theta^*})} + \frac{2}{(1+e^2)^{1/2}(1-e^2)^{1/2}}\tan^{-1}\frac{(1+a_1c_{\theta^*})^{1/2}}{\left(\frac{1-e^2}{1+e^2}\right)^{1/2}}\right]_0^{2\pi} \tag{1.54}$$

with $k = -e(1-e^2)/(1+e^2)$. These integrations must be carried out from perigee to subsequent perigee. This is true because $\Delta\omega$ as computed from Eq. (1.54) modifies the location of the new perigee by a small amount so that, using a common reference direction, small corrections to Δa, Δe, and $\Delta\omega$ must be added to the first-order changes. These corrections are obtained by integrating further between 2π and $\Delta\omega$ by properly accounting for the sign of $\Delta\omega$. From Eq. (1.54), it follows that, after one full revolution from $\theta^* = 0$ to $\theta^* = 2\pi$, $\Delta\omega = 0$, indicating that continuous, constant, tangential acceleration will not induce any motion in the perigee. This will not be the case if shadowing is present, in which case the integration is carried out from 0 to θ_1^* and from θ_2^* to 2π. The change in ω will then be given by

$$\Delta\omega = \frac{2a^2(1-e^2)}{\mu e^2} f_t \left\{ (1+e^2)\left[\frac{(1+a_1c_{\theta_2^*})^{1/2}}{(1+ec_{\theta_2^*})} - \frac{(1+a_1c_{\theta_1^*})^{1/2}}{(1+ec_{\theta_1^*})}\right] + 2(1-e^2)^{-1/2}\left[\tan^{-1}\frac{(1+a_1c_{\theta_2^*})}{\left(\frac{1-e^2}{1+e^2}\right)^{1/2}} - \tan^{-1}\frac{(1+a_1c_{\theta_1^*})}{\left(\frac{1-e^2}{1+e^2}\right)^{1/2}}\right]\right\} \tag{1.55}$$

Both $1 + a_1 c_{\theta_1^*}$ and $1 + a_1 c_{\theta_2^*}$ are positive quantities, as is the $\left[(1-e^2)/(1+e^2)\right]^{1/2}$ term, so that the inverse tangent function will always yield an angle in the first quadrant without ambiguity. It also follows from Eq. (1.55) that, if $\theta_1^* = \theta_2^*$, then $\Delta\omega = 0$ because, in this case, the orbit is in its entirety in sunlight. For a given eccentricity and shadow angle, let θ_1^* vary from 0 to 360 deg. The shadow exit angle θ_2^* is computed simply by adding the shadow angle to θ_1^*. Let $a = 7000$ km, $\mu = 398601.3\ \mathrm{km^3/s^2}$, and $f_t = 3.5 \times 10^{-7}\ \mathrm{km/s^2}$, which corresponds approximately to an acceleration of 3.5×10^{-5} g. In Fig. 1.1, valid for $e = 0.1$, Δa is plotted as a function of θ_1^* for shadow angles of $\theta_2^* - \theta_1^* = 0$, 30, 90, and 140 deg. For a given shadow angle, the variation is due to the relative geometry of the orbit eccentricity vector and the shadow arc. Because θ^* is measured from perigee, a shadow arc centered at perigee has a smaller length than a comparable shadow arc of equal angular opening centered at apogee. In the latter case, the duration of the thrust is less than that for the former case so that there will be comparatively less buildup in the semimajor axis. For small eccentricities, this effect is more important than the velocity efficiency effect, which favors thrusting at perigee where velocity is higher, and therefore, larger changes in the semimajor axis can be achieved for the same thrust duration than at apogee, such as in the impulsive case. This can also be seen from the equation for Δa valid for small intervals dt

$$\Delta a = \int \frac{2Va^2}{\mu} f_t \mathrm{d}t = \frac{2a^2}{\mu} \int V f_t \mathrm{d}t \tag{1.56}$$

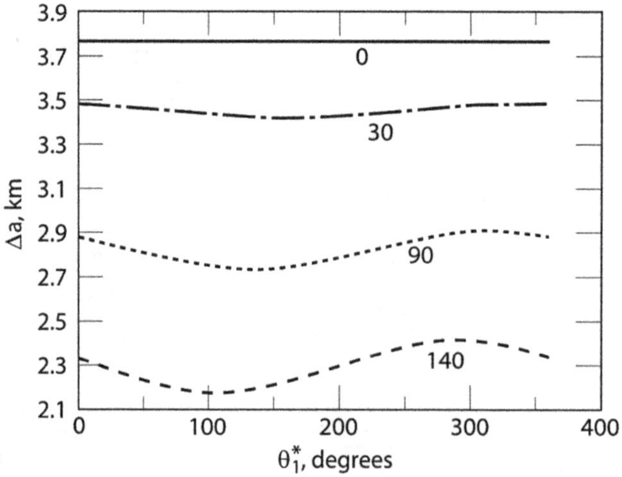

Fig. 1.1 Increase in semimajor axis per revolution vs shadow entry angle for *a* = 7000 km; *e* = 0.1; and total shadow angles of 0, 30, 90, and 140 deg.

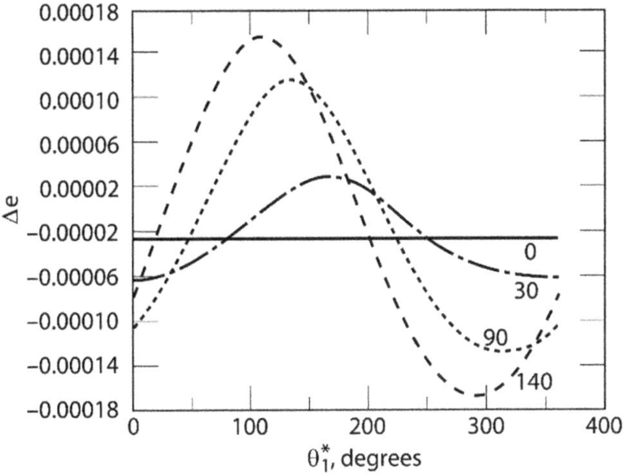

Fig. 1.2 Change in eccentricity per revolution vs shadow entry angle for a = 7000 km; e = 0.1; and total shadow angles of 0, 30, 90, and 140 deg.

The change in semimajor axis is independent of the shadow entry angle θ_1^* only if there is no shadow and Δa is at its constant maximum value. As the shadow angle increases, Δa becomes smaller, and the amplitude of its oscillation becomes greater. The locations of the maximum and minimum values with respect to θ_1^* also shift as the shadow angle varies. As shown in Fig. 1.2, Δe reaches its maximum positive value if the shadow is centered at apogee and its minimum negative value if the shadow is centered at perigee. Δe is constant and negative for the case of no shadow, as was shown in Eq. (1.46), and the amplitude of the oscillations increases with increasing total shadow angle. From Fig. 1.3, one can see that $\Delta\omega = 0$ if the shadow is centered at either perigee or apogee, as expected due to symmetry; $\Delta\omega = 0$ for the no-shadow case, with oscillation amplitude once again increasing with increasing shadow angle. For $e = 0.01$, the Δa oscillations have smaller overall amplitudes, whereas the reverse is true for the Δe and $\Delta\omega$ curves, especially for $\Delta\omega$, which reaches maximum changes on the order of 2 deg. For smaller eccentricities, the $\Delta\omega$ equation will break down as the orbit becomes near-circular and approaches the $e = 0$ singularity, whereas the Δa and Δe equations remain perfectly valid. For $e < 0.01$ down to $e = 0$, it is preferable and convenient to use the following equations, which are valid for tangential thrusting

$$\frac{da}{dt} = \frac{2}{n} f_t \tag{1.57}$$

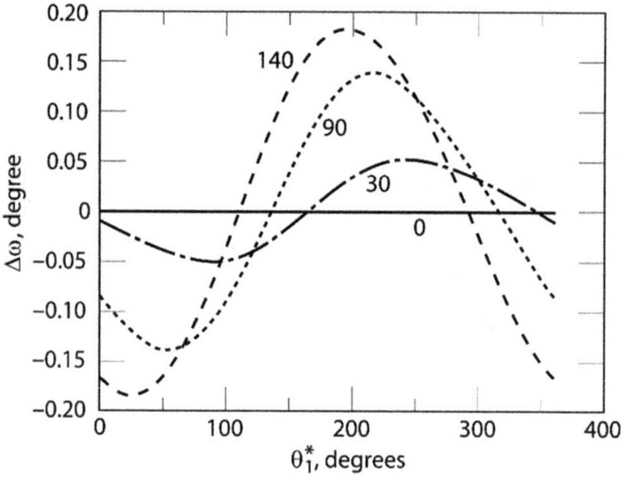

Fig. 1.3 Change in the argument of perigee per revolution vs shadow entry angle for *a* = 7000 km; *e* = 0.1; and total shadow angles of 0, 30, 90, and 140 deg.

$$\frac{de_x}{dt} = \frac{2a^{1/2}}{\mu^{1/2}} c_{nt} f_t \tag{1.58}$$

$$\frac{de_y}{dt} = \frac{2a^{1/2}}{\mu^{1/2}} s_{nt} f_t \tag{1.59}$$

where the x axis is along the line of apsides directed toward perigee, the y axis lies perpendicular to the x axis in the orbit plane, and time is measured from the x axis. Therefore, when e is small, $n\, dt = d\theta$, which is used to convert Eqs. (1.57–1.59) into the form

$$\frac{da}{d\theta} = \frac{2a^3}{\mu} f_t \tag{1.60}$$

$$\frac{de_x}{d\theta} = \frac{2a^2}{\mu} f_t c_\theta \tag{1.61}$$

$$\frac{de_y}{d\theta} = \frac{2a^2}{\mu} f_t s_\theta \tag{1.62}$$

Integrating between zero and θ_1 and between θ_2 and 2π and adding results in

$$\Delta a = \frac{2a^3}{\mu} f_t (2\pi + \theta_1 - \theta_2) \tag{1.63}$$

$$\Delta e_x = \frac{2a^2}{\mu} f_t (s_{\theta_1} - s_{\theta_2}) \tag{1.64}$$

$$\Delta e_y = \frac{-2a^2}{\mu} f_t (c_{\theta_1} - c_{\theta_2}) \tag{1.65}$$

Because perigee is initially along the x axis, the change in the argument of perigee $\Delta\omega$ is simply given by

$$\Delta\omega = \tan^{-1}\left(\frac{\Delta e_y}{e + \Delta e_x}\right) \tag{1.66}$$

which is always well-defined. The change in eccentricity Δe can also be given in this near-circular case by

$$\Delta e = \left[(e + \Delta e_x)^2 + \Delta e_y^2\right]^{1/2} - e \tag{1.67}$$

The secular effect of the second zonal harmonic of Earth's potential J_2 on the elements Ω, ω, and M can be added analytically after each revolution regardless of the existence of a shadow arc.

For the problem of multirevolution orbit prediction in the absence of Earth shadow using continuous, constant-acceleration, tangential thrusting, analytic expressions for the orbit elements as a function of time that are valid for several orbits at the GEO level and several tens of orbits at the LEO level can be developed. This theory is compared in Sec. 1.5 to an exact numerically integrated trajectory involving only the tangential thrust. Because the eccentricity buildup with continuous tangential thrusting is rather small, the analytic integration of the orbital equations linearized about zero eccentricity provides solutions that do not break down rapidly, provided that the mean motion is not held constant during the integration. This analytic integration is further extended by the inclusion of the effects of J_2 during the multirevolution orbit expansion.

1.4 ANALYTIC INTEGRATION FOR NEAR-CIRCULAR ORBITS WITH CONTINUOUS THRUST

The full set of the Gaussian form of the Lagrange planetary equations for near-circular orbits is given by

$$da/dt = 2a(f_t/V) \tag{1.68}$$

$$de_x/dt = 2(f_t/V)\cos\alpha - (f_n/V)\sin\alpha \tag{1.69}$$

$$de_y/dt = 2(f_t/V)\sin\alpha + (f_n/V)\cos\alpha \tag{1.70}$$

$$di/dt = (f_h/V)\cos\alpha \tag{1.71}$$

$$d\Omega/dt = (f_h/V)\sin\alpha/\sin i \tag{1.72}$$

$$d\alpha/dt = n + 2f_n/V - (f_h/V)\sin\alpha/\tan i \tag{1.73}$$

As the classical orbit elements are a, e, i, Ω, ω, and M, they make use of the nonsingular parameters $e_x = e\cos\omega$ and $e_y = e\sin\omega$, as well as $\alpha = \omega + M$, which represents the mean longitude. Furthermore, $n = (\mu/a^3)^{1/2}$ represents the orbit mean motion; $V = na = (\mu/a)^{1/2}$ represents the circular velocity; and f_t, f_n, and f_h correspond to the components of the thrust acceleration vector along the tangent, normal, and out-of-plane directions, respectively, with the normal direction now oriented toward the center of attraction and lying in the instantaneous orbit plane. If we assume only tangential thrusting, then the preceding set of equations reduces to

$$da/dt = \frac{2}{n} f_t \tag{1.74}$$

$$de_x/dt = \frac{2}{na} f_t \cos\alpha \tag{1.75}$$

$$de_y/dt = \frac{2}{na} f_t \sin\alpha \tag{1.76}$$

$$d\alpha/dt = n \tag{1.77}$$

For constant tangential thrust acceleration f_t, Eq. (1.74) for the semimajor axis can be integrated in a straightforward manner to give

$$\int_{a_0}^{a} a^{-3/2} da = \frac{2}{\mu^{1/2}} f_t \int_0^t dt$$

$$a^{-1/2} = a_0^{-1/2} - \frac{f_t}{\mu^{1/2}} t \tag{1.78}$$

which can also be cast into the form

$$a = \frac{1}{At^2 + Bt + C} \tag{1.79}$$

where

$$A = \left(f_t/\mu^{1/2}\right)^2 \tag{1.80}$$

$$B = -2f_t/(a_0\mu)^{1/2} \tag{1.81}$$

$$C = 1/a_0 \tag{1.82}$$

At time $t = 0$, $a = a_0$, and as a approaches infinity, t will tend to

$$t = \mu^{1/2}/\left(f_t a_0^{1/2}\right) \tag{1.83}$$

as can be seen by setting Eq. (1.78) to zero. This result is valid as long as the orbit remains near-circular. Equation (1.77) for the mean longitude can now be integrated because

$$d\alpha/dt = n = \mu^{1/2}a^{-3/2} \tag{1.84}$$

where, from Eq. (1.78)

$$a^{-3/2} = \left(a_0^{-1/2} - \frac{f_t}{\mu^{1/2}}t\right)^3$$

and therefore

$$\int_{\alpha_0}^{\alpha} d\alpha = \int_0^t \mu^{1/2}\left[a_0^{-3/2} - \frac{3}{a_0}\left(\frac{f_t}{\mu^{1/2}}\right)t + 3a_0^{-1/2}\left(\frac{f_t}{\mu^{1/2}}\right)^2 t^2 - \left(\frac{f_t}{\mu^{1/2}}\right)^3 t^3\right] dt$$

resulting in

$$\alpha = \alpha_0 + n_0 t + b' t^2 + c' t^3 + d' t^4 \tag{1.85}$$

with

$$b' = -\frac{3}{2a_0}f_t \tag{1.86}$$

$$c' = a_0^{-1/2} f_t^2/\mu^{1/2} \tag{1.87}$$

$$d' = -f_t^3/(4\mu) \tag{1.88}$$

An expression for α as a function of the semimajor axis can also be obtained by dividing Eq. (1.77) by Eq. (1.74) to obtain $d\alpha/da = \mu a^{-3}/2f_t$, which yields, upon integration

$$\alpha = \alpha_0 - \frac{\mu}{4f_t}(a^{-2} - a_0^{-2}) \tag{1.89}$$

This expression can readily be solved for a in terms of α, giving

$$a = \left\{1\Big/\left[a_0^{-2} - \frac{4f_t}{\mu}(\alpha - \alpha_0)\right]\right\}^{1/2} \tag{1.90}$$

The integration of Eq. (1.75) is best carried out by dividing Eq. (1.75) by Eq. (1.77) such that to give

$$de_x/d\alpha = \left(2f_t a^2/\mu\right)c_\alpha$$

Using a from Eq. (1.90), this derivative can be written as

$$de_x = \left[\frac{2f_t}{\mu}\Big/(b + g\alpha)\right]c_\alpha \, d\alpha$$

with

$$g = \frac{-4f_t}{\mu} \tag{1.91}$$

$$b = a_0^{-2} + \frac{4f_t}{\mu}\alpha_0 \tag{1.92}$$

such that

$$de_x = -\frac{1}{2}\frac{g}{b}\left(1 + \frac{g}{b}\alpha\right)^{-1} c_\alpha \, d\alpha \tag{1.93}$$

The quantity g/b is on the order of 10^{-4} at LEO and 10^{-3} at GEO for a thrust acceleration on the order 10^{-5} g. Therefore, we can expand the term $[1 + (g/b)\alpha]^{-1}$ before integrating. This results in the expression

$$de_x = -\frac{1}{2}\left[\delta - \delta^2\alpha + \delta^3\alpha^2 - \delta^4\alpha^3 + \delta^5\alpha^4 - \delta^6\alpha^5 + \cdots\right] c_\alpha \, d\alpha$$

where δ represents the small quantity

$$\delta = \frac{g}{b} \tag{1.94}$$

Finally, after integration, we obtain

$$\begin{aligned} e_x = e_{x_0} - \frac{1}{2}\{&\delta(s_\alpha - s_{\alpha_0}) - \delta^2[(c_\alpha - c_{\alpha_0}) + (\alpha s_\alpha - \alpha_o s_{\alpha_0})] \\ &+ \delta^3\left[2\alpha c_\alpha - 2\alpha_0 c_{\alpha_0} + (\alpha^2 - 2)s_\alpha - (\alpha_0^2 - 2)s_{\alpha_0}\right] \\ &- \delta^4\left[(3\alpha^2 - 6)c_\alpha - (3\alpha_0^2 - 6)c_{\alpha_0} + (\alpha^3 - 6\alpha)s_\alpha - (\alpha_0^3 - 6\alpha_0)s_{\alpha_0}\right]\} \end{aligned} \tag{1.95}$$

In a similar way, dividing Eq. (1.76) by Eq. (1.77) gives

$$\begin{aligned} de_y &= -\frac{1}{2}\frac{g}{b}\left(1 + \frac{g}{b}\alpha\right)^{-1} s_\alpha \, d\alpha \\ &= -\frac{1}{2}[\delta - \delta^2\alpha + \delta^3\alpha^2 - \delta^4\alpha^3 + \delta^5\alpha^4 - \delta^6\alpha^5 + \cdots] s_\alpha \, d\alpha \end{aligned}$$

which, upon integration, yields

$$\begin{aligned} e_y = e_{y_0} - \frac{1}{2}\{&-\delta(c_\alpha - c_{\alpha_0}) - \delta^2[(s_\alpha - s_{\alpha_0}) - (\alpha c_\alpha - \alpha_0 c_{\alpha_0})] \\ &+ \delta^3\left[2\alpha s_\alpha - 2\alpha_0 s_{\alpha_0} - (\alpha^2 - 2)c_\alpha + (\alpha_0^2 - 2)c_{\alpha_0}\right] \\ &- \delta^4\left[(3\alpha^2 - 6)s_\alpha - (3\alpha_0^2 - 6)s_{\alpha_0} - (\alpha^3 - 6\alpha)c_\alpha + (\alpha_0^3 - 6\alpha_0)c_{\alpha_0}\right]\} \end{aligned} \tag{1.96}$$

The analytic solution of the system of equations (1.74–1.77) is given by Eq. (1.78) for the semimajor axis; Eq. (1.85) for the mean longitude α; and Eqs. (1.95) and (1.96) for the eccentricity vector components e_x and e_y, respectively. The semimajor axis and mean longitude are integrated exactly and are obtained explicitly as functions of time. However, e_x and e_y are given as functions of α and not time and are approximate because they involve expansions. As will be shown later, these expressions are valid for many revolutions and are more accurate than the linearized equations developed earlier because the mean motion n and the semimajor axis are varying in Eqs. (1.74–1.77). On the right-hand sides of the linearized equations, a and n are held constant, and these equations are then integrated readily to yield

$$a = a_0 + \frac{2}{n} f_t t \tag{1.97}$$

$$e_x = e_{x_0} + \frac{2a^{1/2}}{\mu^{1/2}} \frac{f_t}{n} s_{nt} \tag{1.98}$$

$$e_y = e_{y_0} + \frac{2a^{1/2}}{\mu^{1/2}} \frac{f_t}{n} (1 - c_{nt}) \tag{1.99}$$

Of course, $\alpha = nt$ was used to integrate the equations of motion, whereas in Eq. (1.85), α is not linear but quartic in time, which is the exact functional dependence for large t. That is why Eqs. (1.78), (1.85), (1.95), and (1.96) will not break down as fast as the linearized equations (1.97–1.99), which would be accurate for very few revolutions.

A more accurate description of the motion requires the consideration of the effects of J_2 on the various elements, specifically e_x, e_y, Ω, and α, because a and i do not exhibit any secular variations due to J_2. Using the near-circular assumption, the J_2 accelerations can be written as

$$(f_n)_{J_2} = \frac{3\mu J_2 R^2}{2a^4}\left(1 - 3s_i^2 s_\theta^2\right) \tag{1.100}$$

$$(f_t)_{J_2} = \frac{-3\mu J_2 R^2}{a^4} s_i^2 s_\theta c_\theta \tag{1.101}$$

$$(f_h)_{J_2} = \frac{-3\mu J_2 R^2}{a^4} s_i c_i s_\theta \tag{1.102}$$

where $V = \mu^{1/2} a^{-1/2}$, $\alpha = \theta = nt$, and time is measured from the ascending node. The regression of the node Ω is obtained from the variational equation given in Eq. (1.72):

$$\Omega = \Omega_0 - 3\mu^{1/2} J_2 R^2 c_i \int_0^t \left(a_0^{-1/2} - \frac{f_t}{\mu^{1/2}} t\right)^7 s_{nt}^2 \, dt$$

From the relation

$$n = \mu^{1/2} a^{-3/2} = \mu^{1/2} \left(a_0^{-1/2} - \frac{f_t}{\mu^{1/2}} t \right)^3$$

we have $nt = n_0 \, (1 - \varepsilon t)^3 t$, where $n_0 = \mu^{1/2} a_0^{-3/2}$ and $\varepsilon = f_t / (\mu^{1/2} a_0^{-1/2})$. For the LEO example at hand, namely, $a_0 = 7000$ km and $f_t = 3.5 \times 10^{-7}$ km/s^2, ε is on the order of 10^{-8} such that $nt \cong n_0 t(1 - 3\varepsilon t)$. Keeping only first-order terms in ε, we obtain the expressions

$$s_{nt} \cong s_{n_0 t} - 3\varepsilon n_0 t^2 c_{n_0 t} \quad \text{and} \quad s_{nt}^2 \cong s_{n_0 t}^2 - 6\varepsilon n_0 t^2 s_{n_0 t} c_{n_0 t}$$

after making use of the expansion $\sin(x + \varepsilon') \cong s_x + \varepsilon' c_x$, which is valid for small ε'. The preceding integration can be carried out as

$$\begin{aligned} \Omega &= \Omega_0 - 3\mu^{1/2} J_2 R^2 c_i a_0^{-7/2} \int_0^t (1 - 7\varepsilon t) s_{nt}^2 \, \mathrm{d}t \\ &= \Omega_0 - \frac{3}{2} \mu^{1/2} J_2 R^2 c_i a_0^{-7/2} \left[t - \frac{\varepsilon}{2n_0^2} \left(\frac{1}{2} + 7n_0^2 t^2 \right) \right. \\ &\quad \left. - \frac{1}{2n_0} (1 - \varepsilon t) s_{2n_0 t} + \frac{\varepsilon}{2n_0^2} \left(\frac{1}{2} + 6n_0^2 t^2 \right) c_{2n_0 t} \right] \end{aligned} \tag{1.103}$$

In the absence of any thrust, $\varepsilon = 0$, and the orbit will not expand, so that the secular variation

$$\Omega = \Omega_0 - \frac{3}{2} \frac{n_0}{a_0^2} J_2 R^2 c_i t$$

is recovered from Eq. (1.103). In a similar way, Eq. (1.73) for $\dot{\alpha}$ can be integrated using f_n and f_h given by Eqs. (1.100) and (1.102), respectively. Because $b' = (3/2) n_0 \varepsilon$ is first-order in ε whereas $c' = n_0 \varepsilon^2$ and d' in Eq. (1.85) are of higher order, the analytic integration of $\dot{\alpha}$ retaining only first-order terms in ε yields

$$\begin{aligned} \alpha &= (\alpha_0 + n_0 t + b' t^2) + 3\mu^{1/2} J_2 R^2 a_0^{-7/2} \left(t - \frac{7}{2} \varepsilon t^2 \right) \\ &\quad + 3\mu^{1/2} J_2 R^2 a_0^{-7/2} \left(\frac{1 - 4s_i^2}{2} \right) \left[t - \frac{\varepsilon}{2n_0^2} \left(\frac{1}{2} + 7n_0^2 t^2 \right) \right. \\ &\quad \left. - \frac{1}{2n_0} (1 - \varepsilon t) s_{2n_0 t} + \frac{\varepsilon}{2n_0^2} \left(\frac{1}{2} + 6n_0^2 t^2 \right) c_{2n_0 t} \right] \end{aligned} \tag{1.104}$$

with the terms in the first parentheses due to thrust only. In the absence of any thrust (i.e., $\varepsilon = 0$), Eq. (1.104) yields

$$\dot{\alpha} = \dot{\omega} + \dot{M} = n_0 + \frac{3\mu^{1/2}}{2} J_2 R^2 a_0^{-7/2}(3 - 4s_i^2)$$

that is, the combined secular variations of ω and M, namely, ω_S and M_S, are given by

$$\omega_S = \omega_0 + \frac{3}{2} J_2 R^2 \overline{n} a_0^{-2}\left(2 - \frac{5}{2}s_i^2\right)t$$

and $M_S = M_0 + \overline{n}t$, respectively, where

$$\overline{n} = n_0 + \frac{3}{2}\frac{J_2 R^2}{a_0^2}\left(1 - \frac{3}{2}s_i^2\right)n_0$$

Finally, the contributions of J_2 to e_x and e_y are evaluated by direct integration of the rates given in Eqs. (1.69) and (1.70), respectively. Thus

$$\frac{de_x}{dt} = -2(3\mu^{1/2} J_2 R^2 a^{-7/2}) s_i^2 s_\theta c_\theta^2 - \frac{3\mu^{1/2}}{2} J_2 R^2 a^{-7/2}(1 - 3s_i^2 s_\theta^3)$$

Using $s_\theta^3 \cong s_{nt}^3 \approx s_{n_0t}^3 - 9\varepsilon n_0 t^2 s_{n_0t}^2 c_{n_0t}$ and letting $C = \mu^{1/2} J_2 R^2 a_0^{-7/2}$, we have

$$\frac{de_x}{dt} = -\left(\frac{3}{2} + 6s_i^2\right)C(1 - 7\varepsilon t)s_{nt} + \frac{21}{2}Cs_i^2(1 - 7\varepsilon t)s_{nt}^3$$

$$\begin{aligned} e_x = e_{x_0} &- \left(\frac{3}{2} + 6s_i^2\right)C\left[\frac{1}{n_0}(1 - c_{n_0t}) - \frac{\varepsilon}{n_0^2}s_{n_0t} + \frac{\varepsilon t}{n_0}c_{n_0t} - 3\varepsilon t^2 s_{n_0t}\right] \\ &+ \frac{21}{2}Cs_i^2\left[\frac{2}{3n_0}(1 - c_{n_0t}) - \frac{1}{3n_0}c_{n_0t}s_{n_0t}^2 - \frac{9}{4}\varepsilon t^2 s_{n_0t} + \frac{3}{4}\varepsilon t^2 s_{3n_0t}\right. \\ &\left. - \frac{\varepsilon}{12n_0}tc_{3n_0t} + \frac{\varepsilon}{36n_0^2}s_{3n_0t} + \frac{3}{4}\frac{\varepsilon}{n_0}tc_{n_0t} - \frac{3}{4}\frac{\varepsilon}{n_0^2}s_{n_0t}\right] \end{aligned} \tag{1.105}$$

With $a^{-1/2} = a_0^{-1/2} - f_t t\mu^{-1/2}$, the thrust-only contribution can also be obtained as a function of time as

$$(\dot{e}_x)_T = \frac{2f_t}{\mu^{1/2}a^{1/2}}c_\alpha \cong \frac{2f_t}{\mu^{1/2}}a_0^{-1/2}(1 - \varepsilon t)c_{nt}$$

Then, with $c_{nt} = \cos(n_0 t - 3\varepsilon n_0 t^2) \cong c_{n_0 t} + 3\varepsilon n_0 t^2 s_{n_0 t}$ and $C' = 2f_t a_0^{-1/2} \mu^{-1/2}$

$$(e_x)_T = C'\left(\frac{1}{n_0} s_{n_0 t} + \frac{5\varepsilon}{n_0} t s_{n_0 t} - 3\varepsilon t^2 c_{n_0 t} + \frac{5\varepsilon}{n_0^2} c_{n_0 t} - \frac{5\varepsilon}{n_0^2}\right) \tag{1.106}$$

From

$$(\dot{e}_y)_T = \frac{2f_t}{\mu^{1/2} a^{1/2}} s_{nt}$$

the thrust-only contribution to e_y is also obtained as

$$(e_y)_T = C'\left[\frac{1}{n_0}(1 - c_{n_0 t}) - \frac{5\varepsilon}{n_0} t c_{n_0 t} - 3\varepsilon t^2 s_{n_0 t} + \frac{5\varepsilon}{n_0^2} s_{n_0 t}\right] \tag{1.107}$$

Finally, the effects of J_2 on the element e_y can be obtained from Eq. (1.70) as

$$\dot{e}_y = -\frac{21}{2} C s_i^2 (1 - 7\varepsilon t) c_{nt} s_{nt}^2 + \frac{3}{2} C(1 - 7\varepsilon t) c_{nt}$$

with $c_{nt} s_{nt}^2 \cong c_{n_0 t} s_{n_0 t}^2 - 6\varepsilon n_0 t^2 s_{n_0 t} c_{n_0 t}^2 + 3\varepsilon n_0 t^2 s_{n_0 t}^3$, yielding

$$\begin{aligned} e_y = e_{y_0} &- \frac{21}{2} C s_i^2 \left(\frac{1}{3n_0} s_{n_0 t} - \frac{1}{3n_0} s_{n_0 t} c_{n_0 t}^2 - \frac{\varepsilon}{4n_0} t s_{n_0 t} - \frac{3\varepsilon}{4} t^2 c_{n_0 t}\right. \\ &\left. - \frac{\varepsilon}{4n_0^2} c_{n_0 t} + \frac{2\varepsilon}{9n_0^2} + \frac{3}{4}\varepsilon t^2 c_{3n_0 t} + \frac{\varepsilon t}{12 n_0} s_{3n_0 t} + \frac{\varepsilon}{36 n_0^2} c_{3n_0 t}\right) \\ &+ \frac{3}{2} C\left(\frac{1}{n_0} s_{n_0 t} - 3\varepsilon t^2 c_{n_0 t} - \frac{\varepsilon}{n_0^2} c_{n_0 t} - \frac{\varepsilon}{n_0} t s_{n_0 t} + \frac{\varepsilon}{n_0^2}\right) \end{aligned} \tag{1.108}$$

Thus, the combined effects of J_2 and the thrust acceleration yield

$$(e_x)_{\text{tot}} = (e_x)_T + e_x \tag{1.109}$$

$$(e_y)_{\text{tot}} = (e_y)_T + e_y \tag{1.110}$$

where $(e_x)_T$, e_x, $(e_y)_T$, and e_y are given by Eqs. (1.106), (1.105), (1.107), and (1.108), respectively. These equations continue to be accurate as long as the orbit remains near-circular. In this case, the effects of the drag acceleration are taken into account by using an effective thrust acceleration because the drag perturbation will act opposite to the thrust acceleration.

1.5 COMPARISON WITH A NUMERICALLY INTEGRATED EXACT NONSINGULAR SET FOR CONTINUOUS THRUST

The equations of motion for the thrust perturbation case, Eqs. (1.68–1.73), are essentially linearized about a circular orbit. It is for this reason that e_x and e_y will lose accuracy over a long time period. To assess the accuracy of the analytic expressions developed earlier, we need to know the exact behavior of the trajectory, and this can be obtained only by numerically integrating the set of exact first-order differential equations using the nonsingular elements e_x and e_y instead of the classical elements e and ω. These equations are given by

$$\mathrm{d}a/\mathrm{d}t = 2Vf_t/(an^2) \tag{1.111}$$

$$\begin{aligned}\mathrm{d}e_x/\mathrm{d}t = {} & 2f_t(e_x + \cos\alpha_{\theta^*})\Big/V - f_n\left(\frac{r}{a}\sin\alpha_{\theta^*} + 2e_y\right)\Big/V \\ & + e_y r\sin\alpha_{\theta^*} f_h\Big/\left[na^2(1-e^2)^{1/2}\tan i\right]\end{aligned} \tag{1.112}$$

$$\begin{aligned}\mathrm{d}e_y/\mathrm{d}t = {} & 2f_t(e_y + \sin\alpha_{\theta^*})\Big/V + f_n\left(\frac{r}{a}\cos\alpha_{\theta^*} + 2e_x\right)\Big/V \\ & - e_x r\sin\alpha_{\theta^*} f_h\Big/\left[na^2(1-e^2)^{1/2}\tan i\right]\end{aligned} \tag{1.113}$$

$$\mathrm{d}i/\mathrm{d}t = r\cos\alpha_{\theta^*} f_h\Big/\left[na^2(1-e^2)^{1/2}\right] \tag{1.114}$$

$$\mathrm{d}\Omega/\mathrm{d}t = r\sin\alpha_{\theta^*} f_h\Big/\left[na^2(1-e^2)^{1/2}\sin i\right] \tag{1.115}$$

$$\begin{aligned}\mathrm{d}(\omega+M)/\mathrm{d}t = {} & n + [2f_t\sin\theta^*/(Ve)]\left\{1-(1-e^2)^{1/2}\left[1+e^2(1+ec_{\theta^*})^{-1}\right]\right\} \\ & + f_n r\left[2e + \cos\theta^* + e^2\cos\theta^* - (1-e^2)^{3/2}\cos\theta^*\right]\Big/ \\ & \left[Vae(1-e^2)\right] - r\sin\alpha_{\theta^*} f_h\Big/\left[na^2\tan i(1-e^2)^{1/2}\right]\end{aligned} \tag{1.116}$$

Here, θ^* is the true anomaly; $r = a(1-e^2)(1+ec_{\theta^*})^{-1}$ is the radial distance to the spacecraft; $V = \left\{\mu(1+e^2+2ec_{\theta^*})/\left[a(1-e^2)\right]\right\}^{1/2}$ is the orbital velocity on the elliptical orbit; and $\alpha_{\theta^*} = \omega + \theta^* = \alpha + \theta^* - M$ is the mean longitude of the spacecraft measured from the ascending node, where $\alpha = \omega + M$ the mean longitude as before. For coplanar transfers with tangential thrust, $f_n = f_h = 0$ such that the full set of coupled nonlinear differential equations reduces to

$$\mathrm{d}a/\mathrm{d}t = 2Vf_t/(n^2a) \tag{1.117}$$

$$\mathrm{d}e_x/\mathrm{d}t = 2f_t(e_x + \cos\alpha_{\theta^*})/V \tag{1.118}$$

$$\mathrm{d}e_y/\mathrm{d}t = 2f_t(e_y + \sin\alpha_{\theta^*})/V \tag{1.119}$$

$$\mathrm{d}\alpha/\mathrm{d}t = n + 2f_t\sin\theta^*\left[1-(1-e^2)^{1/2}\left(1+\frac{e^2}{1+e\cos\theta^*}\right)\right]\Big/Ve \tag{1.120}$$

Given the values of a, e_x, e_y, and α at time zero, the related quantities needed in the evaluation of the preceding derivatives are computed from $\omega = \tan^{-1}(e_y/e_x)$ and $M = \alpha - \omega$. The eccentric anomaly is calculated from Kepler's equation, $M = E - e \sin E$; the true anomaly is obtained from the equation

$$\tan\left(\frac{\theta^*}{2}\right) = \left(\frac{1+e}{1-e}\right)^{1/2} \tan\left(\frac{E}{2}\right)$$

and finally, $\alpha_{\theta^*} = \alpha + \theta^* - M$. Differential equations (1.117–1.120) are integrated forward in time starting from GEO at $a_0 = 42{,}000$ km, $e_{x_0} = e_{y_0} = 0$, and $\alpha_0 = 0$ with $f_t = 3.5 \times 10^{-7}$ km/s^2. A very long time span corresponding to 50 revolutions or some 1200 h is considered in Figs. 1.4 and 1.5. The analytic expressions in Eqs. (1.78), (1.85), (1.95), and (1.96) for the nonlinear theory, as well as Eqs. (1.97–1.99) for the linearized theory, are plotted against the exact numerical solution of Eqs. (1.117–1.120). In Fig. 1.4, the nonlinear and numerical curves are identical, as expected, because da/dt is insensitive to the eccentricity in both formulations. The evolution of e_x shows that the nonlinear analytic theory matches the exact trajectory for about 10 revolutions in this high-energy orbit and remains in phase with it, whereas the linearized theory breaks down after one or two revolutions in both phase and amplitude.

These observations are valid for the e_y component as well. Figure 1.5 shows the evolution of the eccentricity, displaying excellent agreement between the analytic nonlinear solution and the numerical solution up to $e = 10^{-2}$, after which the orbit is no longer near-circular, so that only the integrated orbit is

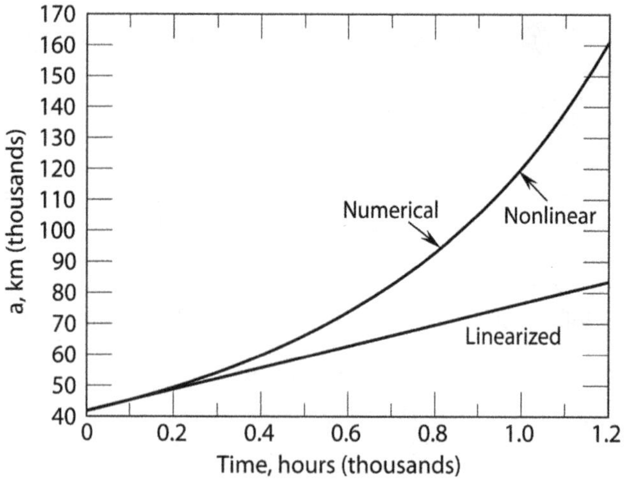

Fig. 1.4 Evolution of the semimajor axis for continuous tangential thrust, with initial $a_0 = 42000$ km and $e_0 = 0$, using numerical, analytic, and linearized methods.

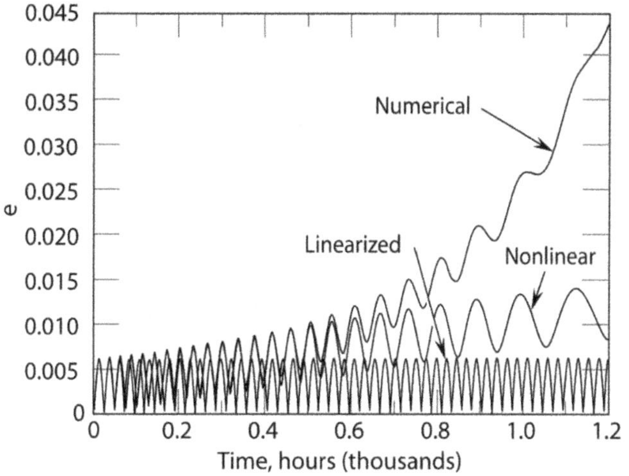

Fig. 1.5 Evolution of the eccentricity magnitude for continuous tangential thrust, with initial $a_0 = 42000$ km and $e_0 = 0$, using numerical, analytic, and linearized methods.

meaningful. However, for lower orbits, the agreement will last much longer, on the order of several tens of revolutions. Orbit prediction can therefore be accomplished over long time periods by means of the analytic expressions developed here as long as the orbit remains near-circular, after which analytic expressions such as those developed earlier and based on the classical elements formulation can be used for larger elliptical orbits by updating the elements one orbit at a time.

1.6 ANALYTIC INTEGRATION FOR NEAR-CIRCULAR ORBITS WITH RESPECT TO THE MEAN MOTION: A SERIES SOLUTION

In this section, we revisit Eqs. (1.74–1.77), and instead of solving them as a function of the mean longitude α, we integrate with respect to the mean motion $n = \mu^{1/2}a^{-3/2}$. Equation (1.74) is integrated as before to yield Eq. (1.78), which reads as $a^{-1/2} = a_0^{-1/2} - f_t t/\mu^{1/2}$. Multiplying both sides by $a_0{}^{-1}\mu^{1/2}/n_0$ yields

$$k_4 n^{1/3} = 1 - k_3 t \tag{1.121}$$

where k_3 and k_4 are constants given by $k_3 = f_t/(a_0 n_0)$ and $k_4 = \mu^{1/3}/(a_0 n_0)$. Equation (1.121) yields

$$\frac{1}{3}k_4 n^{-2/3}\mathrm{d}n = -k_3 \mathrm{d}t$$

which can be used to eliminate time in favor of n, the mean motion. Using this change of variable, Eq. (1.77), namely, $\mathrm{d}\alpha/\mathrm{d}t = n$, yields

$$\frac{\mathrm{d}\alpha}{\mathrm{d}n} = -n^{1/3}\frac{k_4}{3k_3} = k_5 n^{1/3} \tag{1.122}$$

where $k_5 = -k_4/(3k_3)$. Equation (1.122) is now readily integrated as

$$\int_{\alpha_0}^{\alpha} \mathrm{d}\alpha = k_5 \int_{n_0}^{n} n^{1/3}\mathrm{d}n$$

resulting in the following expression for the mean longitude α in terms of the mean motion n

$$\alpha = \alpha_0 + k_6\left(n^{4/3} - n_0^{4/3}\right) \tag{1.123}$$

where $k_6 = 3k_5/4$. Now, from $n = \mu^{1/2}a^{-3/2}$, we have

$$\mathrm{d}n = -\frac{3}{2}\mu^{1/2}a^{-5/2}\mathrm{d}a$$

which allows us to write Eqs. (1.75) and (1.76) in terms of n by first dividing by Eq. (1.74):

$$\frac{\mathrm{d}e_x}{\mathrm{d}t}\Big/\frac{\mathrm{d}a}{\mathrm{d}t} = \frac{\mathrm{d}e_x}{\mathrm{d}a} = \frac{\cos\alpha}{a}$$
$$\frac{\mathrm{d}e_y}{\mathrm{d}t}\Big/\frac{\mathrm{d}a}{\mathrm{d}t} = \frac{\mathrm{d}e_y}{\mathrm{d}a} = \frac{\sin\alpha}{a}$$

Therefore

$$\frac{\mathrm{d}e_x}{\mathrm{d}n} = -\frac{2}{3}\frac{\cos\alpha}{n}$$
$$\frac{\mathrm{d}e_y}{\mathrm{d}n} = -\frac{2}{3}\frac{\sin\alpha}{n}$$

Because α is itself known from Eq. (1.123) as a function of n, these two derivatives can be cast into the following form:

$$\frac{\mathrm{d}e_x}{\mathrm{d}n} = -\frac{2}{3n}\cos\left[\alpha_0 + k_6\left(n^{4/3} - n_0^{4/3}\right)\right]$$
$$\frac{\mathrm{d}e_y}{\mathrm{d}n} = -\frac{2}{3n}\sin\left[\alpha_0 + k_6\left(n^{4/3} - n_0^{4/3}\right)\right]$$

Let us now carry out a change of variable from n to m with $n^{4/3} = m$ and $\mathrm{d}m = (4/3)n^{1/3}\,\mathrm{d}n$. This will yield, after some manipulation

$$\mathrm{d}e_x = -\frac{1}{2}\left(A\frac{\cos k_6 m}{m} - B\frac{\sin k_6 m}{m}\right)\mathrm{d}m \tag{1.124}$$

$$de_y = -\frac{1}{2}\left(B\frac{\cos k_6 m}{m} + A\frac{\sin k_6 m}{m}\right)dm \tag{1.125}$$

with $A = \cos(\alpha_0 - k_6 n_0^{4/3})$ and $B = \sin(\alpha_0 - k_6 n_0^{4/3})$. Integrating Eqs. (1.124) and (1.125) yields

$$\begin{aligned} e_x = e_{x_0} &- \frac{A}{2}\left\{\ln\left(\frac{m}{m_0}\right) + \sum_{\ell=1}^{\infty}\frac{(-1)^\ell (k_6)^{2\ell}}{2\ell(2\ell)!}\left[(m)^{2\ell} - (m_0)^{2\ell}\right]\right\} \\ &+ \frac{B}{2}\left\{\sum_{\ell=0}^{\infty}\frac{(-1)^\ell (k_6)^{2\ell+1}}{(2\ell+1)(2\ell+1)!}\left[(m)^{2\ell+1} - (m_0)^{2\ell+1}\right]\right\} \end{aligned} \tag{1.126}$$

$$\begin{aligned} e_y = e_{y_0} &- \frac{B}{2}\left\{\ln\left(\frac{m}{m_0}\right) + \sum_{\ell=1}^{\infty}\frac{(-1)^\ell (k_6)^{2\ell}}{2\ell(2\ell)!}\left[(m)^{2\ell} - (m_0)^{2\ell}\right]\right\} \\ &- \frac{A}{2}\left\{\sum_{\ell=0}^{\infty}\frac{(-1)^\ell (k_6)^{2\ell+1}}{(2\ell+1)(2\ell+1)!}\left[(m)^{2\ell+1} - (m_0)^{2\ell+1}\right]\right\} \end{aligned} \tag{1.127}$$

Because the semimajor axis is known as a function of time t from Eq. (1.78), the mean longitude α is obtained from Eq. (1.123) as a function of the mean motion. Therefore, the semimajor axis and finally both e_x and e_y can also be evaluated from Eqs. (1.126) and (1.127) in terms of the mean motion after truncating the series at an appropriate value of ℓ. These equations will yield the same medium- to long-term accuracy as the theory used to derive Eqs. (1.78), (1.85), (1.95), and (1.96) in the preceding section.

1.7 CONCLUSION

In this chapter, an orbit prediction capability based on analytic expressions that describe the evolution of the in-plane orbit elements with high accuracy has been presented for the problem of low-thrust tangential thrusting along small- to moderate-eccentricity orbits. The orbit is updated after each revolution if shadow is present to account for the changing shadow geometry, and the secular variation of the node due to J_2 is updated analytically. Analytic expressions for the variations of the pertinent elements due to the combined effects of J_2 and the thrust acceleration are also obtained for the near-circular case in LEO using continuous, constant, low-thrust acceleration along the tangential direction. The analytic modelings of the coplanar phase of a typical LEO-to-GEO transfer allow for straightforward implementation in fast simulation computer programs to support systems design optimization analysis.

REFERENCES

[1] Edelbaum, T. N., "Propulsion requirements for controllable satellites," *ARS Journal*, Vol. 31, No. 8, Aug. 1961, pp. 1079–1089.

[2] Wiesel, W. E., and Alfano, S., "Optimal many-revolution orbit transfer," *AAS/AIAA Astrodynamics Specialist Conference, AAS Paper 83-352*, Lake Placid, NY, Aug. 1983.

[3] Sackett, L. L., Malchow, H., and Edelbaum, T. N., "Solar electric geocentric transfer with attitude constraints: analysis," *Report R901*, Charles Stark Draper Laboratory, Inc., Cambridge, MA, Aug. 1975; also NASA CR-134927.

[4] Kechichian, J. A., "Low-thrust eccentricity-constrained orbit raising," AAS Paper 91-156, AAS/AIAA Spaceflight Mechanics Meeting, Houston, TX, Feb. 1991.

[5] Cass, J. R., "Discontinuous low thrust orbit transfer," M.S. Thesis, Rept. AFIT/GA/AA/83D-1, School of Engineering, U.S. Air Force Institute of Technology, Wright-Patterson AFB, OH, Dec. 1983.

[6] McCann, J. M., "Optimal launch time for a discontinuous low thrust orbit transfer," M.S. Thesis, Rept. AFIT/GA/AA/88D-7, School of Engineering, U.S. Air Force Institute of Technology, Wright-Patterson AFB, OH, Dec. 1988.

[7] Dickey, M. R., Klucz, R. S., Ennix, K. A., and Matuszak, L. M., "Development of the electric vehicle analyzer," Rept. AL-TR-90-006, Astronautics Laboratory, U.S. Air Force Space Technology Center, Edwards AFB, CA, June 1990.

[8] Kechichian, J. A., "Orbit raising with low-thrust tangential acceleration in the presence of Earth shadow," *AAS/AIAA Astrodynamics Specialist Conference, AAS Paper 91-513*, Durango, CO, Aug. 1991.

[9] Kechichian, J. A., "Orbit raising with low-thrust tangential acceleration in presence of Earth shadow," *Journal of Spacecraft and Rockets*, Vol. 35, No. 4, July–Aug. 1998, pp. 516–525.

[10] Danby, J. M. A., *Fundamentals of Celestial Mechanics*, 2nd ed., Willmann-Bell, Inc., Richmond, VA, 1988, pp. 323–337.

CHAPTER 2

Low-Thrust Eccentricity-Constrained Orbit Raising

NOMENCLATURE

a_0, a = initial and current orbit semimajor axes, km
c_θ, s_θ = $\cos\theta$, $\sin\theta$
e_0, e = initial and current orbit eccentricities
e_x, e_y = components of the eccentricity vector along the inertial x and y directions
f_r, f_θ = components of the acceleration vector along the radial and perpendicular directions
f_t, f_n = components of the acceleration vector along the tangential and normal directions
h = orbital angular momentum, $[\mu a(1-e^2)]^{1/2}$
k = thrust acceleration, T/m, km/s^2
m = spacecraft mass, kg
n = orbit mean motion, $(\mu/a^3)^{1/2}$, rad/s
p = orbit parameter, $a(1-e^2)$, km
r = radial distance, $p(1+ec_{\theta^*})^{-1}$, km
$\mathbf{T}$ = thrust vector
t = time, s
v = orbital velocity on a circular orbit, na
δ = shadow angle, deg
θ = angular position of the spacecraft measured from the x axis, deg
θ^* = true anomaly
θ_t = pitch angle of the thrust vector, deg
θ_t', θ_t'' = pitch angles before and after the switch point, respectively, deg
μ = gravitational constant of Earth, 398601.3 km^3/s^2
τ = dimensionless time, nt
τ_1, τ_2 = angular positions of the shadow entry and exit points, deg
τ_c = angular position of the switch point measured from the x axis, deg
ω = argument of perigee

2.1 INTRODUCTION

The problem of zero-eccentricity-constrained orbit raising in a circular orbit in the presence of shadowing is analyzed using both numerical and analytical methods. Given the shadow arc length, piecewise-constant pitch angles are selected analytically, and the location along the orbit where the pitch angle switches is optimized to effect the largest change in semimajor axis per revolution. Two strategies yielding near-optimal performance are presented, differing only in whether a pitch reorientation maneuver is carried out inside or outside the shadow arc. These analytic results are also compared with the exact eccentricity-constrained numerical solutions, which use optimally varying continuous pitch profiles to maximize the change in semimajor axis after each revolution. The suboptimal analytic strategies are almost as effective as the optimal strategy, especially in higher orbits. These analytic strategies remove the need for continuously reorienting the spacecraft attitude in pitch, providing robust real-time onboard guidance capability for EOTV orbit-raising applications. The general problem of circle-to-circle transfer with inclination change using continuous constant acceleration was solved analytically by Edelbaum [1]. The assumption of a constant yaw profile within each revolution, adopted by Edelbaum, was later removed by Wiesel and Alfano [2], who optimized the yaw profile by also allowing the continuous acceleration magnitude to vary as a result of propellant expenditure. Semianalytic solutions of the optimal thrust pitch and yaw profiles for a given transfer were determined by Cass [3] and McCann [4], who also generalized the problem further by considering discontinuous thrust due to eclipsing. Dickey et al. [5] created a computer program that simulates the important effects due to both shadowing or eclipsing and power degradation during transit in the Van Allen radiation belts. To simplify the calculations of the orbit parameters and the shadow entry and exit points, they assumed that the intermediate orbits during the transfer remain circular, even though exact numerical simulations for typical six-month solar-electric transfers from LEO to GEO show that, in the worst case of shadowing geometry, the intermediate orbits could reach eccentricities on the order of 0.2 at most. Analytic solutions for inclination control and orbit raising when shadowing is present have been developed [6, 7] for implementation in efficient codes, such as that of Dickey et al. [5]. In view of the small eccentricity buildup, it is convenient to force eccentricity to remain at zero by developing analytic methods that result in simple but suboptimal steering laws that are easily implementable in mission analysis software and by fully accounting for the varying shadow geometry during the transfer. Thus, orbit raising in the presence of shadowing, which can be difficult to model in LEO, remains complicated if the orbit must be constrained to remain circular during the coplanar transfer. The main advantage of keeping the eccentricity at zero during the transfer is that it simplifies the equations of motion to their simplest form, which then allows the analyst to carry out the optimization problem semianalytically as done by Cass [3]. Another important advantage is the ease of computing the shadow entry and exit points, because they are given in analytic form instead of requiring that a quartic be solved numerically, which would be computationally intensive and therefore may not be of interest for onboard navigation applications. Unlike the approach of Cass [3], we consider constant relative pitch angles that are easily implemented by the onboard attitude control system, which, in this case, must hold the pitch attitude constant relative to the local horizon. The use of piecewise-constant pitch angles allows the spacecraft to pitch at the same rate as the orbital motion, thereby removing the need to continuously maneuver in attitude as was done by Cass [3], who implemented an optimal variable pitch profile instead.

Our approach is based on the theory of function minimization and is purely analytic, requiring minimal computational effort. In Sec. 2.2, we consider that no pitch attitude maneuvers should be implemented in shadow when sun sensors are ineffective (strategy 1). The strategy of Sec. 2.3 (strategy 2) is free of such requirements, and in both cases, the optimal switch location is found such that the largest change in semimajor axis is achieved during one revolution of intermittent thrusting. We also seek solutions with constant pitch angles over a large portion of the sunlit arc such that unique combinations of the pitch angles for a given switch location along the orbit result in no net eccentricity buildup, albeit with a correspondingly degraded semimajor axis buildup when compared with the solution involving the optimal switch location. The analysis in this chapter first appeared in as a conference paper [8] and later as a journal article [9].

2.2 ANALYSIS

Let (x, y) represent an Earth-centered inertial reference frame such that y bisects the shadow arc $O'O$ as in Fig. 2.1. The shadow entry point is O', and the shadow exit point is O. The location of O' is measured from the x axis by the angle τ_1, and because of symmetry, the angular position of O is given by $\tau_2 = \tau - \tau_1$. Thrust is applied only when the spacecraft is in sunlight, and time is measured from the x axis. In a first analysis, starting from point U at time zero, a constant acceleration thrust $\boldsymbol{T}$ is applied along the arc UO' such that the thrust vector is inclined at a constant angle θ_t' with respect to the local horizontal. This simple scheme is easily implemented by the attitude control system of the spacecraft because the vehicle is continuously pitching at the same rate as the orbital mean motion. Furthermore, to prevent any attitude control maneuvers while in shadow, the same constant angle θ_t' is maintained at the exit from the shadow along the arc OVW. At location W, the thrust angle is switched from θ_t' to θ_t'' and held at this constant value from W to U, thereby completing one full revolution. This switch point W is defined by τ_c. The angle τ_1 is always in the first quadrant because the maximum shadow angle $\delta = \tau_2 - \tau_1$ never exceeds 140 deg, which, of course, occurs in LEO. The orbit is in full sunlight if $\tau_2 = \tau_1 = \pi/2$. This problem is best analyzed by making use of the variation-of-parameters

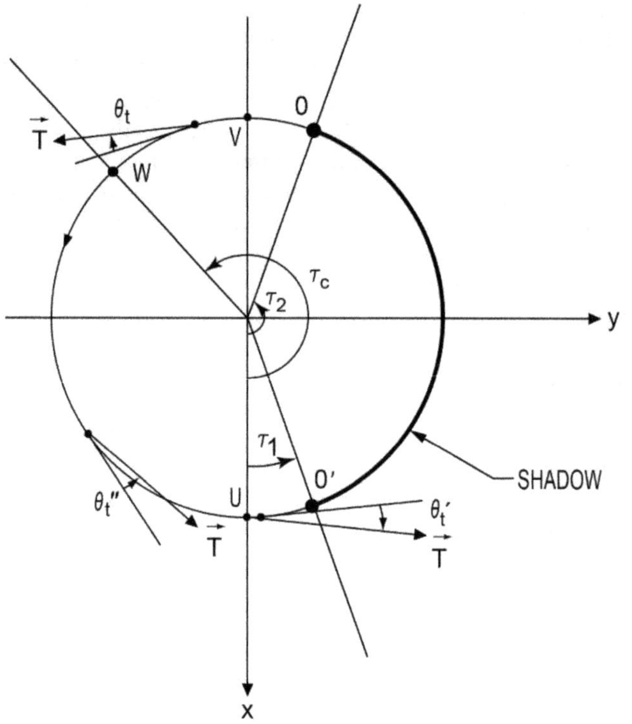

Fig. 2.1 Orbit geometry.

perturbation equations linearized about a reference circular orbit, which we will consider to be the initial circular orbit at time zero. We seek solutions for θ_t', θ_t'', and τ_c such that, after each revolution of Earth, with thrusting only in sunlight, the orbit remains circular with no buildup of eccentricity. Because we are only interested in the coplanar problem, the equations that describe the perturbations of the semimajor axis a and the eccentricity vector $\boldsymbol{e}$ are given by

$$\frac{\mathrm{d}\Delta a}{\mathrm{d}\tau} = \frac{2a_0^3}{\mu} f_\theta \tag{2.1}$$

$$\frac{\mathrm{d}\Delta e_x}{\mathrm{d}\tau} = \frac{a_0^2}{\mu}(2c_\theta f_\theta + s_\theta f_r) \tag{2.2}$$

$$\frac{\mathrm{d}\Delta e_y}{\mathrm{d}\tau} = \frac{a_0^2}{\mu}(2s_\theta f_\theta - c_\theta f_r) \tag{2.3}$$

The rates are with respect to τ. The vector $\boldsymbol{e}$ has been replaced by its components e_x and e_y along the inertial x and y directions such that there are no singularities at $e = 0$ with the use of this formulation. Using $\Delta a = a - a_0$ and the variation-of-parameters equation

$$\dot{a} = \frac{2a_0^2}{\mu} v f_t$$

where $v = na_0$ and $f_t = f_\theta$ for a circular orbit, the rate in Eq. (2.1) is obtained from

$$\frac{\mathrm{d}\Delta a}{\mathrm{d}t} = \frac{2a_0^3}{\mu} n f_\theta$$

Similarly, from $\Delta e_x = e_x - e_{x_0} = ec_\omega - e_0 c_{\omega_0}$ and $\mathrm{d}\Delta e_x/\mathrm{d}t = \dot{e}c_\omega - es_\omega\dot{\omega}$, letting $\omega = 0$ and using the variation-of-parameters equation for the eccentricity [10], we obtain

$$\dot{e} = \frac{r}{h}\left[s_{\theta^*}(1 + ec_{\theta^*})f_r + (e + 2c_{\theta^*} + ec_{\theta*}^2)f_\theta\right]$$

where $r = a_0$, $h = (\mu a_0)^{1/2}$, and $\theta^* = \theta - \omega$ with $\omega = 0$. Finally, for $e = 0$, the rate in Eq. (2.2) is obtained. Now, using $\Delta e_y = e_y - e_{y_0}$ and $\mathrm{d}\Delta e_y/\mathrm{d}t = \dot{e}s_\omega + ec_\omega\dot{\omega}$, where [10]

$$e\dot{\omega} = \frac{r}{h}\left[-c_{\theta^*}(1 + ec_{\theta^*})f_r + s_{\theta^*}(2 + ec_{\theta^*})f_\theta\right]$$

and making the same substitutions as for Δe_x, Eq. (2.3) is obtained. If k represents the magnitude of the constant acceleration imparted to the spacecraft and m is updated if needed from orbit to orbit, then

$$f_r = ks_{\theta_t} \tag{2.4}$$

$$f_\theta = kc_{\theta_t} \tag{2.5}$$

Then, because the orbit is to remain circular at the end of each revolution and near-circular between these points, $\theta \cong nt = \tau$ is used in Eqs. (2.1–2.3), yielding

$$\frac{\mathrm{d}\Delta a}{\mathrm{d}\tau} = \frac{2}{n^2} kc_{\theta_t} \tag{2.6}$$

$$\frac{\mathrm{d}\Delta e_x}{\mathrm{d}\tau} = \frac{k}{a_0 n^2}(2c_{\theta_t}c_\tau + s_{\theta_t}s_\tau) \tag{2.7}$$

$$\frac{\mathrm{d}\Delta e_y}{\mathrm{d}\tau} = \frac{k}{a_0 n^2}(2c_{\theta_t}s_\tau - s_{\theta_t}c_\tau) \tag{2.8}$$

Let us examine first the case of pure tangential thrust where $\theta_t = 0$. In this case, the preceding system of equations is integrated between 0 and τ_1 and between τ_2 and 2π, and after the two segments are added, the changes in a, e_x, and

e_y are obtained as

$$\Delta a = \frac{2k}{n^2}(\tau_1 - \tau_2 + 2\pi)$$
$$\Delta e_x = \frac{2k}{a_0 n^2}(s_{\tau_1} - s_{\tau_2})$$
$$\Delta e_y = -\frac{2k}{a_0 n^2}(c_{\tau_1} - c_{\tau_2})$$

These expressions show that Δe_x and Δe_y are equal to zero only for $\tau_1 = \tau_2$, which occurs when $\tau_1 = \pi/2$, that is, in the case of no shadow. It is therefore impossible to thrust tangentially along the velocity vector without building up net eccentricity. In fact, because $\tau_2 = \pi - \tau_1$, the preceding expressions reduce to

$$\Delta a = \frac{2k}{n^2}(\pi + 2\tau_1)$$
$$\Delta e_x = 0$$
$$\Delta e_y = \frac{-4k}{a_0 n^2} c_{\tau_1}$$

with $\Delta a > 0$ and $\Delta e_y \leq 0$, because τ_1 is always such that $0 < \tau_1 \leq \pi/2$. If we integrate the equations of motion with $\theta_t = \theta_t'$ for $0 \leq \tau \leq \tau_1$ and $\tau_2 \leq \tau \leq \tau_c$ and with $\theta_t = \theta_t''$ for $\tau_c \leq \tau \leq 2\pi$, then we obtain

$$\Delta a = \frac{2k}{n^2}[(\tau_1 + \tau_c - \tau_2)c_{\theta_t'} + (2\pi - \tau_c)c_{\theta_t''}] \tag{2.9}$$

$$\Delta e_x = \frac{k}{a_0 n^2}\Big[2(s_{\tau_1} + s_{\tau_c} - s_{\tau_2})c_{\theta_t'} - 2s_{\tau_c}c_{\theta_t''} - (c_{\tau_1} - c_{\tau_2} - 1 + c_{\tau_c})s_{\theta_t'} - (1 - c_{\tau_c})s_{\theta_t''}\Big] \tag{2.10}$$

$$\Delta e_y = -\frac{k}{a_0 n^2}\Big[2(c_{\tau_1} - 1 + c_{\tau_c} - c_{\tau_2})c_{\theta_t'} + 2(1 - c_{\tau_c})c_{\theta_t''} + (s_{\tau_1} + s_{\tau_c} - s_{\tau_2})s_{\theta_t'} - s_{\tau_c}s_{\theta_t''}\Big] \tag{2.11}$$

Let us consider purely tangential deceleration/acceleration programs. If we choose $\theta_t' = \pi$ and $\theta_t'' = 0$, then we have a deceleration/acceleration program such that the preceding expressions reduce to

$$\Delta a = \frac{2k}{n^2}(3\pi - 2\tau_c - 2\tau_1) \tag{2.12}$$

$$\Delta e_x = -\frac{4k}{a_0 n^2} s_{\tau_c} \tag{2.13}$$

$$\Delta e_y = \frac{4k}{a_0 n^2}(c_{\tau_1} + c_{\tau_c} - 1) \tag{2.14}$$

It is seen here that $\Delta e_x = 0$ if $\tau_c = \pi$ or 2π. For $\tau_c = \pi$, $\Delta e_y = 0$ if $c_{\tau_1} - 2 = 0$, which is clearly impossible. For $\tau_c = 2\pi$, Δe_y vanishes only if $c_{\tau_1} = 0$ or $\tau_1 = \pi/2$, indicating the case of no shadow. Therefore, this deceleration/acceleration program cannot yield $\Delta e_x = 0$ and $\Delta e_y = 0$ simultaneously as long as there is a shadow arc along the orbit.

For $\tau_1 = \pi/2$ and $\tau_c = 2\pi$, we have $\Delta a = -4\pi k/n^2 < 0$, which would shrink the orbit. The same discussion will hold true if we consider an acceleration/deceleration program with $\theta_t' = 0$ and $\theta_t'' = \pi$. In this case, Eqs. (2.9–2.11) yield the exact negative expressions of Eqs. (2.12–2.14). In these expressions, $\Delta e_x = 0$ for $\tau_c = \pi$ or 2π. If $\tau_c = \pi$, $\Delta e_y = 0$ requires $2 - c_{\tau_1} = 0$, which is impossible, and if $\tau_c = 2\pi$, the condition $\Delta e_y = 0$ requires $c_{\tau_1} = 0$ or $\tau_1 = \pi/2$, where no shadow arc exists. In this case, $\Delta a = 4\pi k/n^2 > 0$, which is strictly positive, as expected. If one uses the first strategy (deceleration/acceleration) during one orbit and the second strategy (acceleration/deceleration) during the following orbit or vice versa, it is possible to keep $\Delta e_y = 0$ if τ_c is selected such that $c_{\tau_c} = 1 - c_{\tau_1}$. In this case, Δe_x from the first revolution will cancel out Δe_x from the subsequent revolution so that both Δe_x and Δe_y will remain at zero after each set of two orbits. However, this strategy will also yield a net change in the semimajor axis Δa of zero, which is not desirable. Let us consider one final strategy in which $\theta_t = \theta_t'$ for $0 \le \tau \le \tau_1$, $\theta_t = \theta_t''$ for $\tau_2 \le \tau \le \tau_c$, and $\theta_t = \theta_t'''$ for $\tau_c \le \tau \le 2\pi$. Then, the integration of the linearized differential equations will yield

$$\Delta a = \frac{2k}{n^2}\left[\tau_1 c_{\theta_t'} + (\tau_c - \tau_2) c_{\theta_t''} + (2\pi - \tau_c) c_{\theta_t'''}\right] \tag{2.15}$$

$$\begin{aligned}\Delta e_x = &\frac{k}{a_0 n^2}\Big[2 s_{\tau_1} c_{\theta_t'} + (1 - c_{\tau_1}) s_{\theta_t'} + 2(s_{\tau_c} - s_{\tau_2}) c_{\theta_t''} \\ &- (c_{\tau_c} - c_{\tau_2}) s_{\theta_t''} - 2 s_{\tau_c} c_{\theta_t'''} - (1 - c_{\tau_c}) s_{\theta_t'''}\Big]\end{aligned} \tag{2.16}$$

$$\begin{aligned}\Delta e_y = -&\frac{k}{a_0 n^2}\Big[-2(1 - c_{\tau_1}) c_{\theta_t'} + s_{\tau_1} s_{\theta_t'} + 2(c_{\tau_c} - c_{\tau_2}) c_{\theta_t''} \\ &+ (s_{\tau_c} - s_{\tau_2}) s_{\theta_t''} + 2(1 - c_{\tau_c}) c_{\theta_t'''} - s_{\tau_c} s_{\theta_t'''}\Big]\end{aligned} \tag{2.17}$$

If we consider now an acceleration/deceleration/acceleration program with $\theta_t' = 0$, $\theta_t'' = \pi$, and $\theta_t''' = 0$, then these equations reduce to

$$\Delta a = \frac{2k}{n^2}(3\pi - 2\tau_c) \tag{2.18}$$

$$\Delta e_x = \frac{4k}{a_0 n^2}(s_{\tau_1} - s_{\tau_c}) \tag{2.19}$$

$$\Delta e_y = \frac{4k}{a_0 n^2} c_{\tau_c} \tag{2.20}$$

Clearly, both Δe_x and Δe_y cannot be equal to zero regardless of the value of τ_c because $\Delta e_x = 0$ for $\tau_c = \pi - \tau_1 = \tau_2$ and $\Delta e_y = 0$ for $\tau_c = 3\pi/2$. In a similar manner, if we consider a deceleration/acceleration/deceleration program with $\theta_t' = \pi$, $\theta_t'' = 0$, and $\theta_t''' = \pi$, then the variations are opposite in sign to Eqs. (2.18–2.20), and for the same reason, it is impossible to satisfy $\Delta e_x = \Delta e_y = 0$ simultaneously. From these simple examples, it is seen that purely tangential thrusting in the presence of shadow arcs will not satisfy the $e = 0$ constraint. Therefore, we must investigate strategies with $\theta_t \neq 0$ such that we can control the eccentricity buildup through the radial components of the acceleration vectors when shadowing is present. Returning to Eqs. (2.9–2.11), the conditions $\Delta e_x = 0$ and $\Delta e_y = 0$ yield the system of equations given by

$$2Ac_{\theta_t'} - Bs_{\theta_t'} - 2Cc_{\theta_t''} - Ds_{\theta_t''} = 0 \tag{2.21}$$

$$2Bc_{\theta_t'} + As_{\theta_t'} + 2Dc_{\theta_t''} - Cs_{\theta_t''} = 0 \tag{2.22}$$

with $A = s_{\tau_1} - s_{\tau_2} + s_{\tau_c} = s_{\tau_c}$, $B = c_{\tau_1} - c_{\tau_2} - 1 + c_{\tau_c} = 2c_{\tau_1} - 1 + c_{\tau_c}$, $C = s_{\tau_c}$, and $D = 1 - c_{\tau_c}$. For given τ_1 and τ_c, Eqs. (2.21) and (2.22) can be solved to give $s_{\theta_t''}$ and $c_{\theta_t''}$ in terms of $s_{\theta_t'}$ and $c_{\theta_t'}$ as follows:

$$s_{\theta_t''} = \frac{2(AD + BC)c_{\theta_t'} + (AC - BD)s_{\theta_t'}}{(C^2 + D^2)}$$

$$c_{\theta_t''} = \frac{2(AC - BD)c_{\theta_t'} - (BC + AD)s_{\theta_t'}}{2(C^2 + D^2)}$$

However, from the preceding definitions, $B + D = 2c_{\tau_1}$, $A^2 + D^2 = C^2 + D^2 = 2D = 2(1 - c_{\tau_c})$, and $A^2 - BD = 2(1 - c_{\tau_c})(1 - c_{\tau_1})$ such that

$$s_{\theta_t''} = \frac{2s_{\tau_c}c_{\tau_1}c_{\theta_t'} + (1 - c_{\tau_c})(1 - c_{\tau_1})s_{\theta_t'}}{(1 - c_{\tau_c})} \tag{2.23}$$

$$c_{\theta_t''} = \frac{2(1 - c_{\tau_1})(1 - c_{\tau_c})c_{\theta_t'} - s_{\tau_c}c_{\tau_1}s_{\theta_t'}}{2(1 - c_{\tau_c})} \tag{2.24}$$

These expressions will be useful when we evaluate θ_t'' once θ_t' is known. Let us now rewrite Eqs. (2.9–2.11) as

$$\Delta a = \frac{2k}{n^2}\left[(\tau_c + 2\tau_1 - \pi)c_{\theta_t'} + (2\pi - \tau_c)c_{\theta_t''}\right] \tag{2.25}$$

$$\Delta e_x = \frac{k}{a_0 n^2}\left[2s_{\tau_c}c_{\theta_t'} - (2c_{\tau_1} - 1 + c_{\tau_c})s_{\theta_t'} - 2s_{\tau_c}c_{\theta_t''} - (1 - c_{\tau_c})s_{\theta_t''}\right] \tag{2.26}$$

$$\Delta e_y = \frac{-k}{a_0 n^2}\left[2(2c_{\tau_1} - 1 + c_{\tau_c})c_{\theta_t'} + s_{\tau_c}s_{\theta_t'} + 2(1 - c_{\tau_c})c_{\theta_t''} - s_{\tau_c}s_{\theta_t''}\right] \tag{2.27}$$

The last two expressions can be written in terms of $s_{\theta'_t}$ and $s_{\theta''_t}$, and because we require $\Delta e_x = \Delta e_y = 0$, they reduce to

$$2s_{\tau_c}(\pm)\left(1 - s^2_{\theta'_t}\right)^{1/2} - (2c_{\tau_1} - 1 + c_{\tau_c})s_{\theta'_t} - 2s_{\tau_c}(\pm) \times \left(1 - s^2_{\theta''_t}\right)^{1/2} - (1 - c_{\tau_c})s_{\theta''_t} = 0 \tag{2.28}$$

$$2(2c_{\tau_1} - 1 + c_{\tau_c})(\pm)\left(1 - s^2_{\theta'_t}\right)^{1/2} + s_{\tau_c}s_{\theta'_t} + 2(1 - c_{\tau_c})(\pm)\left(1 - s^2_{\theta''_t}\right)^{1/2} - s_{\tau_c}s_{\theta''_t} = 0 \tag{2.29}$$

From Eqs. (2.28) and (2.29), we can write

$$s_{\theta''_t} = \frac{2s_{\tau_c}c_{\tau_1}(\pm)\left(1 - s^2_{\theta'_t}\right)^{1/2} + (1 - c_{\tau_c})(1 - c_{\tau_1})s_{\theta'_t}}{(1 - c_{\tau_c})} \tag{2.30}$$

If we square this expression for $s_{\theta''_t}$, substitute it into Eq. (2.28), and regroup terms, we obtain

$$\pm\left(1 - s^2_{\theta'_t}\right)^{1/2} K_1 - s_{\theta'_t}K_2 = \pm 2s_{\tau_c}\left(1 - s^2_{\theta''_t}\right)^{1/2} \tag{2.31}$$

where the right-hand side is written in terms of Eq. (2.30) and where $K_1 = 2s_{\tau_c}(1 - c_{\tau_1})$ and $K_2 = c_{\tau_1}(1 + c_{\tau_c})$. Squaring Eq. (2.31) once more yields

$$L_1 + s^2_{\theta'_t}L_2 = \pm s_{\theta'_t}\left(1 - s^2_{\theta'_t}\right)^{1/2} L_3 \tag{2.32}$$

with

$$L_1 = 4s^2_{\tau_c}c_{\tau_1}\left[(c_{\tau_1} - 2) + \frac{4s^2_{\tau_c}c_{\tau_1}}{(1 - c_{\tau_c})^2}\right] \tag{2.33}$$

$$L_2 = -15c^2_{\tau_1}(1 + c_{\tau_c})^2 \tag{2.34}$$

$$L_3 = \frac{-12s^3_{\tau_1}c_{\tau_1}(1 - c_{\tau_1})}{(1 - c_{\tau_c})} \tag{2.35}$$

The term L_1 can also be written as

$$L_1 = 4(1 + c_{\tau_c})c_{\tau_1}[c_{\tau_c}(3c_{\tau_1} + 2) + (5c_{\tau_1} - 2)] \tag{2.36}$$

We square Eq. (2.32) once again to obtain the following expression in $s_{\theta'_t}$:

$$s^4_{\theta'_t}(L_2^2 + L_3^2) + (2L_1L_2 - L_3^2)s^2_{\theta'_t} + L_1^2 = 0$$

Because this expression is a quadratic in $s^2_{\theta'_t}$, its solutions are given by

$$s^2_{\theta'_t} = \frac{-2\left(L_1L_2 - L_3^2\right) \pm \left[\left(2L_1L_2 - L_3^2\right)^2 - 4\left(L_2^2 + L_3^2\right)L_1^2\right]^{1/2}}{2\left(L_2^2 + L_3^2\right)} = (S_1, S_2) \tag{2.37}$$

where S_1 corresponds to the plus sign and S_2 to the minus sign in Eq. (2.37). Therefore, the angle θ'_t is such that

$$s_{\theta'_t} = \pm(S_1)^{1/2}, \quad \pm(S_2)^{1/2} \tag{2.38}$$

providing no fewer than eight solutions because of the double squaring operations just carried out. Four of these solutions can be eliminated if we require that $c_{\theta'_t} > 0$ or $-\pi/2 \le \theta'_t \le \pi/2$, so that the $c_{\theta'_t}$ term in the expression for Δa in Eq. (2.25) contributes to a positive Δa for orbit raising. From Eq. (2.25), $(\tau_1 - \tau_2 + \tau_c) > 0$ and $(\tau_c - 2\pi) < 0$ so that $\Delta a > 0$ would require $c_{\theta'_t}(\tau_1 + \tau_c - \tau_2) > -c_{\theta''_t}(2\pi - \tau_c)$. If $c_{\theta'_t} < 0$, then we must have $c_{\theta''_t} > 0$, and to have $\Delta a > 0$, we must additionally satisfy $\left|c_{\theta''_t}(2\pi - \tau_c)\right| > \left|c_{\theta'_t}(\tau_1 + \tau_c - \tau_2)\right|$, which requires, in turn, that

$$\tau_c < \frac{2\pi\left|c_{\theta''_t}\right| + (\tau_2 - \tau_1)\left|c_{\theta'_t}\right|}{\left|c_{\theta'_t}\right| + \left|c_{\theta''_t}\right|}$$

If $c_{\theta'_t} > 0$, then $c_{\theta''_t} > 0$ always satisfies the inequality $c_{\theta'_t}(\tau_1 + \tau_c - \tau_2) > -c_{\theta''_t}(2\pi - \tau_c)$, and $\Delta a > 0$ is always satisfied. However, if $c_{\theta''_t} < 0$, then $\Delta a > 0$ requires the satisfaction of $\left|c_{\theta'_t}(\tau_1 + \tau_c - \tau_2)\right| > \left|c_{\theta''_t}(2\pi - \tau_c)\right|$, which is possible for

$$\tau_c > \frac{2\pi\left|c_{\theta''_t}\right| + (\tau_2 - \tau_1)\left|c_{\theta'_t}\right|}{\left|c_{\theta'_t}\right| + \left|c_{\theta''_t}\right|}$$

Let us consider only accelerating solutions of the type with both $c_{\theta'_t} > 0$ and $c_{\theta''_t} > 0$. Equation (2.34) for L_2 is such that $L_2 < 0$ is always satisfied for any values of τ_1 and τ_c. The expression for L_3 in Eq. (2.35) is such that

$$L_3 \le 0 \quad \text{if} \quad \tau_2 < \tau_c \le \pi \quad \text{or} \quad s_{\tau_c} \ge 0$$

$$L_3 \ge 0 \quad \text{if} \quad \pi \le \tau_c \le 2\pi \quad \text{or} \quad s_{\tau_c} \le 0$$

From Eq. (2.36), it is seen that $L_1 > 0$ if $c_{\tau_c} > (2 - 5c_{\tau_1})/(2 + 3c_{\tau_1})$ because $(1 + c_{\tau_c})$ and c_{τ_1} are always positive. Furthermore, $\tau_c \ge \tau_2$ must hold, such that

$L_1 < 0$ for $\tau_2 \le \tau_c \le \tau'_c$, where τ'_c is obtained from

$$c_{\tau'_c} = \frac{(2 - 5c_{\tau_1})}{(2 + 3c_{\tau_1})} \tag{2.39}$$

At $\tau_c = \tau'_c$, $L_1 = 0$, and for $\tau'_c < \tau \le 2\pi$, $L_1 > 0$. From Eq. (2.37), it is clear that $(2L_1L_2 - L_3^2)$ must be negative because $4(L_2^2 + L_3^2)L_1^2 \ge 0$ and $L_2^2 + L_3^2 \ge 0$; otherwise, $s^2_{\theta'_t}$ would be negative, which is impossible. Therefore, we can write the condition

$$M_1 = 2L_1L_2 - L_3^2 < 0 \tag{2.40}$$

Upon substitution from Eqs. (2.34–2.36), this inequality reduces to

$$M_1 = -24(1 + c_{\tau_c})^3 c^2_{\tau_1}\left[c_{\tau_c}\left(9c^2_{\tau_1} + 22c_{\tau_1} - 6\right) + \left(31c^2_{\tau_1} - 22c_{\tau_1} + 6\right)\right] < 0$$

which, in view of the minus sign and the fact that $(1 + c_{\tau_c})^3 c^2_{\tau_1} > 0$ is always satisfied, translates to the condition

$$c_{\tau_c}\left(9c^2_{\tau_1} + 22c_{\tau_1} - 6\right) + \left(31c^2_{\tau_1} - 22c_{\tau_1} + 6\right) > 0 \tag{2.41}$$

For a given value of τ_1, this inequality is satisfied if τ_c is such that

$$c_{\tau_c} > \frac{\left(-31c^2_{\tau_1} + 22c_{\tau_1} - 6\right)}{\left(9c^2_{\tau_1} + 22c_{\tau_1} - 6\right)} \tag{2.42}$$

The quadratic functions appearing in Eq. (2.41) are plotted in Fig. 2.2. At $\tau_1 = 70.985$ deg, they intersect, and Eq. (2.42) becomes $c_{\tau_c} > -1$, which is satisfied for every τ_c. The values of τ_c that are the switch points satisfying Eq. (2.41), or $M_1 < 0$, are shown in Figs. 2.3 and 2.4 for various values of τ_1. For $\tau_1 = 0$ deg, $M_1 < 0$ is satisfied for 233.13 deg $< \tau_c \le 2\pi$. This case of maximum shadow angle $\tau_2 - \tau_1 = \pi$ is for an orbit with zero altitude. For Earth orbiters in LEO, τ_1 is always larger than 20 deg, and the locus of τ_c that satisfies $M_1 < 0$ is given by 215.27 deg $< \tau_c \le 2\pi$. The switch points are, of course, obtained from

$$c_{\tau_c} = \frac{-31c^2_{\tau_1} + 22c_{\tau_1} - 6}{9c^2_{\tau_1} + 22c_{\tau_1} - 6} \tag{2.43}$$

with the additional condition that $\tau_c \ge \tau_2$. Figure 2.4 shows that, as we approach $\tau_1 = 70.98$ deg, there are two regions along the orbit for which $M_1 < 0$ is satisfied; these two regions eventually merge such that, for $\tau_1 > 70.98$ deg, all values of $\tau_c \ge \tau_2$ satisfy $M_1 < 0$. Another condition that the switch point τ_c must satisfy is given by

$$\left(2L_1L_2 - L_3^2\right)^2 - 4\left(L_2^2 + L_3^2\right)L_1^2 \ge 0 \tag{2.44}$$

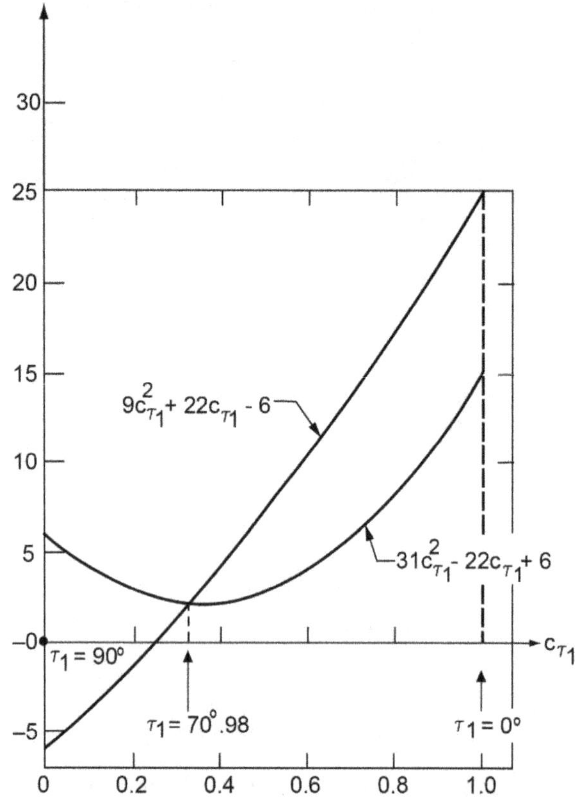

Fig. 2.2 Switch point conditions for the full range of shadow entry angles.

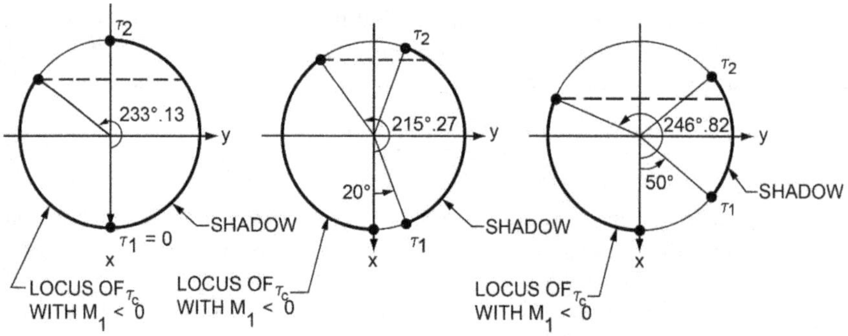

Fig. 2.3 Feasible regions for τ_c satisfying the condition $M_1 < 0$ for three values of $\tau_1 = 0$, 20, and 50 deg.

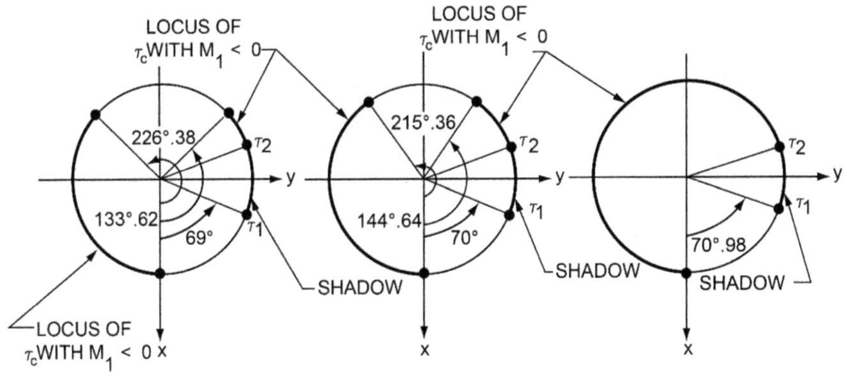

Fig. 2.4 Feasible regions for τ_c satisfying the condition $M_1 < 0$ for larger values of $\tau_1 = 69$, 70, and 70.98 deg.

The preceding equation is the discriminant in Eq. (2.37), and its satisfaction ensures that the roots $s^2_{\theta_t'}$ do indeed exist. This condition is equivalent to

$$N_1 = L_3^2 - 4L_1(L_1 + L_2) \geq 0 \tag{2.45}$$

Moreover, from Eqs. (2.34–2.36)

$$L_1 + L_2 = c_{\tau_1}(1 + c_{\tau_c})[c_{\tau_1}(5 - 3c_{\tau_c}) - 8(1 - c_{\tau_c})]$$
$$L_3^2 - 4L_1(L_1 + L_2) \geq 0$$

so the condition in Eq. (2.45) is reduced, after some algebraic manipulation, to the following quadratic form in c_{τ_c}

$$N_2 = -25c_{\tau_c}^2 + 32(1 - c_{\tau_1})c_{\tau_c} - [7 + 16c_{\tau_1}(c_{\tau_1} - 2)] \geq 0 \tag{2.46}$$

whose roots are given by

$$c_{\tau_c} = \frac{16(1 - c_{\tau_1}) \pm \left(-144c_{\tau_1}^2 + 288c_{\tau_1} + 81\right)^{1/2}}{25} \tag{2.47}$$

Here, the discriminant is positive semidefinite for $c_{\tau_1} \geq -0.250$, so that, for $0 \leq \tau_1 \leq 90$, it always holds, and therefore, c_{τ_c} always exists for these values of τ_1. In fact, the condition in Eq. (2.46) is always satisfied by τ_c values for which c_{τ_c} is between the two roots of Eq. (2.47). This condition is shown in Fig. 2.5 for three values of τ_1, namely, $\tau_1 = 20$, 50, and 90 deg. As τ_1 varies, the τ_c arc for which $N_2 > 0$ maintains a constant length of 73.74 deg while shifting toward the x axis, which is reached for $\tau_1 = 90$ deg. Because both conditions $M_1 < 0$ and $N_2 > 0$ must be satisfied by τ_c for any given τ_1, the intersection of

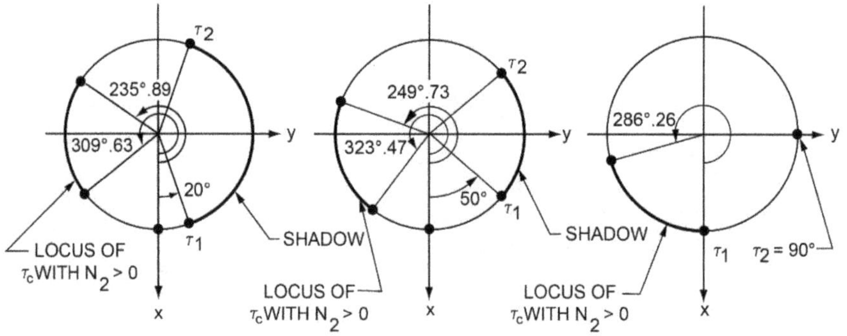

Fig. 2.5 Feasible regions for τ_c satisfying the condition $N_2 > 0$ for $\tau_1 = 20$, 50, and 90 deg.

the two regions defines the locus of feasible τ_c values. It is found that the $N_2 > 0$ region is always contained within the larger $M_1 < 0$ region, with the lower boundaries being as close as 0.1 deg to one another for $\tau_1 = 24$ deg. Therefore, the satisfaction of $N_2 > 0$ guarantees the satisfaction of $M_1 < 0$ for all values of τ_1 in the range of interest $0 \le \tau_1 \le 90$ deg. Now, the same manipulations as led to the generation of Eq. (2.32) can be repeated to obtain an expression in terms of $c_{\theta_t'}$ instead of $s_{\theta_t'}$. Without going through the details of the derivations, we can obtain an expression for $s_{\theta_t''}$ from Eqs. (2.26) and (2.27) identical to Eq. (2.30), except that $s_{\theta_t''}$ is now expressed in terms of $c_{\theta_t'}$ instead of $s_{\theta_t'}$ as in that case. Squaring the resulting expression yields

$$L_0 - L_2 c^2_{\theta_t'} = \pm c_{\theta_t'}\left(1 - c^2_{\theta_t'}\right)^{1/2} L_3 \tag{2.48}$$

with L_0 given by

$$L_0 = 4s^2_{\tau_c} c_{\tau_1}(c_{\tau_1} - 2) + c^2_{\tau_1}(1 + c_{\tau_c})^2 \tag{2.49}$$

and with L_2 and L_3 as before in Eqs. (2.34) and (2.35), respectively. Equation (2.46) then reduces, after squaring and substitution, to the form

$$c^4_{\theta_t'}(L_2^2 + L_3^2) + (-2L_0L_2 - L_3^2)c^2_{\theta_t'} + L_0^2 = 0$$

whose solutions are given by

$$c^2_{\theta_t'} = \frac{-(-2L_0L_2 - L_3^2) \pm \left[(-2L_0L_2 - L_3^2)^2 - 4(L_2^2 + L_3^2)L_0^2\right]^{1/2}}{2(L_2^2 + L_3^2)} \tag{2.50}$$

Letting $M_2 = L_2^2 + L_3^2$ and $M_3 = -2L_0L_2 - L_3^2$, we obtain

$$c_{\theta_t'}^2 = \frac{-M_3 + \left(M_3^2 - 4M_2L_0^2\right)^{1/2}}{2M_2} = S_3$$

$$c_{\theta_t'}^2 = \frac{-M_3 - \left(M_3^2 - 4M_2L_0^2\right)^{1/2}}{2M_2} = S_4$$

so that the solutions ultimately reduce to

$$c_{\theta_t'} = \pm(S_3)^{1/2} \tag{2.51}$$

$$c_{\theta_t'} = \pm(S_4)^{1/2} \tag{2.52}$$

It is easily verified that $L_0 = L_1 + L_2$, where L_1 is as given in Eq. (2.36). We could have also obtained $c_{\theta_t'}^2$ directly from $c_{\theta_t'}^2 = 1 - s_{\theta_t'}^2$ such that

$$c_{\theta_t'}^2 = \frac{\left(2L_1L_2 + 2L_2^2 + L_3^2\right) \mp \left[\left(2L_1L_2 - L_3^2\right)^2 - 4\left(L_2^2 + L_3^2\right)L_1^2\right]^{1/2}}{2\left(L_2^2 + L_3^2\right)}$$

We obtain the following four solutions for the thrust angle θ_t' by observing that from Eqs. (2.38), (2.51), and (2.52) the S_1 solution for $s_{\theta_t'}$ corresponds to the S_4 solution for $c_{\theta_t'}$, and similarly for S_2 and S_3, such that the four solutions that satisfy $\Delta e = 0$ are given by

$$\theta_t' = \tan^{-1}\left[\frac{(S_1)^{1/2}}{-(S_4)^{1/2}}\right] \tag{2.53}$$

$$\theta_t' = \tan^{-1}\left[\frac{-(S_1)^{1/2}}{(S_4)^{1/2}}\right] \tag{2.54}$$

$$\theta_t' = \tan^{-1}\left[\frac{(S_2)^{1/2}}{(S_3)^{1/2}}\right] \tag{2.55}$$

$$\theta_t' = \tan^{-1}\left[\frac{-(S_2)^{1/2}}{-(S_3)^{1/2}}\right] \tag{2.56}$$

We could also have obtained an expression for $c_{\theta_t''}$ in terms of $c_{\theta_t'}$ from $\Delta e_x = \Delta e_y = 0$ in. Eqs. (2.26) and (2.27) , which would have led to

$$L_1' + c_{\theta_t'}^2 L_2' = \pm c_{\theta_t'}\left(1 - c_{\theta_t'}^2\right)^{1/2} L_3'$$

This equation, when squared, yields the quadratic form in $c_{\theta_t'}^2$

$$c_{\theta_t'}^4\left(L_2'^2 + L_3'^2\right) + \left(2L_1'L_2' - L_3'^2\right)c_{\theta_t'}^2 + L_1'^2 = 0$$

whose solutions are given by

$$c^2_{\theta'_t} = \frac{-M'_1 \pm \left(M'^2_1 - 4M'_2 L'^2_1\right)^{1/2}}{2M'_2} \tag{2.57}$$

where

$$M'_1 = 2L'_1 L'_2 - L'^2_3$$
$$M'_2 = L'^2_2 + L'^2_3$$
$$L'_1 = \frac{(1 + c_{\mathcal{T}_c})c_{\mathcal{T}_1}}{4}[5c_{\mathcal{T}_1} - 3c_{\mathcal{T}_1}c_{\mathcal{T}_c} - 8(1 - c_{\mathcal{T}_c})]$$
$$L'_2 = \frac{15}{4}c^2_{\mathcal{T}_1}(1 + c_{\mathcal{T}_c})^2 = \frac{-L_2}{4}$$
$$L'_3 = 3s_{\mathcal{T}_c}c_{\mathcal{T}_1}(1 + c_{\mathcal{T}_c})(1 - c_{\mathcal{T}_1}) = \frac{-L_3}{4}$$

The solutions are given by

$$\left.\begin{aligned} c_{\theta'_t} &= \pm(S'_3)^{1/2} \\ c_{\theta'_t} &= \pm(S'_4)^{1/2} \end{aligned}\right\} \tag{2.58}$$

where S'_3 corresponds to the plus sign in Eq. (2.57) and S'_4 corresponds to the minus sign. We can also write the solutions for θ'_t in terms S_1, S_2, S'_3, and S'_4 as

$$\theta'_t = \tan^{-1}\left[\frac{(S_1)^{1/2}}{-(S'_4)^{1/2}}\right] \tag{2.59}$$

$$\theta'_t = \tan^{-1}\left[\frac{-(S_1)^{1/2}}{(S'_4)^{1/2}}\right] \tag{2.60}$$

$$\theta'_t = \tan^{-1}\left[\frac{(S_2)^{1/2}}{(S'_3)^{1/2}}\right] \tag{2.61}$$

$$\theta'_t = \tan^{-1}\left[\frac{-(S_2)^{1/2}}{-(S'_3)^{1/2}}\right] \tag{2.62}$$

Once the four solutions are found for θ'_t, the solutions for θ''_t are obtained from Eqs. (2.23) and (2.24), which express $s_{\theta''_t}$ and $c_{\theta''_t}$, respectively, in terms of θ'_t. Next, Δa in kilometers and Δe_x and Δe_y are also evaluated from Eqs. (2.25–2.27). It is thus found that the preceding four solutions are symmetrical in pairs, meaning that two solutions provide positive Δa and the other two provide mirror values of Δa that are negative. The positive solutions are to be used for orbit raising, whereas the negative solutions will be used for the return leg to LEO. The obvious choice is, of course, the solution that provides the largest value of Δa, whether positive or negative. The solution given in Eq. (2.61) provides the largest $\Delta a > 0$, and the solution given in Eq. (2.62) provides the largest $\Delta a < 0$. Let $a_0 = 40{,}000$ km, $\mu = 398{,}601.3\ \text{km}^3/\text{s}^2$, $k = 3.5 \times 10^{-7}$ km/s^2, and $\tau_1 = 80$ deg. Starting from $\tau_c = \tau_2 = \pi - \tau_1 = 100$ deg, Δa, Δe_x, and Δe_y are evaluated from Eqs. (2.25–2.27) after computing the four solutions θ_t' that satisfy $\Delta e_x = \Delta e_y = 0$ from Eqs. (2.59–2.62) and evaluating the corresponding angles θ_t'' from Eqs. (2.23) and (2.24). For each value of τ_c, the discriminants in Eqs. (2.44), (2.50), and (2.57) are evaluated to determine whether they are positive or negative. If they are negative, then no solution is possible, and a larger value of τ_c is selected until the discriminants are positive. Once this condition is satisfied, the four θ_t' angles are computed, followed by the four corresponding θ_t'' values.

Initially, only the first two solutions satisfy $\Delta e_x = \Delta e_y = 0$, and as τ_c is increased further, all four solutions become acceptable, whereas the last τ_c value used in this example will result in the satisfaction of only the last two solutions. Clearly, there exists a value of τ_c for which Δa reaches a maximum, which, in this case, is about 640 km. It is very easy to automate this algorithm and step τ_c one degree at a time, or less if so desired, and select the solution that provides the largest value of Δa using a search-and-sort routine. Incidentally, only those values of τ_c in Fig. 2.5 that satisfy the condition $N_2 > 0$ produce any feasible solutions, as expected.

2.3 MODIFIED STRATEGY FOR ECCENTRICITY CONTROL

In the current analysis, the x axis is along the shadow exit point, as indicated by point O in Fig. 2.6. The thrust $\boldsymbol{T} = \boldsymbol{f}$ is again applied at a constant pitch angle θ_t' with respect to the local horizon from shadow exit to point W defined by the angle τ_c. At this point, the pitch angle is changed to the constant value of θ_t'' until shadow entry, defined by point O', where thrust is interrupted until shadow exit.

Unlike the strategy in Sec. 2.2, this strategy requires that the vehicle perform an attitude-adjust maneuver in shadow so that the pitch angle θ_t' is recovered at shadow exit for another revolution of intermittent low thrust around Earth. In this analysis, the shadow entry point O' is defined by the angular position of τ_f, and the shadow arc length is defined by the angle $\delta = 2\alpha$ such that $\tau_f = 2\pi - \delta$. Returning to the linearized perturbation equations (2.6–2.8) and integrating between 0 and τ_c with a pitch angle of θ_t' and then from τ_c to τ_f with a pitch angle of θ_t'', we obtain

$$\Delta a = \frac{2k}{n^2}\left[c_{\theta_t'}\tau_c + c_{\theta_t''}(\tau_f - \tau_c)\right] \tag{2.63}$$

$$\Delta e_x = \frac{k}{a_0 n^2}\left[2c_{\theta_t'}s_{\tau_c} + s_{\theta_t'}(1 - c_{\tau_c}) + 2c_{\theta_t''}(s_{\tau_f} - s_{\tau_c}) - s_{\theta_t''}(c_{\tau_f} - c_{\tau_c})\right] \tag{2.64}$$

$$\Delta e_y = \frac{-k}{a_0 n^2}\left[-2c_{\theta_t'}(1 - c_{\tau_c}) + s_{\theta_t'}s_{\tau_c} + 2c_{\theta_t''}(c_{\tau_f} - c_{\tau_c}) + s_{\theta_t''}(s_{\tau_f} - s_{\tau_c})\right] \tag{2.65}$$

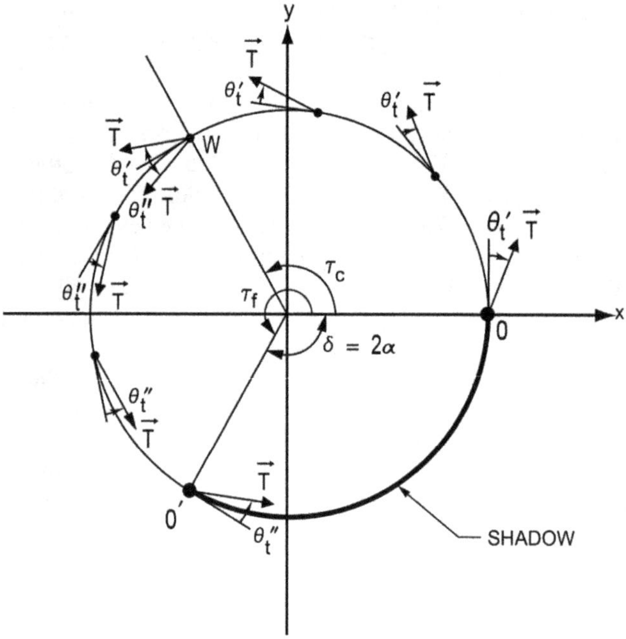

Fig. 2.6 Transfer geometry with pitch reorientation maneuver in shadow.

Once again, we require $\Delta e_x = \Delta e_y = 0$ such that expressions for $s_{\theta_t''}$ and $c_{\theta_t''}$ in terms of $s_{\theta_t'}$ and $c_{\theta_t'}$ can be obtained from Eqs. (2.64) and (2.65)

$$s_{\theta_t''} = \frac{2(A'D' + B'C')c_{\theta_t'} + (B'D' - A'C')s_{\theta_t'}}{\left(C'^2 + D'^2\right)} \tag{2.66}$$

$$c_{\theta_t''} = \frac{2(B'D' - A'C')c_{\theta_t'} - (B'C' + A'D')s_{\theta_t'}}{2\left(C'^2 + D'^2\right)} \tag{2.67}$$

where

$$A' = s_{\tau_c}$$

$$B' = 1 - c_{\tau_c}$$

$$C' = s_{\tau_f} - s_{\tau_c}$$

$$D' = c_{\tau_f} - c_{\tau_c}$$

and where

$$A'D' + B'C' = s_{T_f}(1 - c_{T_c}) - s_{T_c}(1 - c_{T_f})$$
$$\left(C'^2 + D'^2\right) = 2\left[1 - s_{T_c}s_{T_f} - c_{T_c}c_{T_f}\right]$$
$$B'D' - A'C' = 1 + c_{T_f} - c_{T_c} - s_{T_f}s_{T_c} - c_{T_f}c_{T_c}$$

To solve for the pitch angle θ_t', Eqs. (2.64) and (2.65) are written in terms of the sine functions only:

$$\pm 2\left(1 - s_{\theta_t'}^2\right)^{1/2} A' + s_{\theta_t'}B' \pm 2\left(1 - s_{\theta_t''}^2\right)^{1/2} C' - s_{\theta_t''}D' = 0 \tag{2.68}$$

$$-(\pm)2\left(1 - s_{\theta_t'}^2\right)^{1/2} B' + s_{\theta_t'}A' \pm 2\left(1 - s_{\theta_t''}^2\right)^{1/2} D' + s_{\theta_t''}C' = 0 \tag{2.69}$$

Let us solve for $s_{\theta_t''}$ in terms of $s_{\theta_t'}$ from these two expressions. Then

$$s_{\theta_t''} = \frac{\pm 2(A'D' + B'C')\left(1 - s_{\theta_t'}^2\right)^{1/2} - (A'C' - B'D')s_{\theta_t'}}{\left(C'^2 + D'^2\right)}$$

If we substitute this expression back into Eq. (2.68) and square the resulting expression, we obtain

$$L_1'' + s_{\theta_t'}^2 L_2'' = -s_{\theta_t'}(\pm)\left(1 - s_{\theta_t'}^2\right)^{1/2} L_3''$$

which is squared one more time to yield the quadratic form in $s_{\theta_t'}^2$

$$s_{\theta_t'^4}\left(L_2''^2 + L_3''^2\right) + \left(2L_1''L_2'' - L_3''^2\right)s_{\theta_t'}^2 + L_1''^2 = 0 \tag{2.70}$$

with

$$L_1'' = K_1'^2 - 4C'^2 + \frac{16C'^2(A'D' + B'C')^2}{\left(C'^2 + D'^2\right)^2}$$

$$L_2'' = K_2'^2 - K_1'^2 - \frac{4C'^2}{\left(C'^2 + D'^2\right)^2}\left[4(A'D' + B'C')^2 - (A'C' - B'D')^2\right]$$

$$L_3'' = 2K_1'K_2' - \frac{16C'^2}{\left(C'^2 + D'^2\right)^2}(A'D' + B'C')(A'C' - B'D')$$

$$K_1' = 2\left[A' - \frac{D'(A'D' + B'C')}{\left(C'^2 + D'^2\right)}\right]$$

$$K'_2 = B' + \frac{D'(A'C' - B'D')}{\left(C'^2 + D'^2\right)}$$

The solutions of Eq. (2.70) are given by

$$s^2_{\theta'_t} = \frac{-M''_1 \pm \left(M''^2_1 - 4M''_2 L''^2_1\right)^{1/2}}{2M''_2} = (S_1, S_2) \tag{2.71}$$

such that

$$s_{\theta'_t} = \pm(S_1)^{1/2} \tag{2.72}$$

$$s_{\theta'_t} = \pm(S_2)^{1/2} \tag{2.73}$$

where S_1 corresponds to the plus sign in front of the radical and S_2 corresponds to the minus sign. Finally, from $c^2_{\theta'_t} = 1 - s^2_{\theta'_t}$, we obtain

$$c^2_{\theta'_t} = \frac{(2M''_2 + M''_1) \mp \left(M''^2_1 - 4M''_2 L''^2_1\right)^{1/2}}{2M''_2} = (S_4, S_3) \tag{2.74}$$

$$c_{\theta'_t} = \pm(S_4)^{1/2} \tag{2.75}$$

$$c_{\theta'_t} = \pm(S_3)^{1/2} \tag{2.76}$$

The solution S_4 corresponds to the minus sign in Eq. (2.74), and S_3 corresponds to the plus sign. Starting from $\tau_c = 0$ deg, the radical in Eqs. (2.71) and (2.74) is evaluated to determine whether it is positive or negative. If it is negative, no solution is possible, and so τ_c is increased and the test is carried out again until radical satisfies the positive semidefinite property. Then, the four solutions of interest are obtained from

$$\theta'_t = \tan^{-1}\left[\frac{(S_1)^{1/2}}{-(S_4)^{1/2}}\right] \tag{2.77}$$

$$\theta'_t = \tan^{-1}\left[\frac{-(S_1)^{1/2}}{(S_4)^{1/2}}\right] \tag{2.78}$$

$$\theta'_t = \tan^{-1}\left[\frac{(S_2)^{1/2}}{(S_3)^{1/2}}\right] \tag{2.79}$$

$$\theta'_t = \tan^{-1}\left[\frac{-(S_2)^{1/2}}{-(S_3)^{1/2}}\right] \tag{2.80}$$

after which the corresponding solutions for θ_t'' are obtained from Eqs. (2.66) and (2.67). Then, Δa is evaluated from Eq. (2.63), and both Δe_x and Δe_y are computed from Eqs. (2.64) and (2.65), respectively, to ensure that they both vanish. The solutions in Eqs. (2.77) and (2.80) provide negative strategies for Δa, whereas their mirror solutions obtained from Eqs. (2.78) and (2.79), respectively, provide equal but positive Δa values. The largest Δa values are the ones of interest, and they are found to correspond to Eq. (2.79) for $\Delta a > 0$ and to Eq. (2.80) for $\Delta a < 0$. As τ_c is increased, the largest change in semimajor axis a is reached when $\tau_c = \tau_f/2$, meaning that the optimal switch point is located at the midpoint of the sunlit arc OO'. Let $\tau_f = 280$ deg, $a_0 = 10{,}000$ km, $\mu = 398{,}601.3$ km^3/s^2, and $k = 3.5 \times 10^{-7}$ km/s^2. Starting from $\tau_c = 10$ deg and increasing τ_c in 10-deg increments thereafter, the radical in Eq. (2.71) is found to be negative until $\tau_c = 80$ deg. Next, the four solutions for θ_t' in Eqs. (2.77–2.80) are evaluated, followed by the four corresponding angles θ_t''. Finally, the values of Δa, Δe_x, and Δe_y corresponding to each of the preceding solutions are computed. As was the case for the analysis in the previous section (strategy 1), not all four solutions pass the $\Delta e_x = \Delta e_y = 0$ condition initially. After $\tau_c = 170$ deg, only the last two solutions are valid. At $\tau_c = 210$ deg, the radical fails the positive semidefinite test, and at $\tau_c = 260$ deg, it is exactly equal to zero, but all four solutions are invalid because Δe_x and Δe_y are not driven to zero (in practice, 10^{-19} or less). The reason for this is that we have not enforced either the $M_1'' < 0$ condition as in Sec. 2.2 or the more relevant condition $s^2_{\theta_t'} < 1$, which would have eliminated all invalid solutions. Nonetheless, the numerical scheme derived earlier is simple enough that there is no need for the enforcement of additional conditions. At $\tau_c = 140$ deg, the largest value of Δa, namely, 6.938 km, is achieved with $\theta_t' = 36.05$ deg and $\theta_t'' = -36.05$ deg.

It is then seen that, for $\tau_c = \tau_f/2$, the pitch angles are equal in magnitude but opposite in sign. As was conjectured in the analysis of Sec. 2.2, the other solution that satisfies $\Delta e_x = \Delta e_y = 0$ and provides the positive smaller Δa value is obtained with $\theta_t' = 100.31$ deg and $\theta_t'' = 79.68$ deg for the same value of $\tau_c = 140$ deg, yielding $\Delta a = 0$ km in this case. This type of solution almost always involves a deceleration leg, whereas the Δa-maximizing solutions are always either accelerating or decelerating, depending on whether positive or negative Δa is desired. Finally, in Fig. 2.7, the results of a series of runs for various circular orbits from LEO to GEO are depicted for the worst case of shadowing, which occurs when the sun–Earth line is contained within the spacecraft orbit plane. The total maximum shadow angle δ in this case is obtained simply from

$$\delta = \pi - 2\cos^{-1}\left(\frac{R}{a}\right) \tag{2.81}$$

where R is the radius of Earth. Then

$$\tau_f = \pi + 2\cos^{-1}\left(\frac{R}{a}\right) \tag{2.82}$$

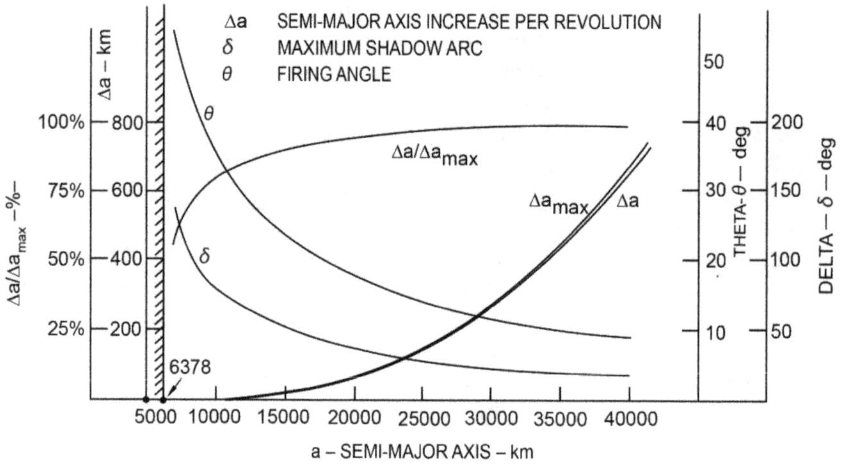

Fig. 2.7 Relative performance characteristics vs orbit semimajor axis.

and the Δa-maximizing switch point is

$$\tau_c = \frac{\pi}{2} + \cos^{-1}\left(\frac{R}{a}\right) \tag{2.83}$$

Table 2.1 and Fig. 2.7 show, as functions of the semimajor axis, the corresponding values of τ_f; $\Delta a_{\max}$ (obtained through tangential thrusting with corresponding eccentricity buildup); δ along with τ_1 from the preceding section and strategy; and finally $\theta = |\theta_t'| = |\theta_t''|$, Δa [from Eq. (2.63)], and $\Delta a/\Delta a_{\max}$. The numbers in parentheses in Table 2.1 are for τ_1 and the maximized Δa value from strategy 1. It is clear that, as the shadow arc becomes smaller with altitude, the pitch angle decreases accordingly, and the Δa value achieved after one revolution of intermittent thrusting becomes comparable to $\Delta a_{\max}$ achieved by tangential thrust, which would, by definition, be the absolute maximum value. It is also seen that strategy 1 is very nearly optimal because it provides Δa values slightly less than the optimal values possible with strategy 2. Letting α now be the pitch angle of the thrust acceleration measured from the radial direction, the optimal α profile that maximizes Δa while constraining Δe to zero can also be determined. The variation-of-parameters equations in terms of f_r and f_θ for the general elliptical case that are written as

$$\dot{a} = \frac{2a^2}{h}\left(es_{\theta^*}f_r + \frac{p}{r}f_\theta\right)$$

$$\dot{e} = \frac{1}{h}\{ps_{\theta^*}f_r + [(p+r)c_{\theta^*} + re]f_\theta\}$$

TABLE 2.1 RELATIVE PERFORMANCES OF STRATEGIES 1 AND 2[a]

a, km	7000	10,000	20,000	30,000	40,000
τ_f, deg	220	280	322	335	341
Δa_{max}, km	2.31	8.58	78.9	277	669
δ, deg (τ_1, deg)	140 (20)	80 (50)	38 (70)	25 (77)	19 (80)
θ'_t, deg	54	36	18	12	9.4
θ''_t, deg	−54	−36	−18	−12	−9.4
Δa, km	1.34 (1.18)	6.93 (6.56)	74.8 (71.3)	270 (263)	659 (639)
$\Delta a/\Delta a_{max}$, %	58	71	80	97	98
λ	5823.049823	8379.265958	13034.177567	14569.241955	15548.465162
Δa_{opt}, km	2.107	7.660	75.856	272.102	661.563
$\Delta a/\Delta a_{opt}$, %	63	90	98.6	99.2	99.6

[a]Data for strategy 1 in parentheses.

can also be cast as [3]

$$\dot{a} = \frac{2es_{\theta^*}}{n(1-e^2)^{1/2}} f_r + \frac{2(1+ec_{\theta^*})}{n(1-e^2)^{1/2}} f_\theta \tag{2.84}$$

$$\dot{e} = \frac{(1-e^2)^{1/2} s_{\theta^*}}{na} f_r + \frac{(1-e^2)^{1/2}}{nae}\left[1+ec_{\theta^*} - \frac{(1-e^2)}{1+ec_{\theta^*}}\right] f_\theta \tag{2.85}$$

Measuring θ^* from the shadow exit point and setting $e = 0$ in Eqs. (2.84) and (2.85) gives

$$\dot{a} = \frac{2f_\theta}{n} \tag{2.86}$$

$$\dot{e} = \frac{s_{\theta^*}}{na} f_r + \frac{2c_{\theta^*}}{na} f_\theta \tag{2.87}$$

which can be further reduced by using $f_r = kc_\alpha$ and $f_\theta = ks_\alpha$ and integrating between 0 and τ_f (Fig. 2.6), yielding the changes in a and e as given by Cass [3]

$$\Delta a = \frac{2ka^3}{\mu}\int_0^{\tau_f} s_\alpha \, \mathrm{d}\theta^* \tag{2.88}$$

$$\Delta e = \frac{ka^2}{\mu}\int_0^{\tau_f} (c_\alpha s_{\theta^*} + 2s_\alpha c_{\theta^*}) \, \mathrm{d}\theta^* \tag{2.89}$$

The maximization of Δa subject to the condition $\Delta e = 0$ leads to the adoption of the following performance index

$$I(\alpha) = \int_0^{\tau_f} \frac{2ka^3}{\mu} s_\alpha \, \mathrm{d}\theta^* + \lambda\left[\int_0^{\tau_f} \frac{ka^2}{\mu}(c_\alpha s_{\theta^*} + 2s_\alpha c_{\theta^*}) \, \mathrm{d}\theta^*\right]$$

$$I(\alpha) = \int_0^{\tau_f} \left\{\frac{ka^2}{\mu}[2as_\alpha + \lambda(c_\alpha s_{\theta^*} + 2s_\alpha c_{\theta^*})]\right\} \mathrm{d}\theta^* \tag{2.90}$$

where λ is the Lagrange multiplier for the $\Delta e = 0$ constraint. The optimal control law for α is obtained from Euler's equation, which reduces to $\partial F/\partial\alpha = 0$, where F is the integrand in Eq. (2.90), resulting in

$$\tan\alpha = \frac{2(a + \lambda c_{\theta^*})}{\lambda s_{\theta^*}} \tag{2.91}$$

such that

$$s_\alpha^2 = \frac{4(a+\lambda c_{\theta^*})^2}{\lambda^2 s_{\theta^*}^2 + 4(a+\lambda c_{\theta^*})^2}$$

$$c_{\alpha}^2 = \frac{\lambda^2 s_{\theta^*}^2}{\lambda^2 s_{\theta^*}^2 + 4(a + \lambda c_{\theta^*})^2}$$

Substituting these expressions into Eq. (2.89) gives

$$\Delta e = \frac{ka^2}{\mu}\int_0^{\tau_f} \frac{\left(4ac_{\theta^*} + \lambda + 3\lambda c_{\theta^*}^2\right)}{\left[4a^2 + 4a\lambda c_{\theta^*} + \lambda\left(\lambda + 3\lambda c_{\theta^*}^2 + 4ac_{\theta^*}\right)\right]^{1/2}}\,\mathrm{d}\theta^* = 0 \qquad (2.92)$$

The value of λ is determined numerically by performing a 10-point Gauss–Legendre quadrature of the preceding integral and searching on λ to establish the zero crossing of the Δe function, after which the van Wijngaarden–Dekker–Brent method, which combines root bracketing, bisection, and inverse quadratic interpolation, is applied to converge on the exact value of λ from the neighborhood of the zero crossing. Once λ is thus determined, s_α and c_α are readily obtained from the control law in Eq. (2.91), and the expression for Δa in Eq. (2.88) is evaluated also by numerical quadrature. The last three columns of Table 2.1 report, for the same constant acceleration of $k = 3.5 \times 10^{-7}$ km/s^2, the achieved optimal values of Δa for the five orbit radii with maximum shadowing. As expected, the optimal value Δa_{opt} falls between the maximum value Δa_{max} for pure tangential thrusting and the value of Δa obtained using strategy 2 of this section. Constraining the eccentricity at the zero value, even in the worst case of maximum shadowing, results in a loss of only 10% in the achieved Δa value compared to the unconstrained tangential thrust case in LEO and negligible loss in higher orbits. The purely analytic scheme using piecewise-constant pitch profiles is also very effective in orbit raising, being less effective only in low orbit, as shown in Table 2.1.

2.4 CONCLUSION

In this chapter, two coplanar orbit-raising strategies with near-optimal performance have been analyzed. The orbit is kept circular during the transfer even though thrusting is intermittent because of the presence of Earth shadow. From a practical point of view, it is better to use strategy 1 initially in LEO and to later switch to the other strategy when the period of the orbit becomes larger, thus allowing a thrust attitude reorientation maneuver to take place just before entry into or immediately after exit from shadow with negligible penalty in overall performance. The solutions presented here are very easy to generate and robust inasmuch as a table of the achieved change in semimajor axis vs the switch point location is established and the largest change is selected by means of a search-and-sort routine. These results are further compared with the optimal eccentricity-constrained solutions, which require integral evaluations through numerical quadrature and Lagrange multiplier determination through numerical search, and shown to be near-optimal in LEO and almost equivalent to the optimal solution in higher orbits.

REFERENCES

[1] Edelbaum, T. N., "Propulsion requirements for controllable satellites," *ARS Journal*, Vol. 31, No. 8, Aug. 1961, pp. 1079–1089.

[2] Wiesel, W. E., and Alfano, S., "Optimal many-revolution orbit transfer," *AAS/AIAA Astrodynamics Specialist Conference*, AAS Paper 83-352, Lake Placid, NY, Aug. 1983.

[3] Cass, J. R., "Discontinuous low thrust orbit transfer," *M.S. Thesis, Rept. AFIT/GA/AA/83D-1,* School of Engineering, U.S. Air Force Institute of Technology, Wright-Patterson AFB, OH, Dec. 1983.

[4] McCann, J. M., "Optimal launch time for a discontinuous low thrust orbit transfer," *M.S. Thesis, Rept. AFIT/GA/AA/88D-7,* School of Engineering, U.S. Air Force Institute of Technology, Wright-Patterson AFB, OH, Dec. 1988.

[5] Dickey, M. R., Klucz, R. S., Ennix, K. A., and Matuszak, L. M., "Development of the electric vehicle analyzer," *Rept. AL-TR-90-006,* Astronautics Laboratory, U.S. Air Force Space Technology Center, Edwards AFB, CA, June 1990.

[6] Kechichian, J. A., "Low-thrust inclination control in the presence of Earth shadow," *AAS Paper 91-157*, AAS/AIAA Spaceflight Mechanics Meeting, Houston, TX, Feb. 1991.

[7] Kechichian, J. A., "Orbit raising with low-thrust tangential acceleration in the presence of Earth shadow," *AAS/AIAA Astrodynamics Specialist Conference*, AAS Paper 91-513, Durango, CO, Aug. 1991.

[8] Kechichian, J. A., "Low-thrust eccentricity-constrained orbit raising," *AAS Paper 91-156*, AAS/AIAA Spaceflight Mechanics Meeting, Houston, TX, Feb. 1991.

[9] Kechichian, J. A., "Low-thrust eccentricity-constrained orbit raising," *Journal of Spacecraft and Rockets*, Vol. 35, No. 3, May 1998, pp. 327–335.

[10] Kaplan, M. H., *Modern Spacecraft Dynamics and Control*, John Wiley & Sons, Inc., 1976, pp. 348–353.

CHAPTER 3

Low-Thrust Inclination Control in the Presence of Earth Shadow

NOMENCLATURE

a_0 = semimajor axis of the reference orbit, km
f = thrust magnitude, N
$\boldsymbol{f}$ = acceleration vector due to thrust
f_r, f_θ, f_h = components of the acceleration vector along the $\hat{\boldsymbol{r}}$, $\hat{\boldsymbol{\theta}}$, and $\hat{\boldsymbol{h}}$ Euler–Hill directions
h = orbital angular momentum, km^2/s
$\hat{\boldsymbol{h}}$ = unit vector along the instantaneous angular momentum vector $\boldsymbol{h}$
m = spacecraft mass
n = mean motion of the reference orbit, $(\mu/a_0^3)^{1/2}$, rad/s
$\boldsymbol{r}$ = spacecraft position vector, Earth-centered
$\hat{\boldsymbol{r}}$ = unit vector along the instantaneous radius vector $\boldsymbol{r}$
$\boldsymbol{T}$ = thrust vector
$\hat{\boldsymbol{x}},\hat{\boldsymbol{y}}$ = unit vectors along the inertial x and y directions
β_s = sun look angle
$\hat{\boldsymbol{\theta}}$ = unit vector in the instantaneous orbit plane, perpendicular to $\hat{\boldsymbol{r}}$
θ' = spacecraft angular position at time t measured from the x axis, deg
θ_h = out-of-plane or thrust yaw angle, deg
θ_t = thrust pitch angle, deg
μ = gravitational constant of Earth, 398,601.3 km^3/s^2
τ = dimensionless time
Ω = right ascension of ascending node

3.1 INTRODUCTION

The problem of inclination control in a near-circular orbit using intermittent, low-thrust solar-electric propulsion is analyzed in this chapter. Given the shadow arc length and the line of nodes of the initial and final orbits, piecewise-constant yaw angles are selected, and the location along the orbit where the yaw angle switches is optimized to carry out the largest change in inclination per revolution for that particular geometry. Single-switch and two-switch strategies are analyzed, and several algorithms of varying complexity are described and numerically tested for their relative performance. This approach will yield suboptimal but robust and real-time autonomous onboard guidance software for EOTV orbit-transfer applications. The problem of low-thrust transfer between inclined circular orbits has benefited from the contributions of Edelbaum [1] and Wiesel and Alfano [2], who provided analytic solutions for executing minimum-time transfers using continuous constant or variable acceleration and a constant or optimized yaw profile within each revolution. Cass [3] and McCann [4] provided semianalytic solutions of the optimal thrust pitch and yaw profiles for transfers using intermittent thrusting due to shadowing. Dickey et al. developed a simulation tool [5] using simplifying assumptions to carry out preliminary parametric studies, as well as spacecraft systems design and optimization assessments. Analytic solutions leading to suboptimal transfers were developed [6, 7] for possible implementation in fast computer programs similar to that of Dickey et al. [5]. For ease of calculation of various dynamic and geometric parameters during the generation of the transfer solution, many of these analytic methods assume that the orbit remains circular during the transfer. Higher-fidelity simulations using averaging techniques with optimized pitch and yaw profiles and intermittent thrusting due to shadowing have also been produced [8–10]. These simulations show that the eccentricities of the intermediate orbits remain below 0.2 for the worst case of shadowing geometry for typical LEO-to-GEO transfers. However, these numerically generated transfers are difficult to obtain and are thus not suitable for parametric studies, which require a great number of iterations. In view of the small eccentricity buildup and the need to use simple but efficient control strategies that are easily implemented in analytic-type codes, the control problem for inclination change in the presence of Earth shadow is addressed in this chapter from a purely analytic but suboptimal point of view. Marec [11] reported several optimal thrust acceleration laws for various low-thrust systems applicable in general elliptical orbits with no restrictions as to the location of the thrust arcs. In our problem, the acceleration magnitude is assumed to remain constant, and the orbit is assumed to remain circular.

Thus, a pure inclination change in near-circular orbit using low thrust is analyzed for solar-electric propulsion systems using intermittent thrust along the orbit. The continuous-thrust solutions are such that the out-of-plane thrust angle or the yaw angle switches between ± 90 deg every half orbit, with the switch points located at the antinodes. These results are extended to the case in which an eclipse or shadow arc restricts thrusting to sunlight only, so that thrusting is now intermittent during each orbit. Given an arbitrarily selected line of nodes between the initial and final orbits and the length of the shadow arc, simple but robust algorithms are presented that effect the largest change in inclination during the revolution by thrusting only in sunlight. The geometry of the

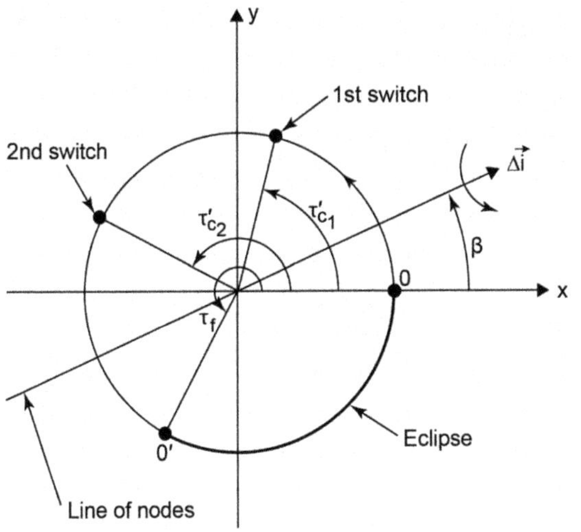

Fig. 3.1 Orbit transfer and switch geometry.

problem is illustrated in Fig. 3.1, where the Earth-centered inertial xy frame is such that x is pointing toward the point on the orbit corresponding to exit from shadow. The line of nodes is inclined with respect to the x axis by an angle β, where β can be any value between 0 and 360 deg. Furthermore, the arc OO' is the sunlit arc where thrusting is allowed, and the arc $O'O$ is the shadowed arc where thrusting is not possible. The shadow geometry is easily calculated once the sun look angle β_s is evaluated from the knowledge of the solar right ascension and declination and the spacecraft orbit equatorial inclination and ascending node. The node is allowed to regress because of J_2, and the amount by which Ω changes is calculated from the analytic expression defining the regression rate. This rate is dependent on both the orbital equatorial inclination and the orbit semimajor axis. Therefore, the angle β_s must be updated from revolution to subsequent revolution, and the x axis must be repositioned to point toward the current exit from shadow to perform the orbit rotation calculations during that particular revolution, in accordance with the algorithms presented in the following sections.

Here, we analyze both single-switch and two-switch strategies while holding the yaw angle constant between the two switches. The analysis in this chapter first appeared as a conference paper [12] and later as a journal article [13].

3.2 ANALYSIS

The variation-of-parameters equations linearized about a reference circular orbit, representing the initial circular orbit, are a complete set of first-order differential equations that describe the motion of the thrusting spacecraft. The pertinent equations that describe the out-of-plane motion are

$$\frac{\mathrm{d}\Delta i_x}{\mathrm{d}\tau} = \frac{a_0^2}{\mu} c_{\theta'} f_h \tag{3.1}$$

$$\frac{\mathrm{d}\Delta i_y}{\mathrm{d}\tau} = \frac{a_0^2}{\mu} s_{\theta'} f_h \tag{3.2}$$

The change in the inclination $\mathbf{\Delta i}$ along the line of nodes is a vector with components Δi_x and Δi_y along the x and y directions, respectively, as in Fig. 3.1. The rates are with respect to the dimensionless time $\tau = nt$, where n is the reference-orbit mean motion corresponding to the semimajor axis a_0. Given the $\hat{\boldsymbol{x}} - \hat{\boldsymbol{y}}$ plane of the initial reference circular orbit of semimajor axis a_0 and mean motion n, we can write $\mathbf{\Delta i} = \Delta i_x \hat{\boldsymbol{x}} + \Delta i_y \hat{\boldsymbol{y}}$, where $\Delta i_x = \Delta i\, c_{\theta'}$ and $\Delta i_y = \Delta i\, s_{\theta'}$. Here, the magnitude Δi is obtained from the variation-of-parameters equation $\mathrm{d}i/\mathrm{d}t = (rf_h/h)c_\theta$, where θ represents the angular position of the spacecraft measured from the ascending node. Because the spacecraft is initially on the reference circular orbit, $\theta = 0$, $r = a_0$, and $h = na_0^2$, such that Δi due to a small impulse $\mathbf{\Delta V}$ applied along the direction $\boldsymbol{\alpha} = (\alpha_r, \alpha_\theta, \alpha_h)$ is given by $\Delta i = (\Delta V/na_0)\alpha_h$ because $\Delta t f_h = \Delta V \alpha_h$. Finally, $\mathbf{\Delta i} = (\Delta V/na_0)\alpha_h \hat{\boldsymbol{r}}$, $\Delta i_x = (\Delta V/na_0)\alpha_h c_{\theta'}$, and $\Delta i_y = (\Delta V/na_0)\alpha_h s_{\theta'}$. The dimensionless rates in Eqs. (3.1) and (3.2) can now be obtained because $h^2/\mu = a_0$ for a nominally circular orbit and $\theta' \cong nt = \tau$. Figure 3.2 shows the spacecraft at position $\boldsymbol{r}$ with

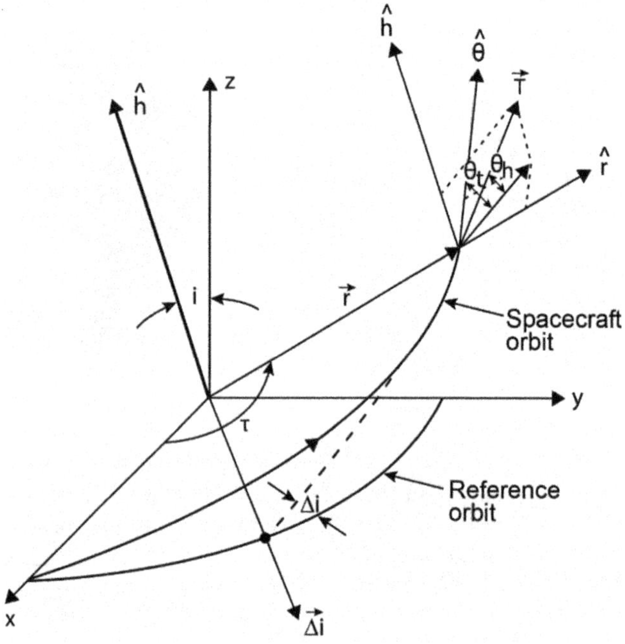

Fig. 3.2 Euler–Hill frame and thrust geometry.

respect to the center of the Earth, with $\boldsymbol{T}$ such that θ_h is the angle between this thrust vector and its projection on the $\hat{\boldsymbol{r}} - \hat{\boldsymbol{\theta}}$ plane and θ_t is the angle between this projection and the $\hat{\boldsymbol{\theta}}$ direction. Assuming that the instantaneous osculating orbit never departs from the near-circular shape, it is then possible to use the approximation mentioned earlier, namely, $\theta' \simeq nt = \tau$, $c_{\theta'} \simeq c_\tau$, and $s_{\theta'} \simeq s_\tau$, such that, with $k = f/m$, Eqs. (3.1) and (3.2) can now be written in linearized form as

$$\frac{\mathrm{d}\Delta i_x}{\mathrm{d}\tau} = \frac{k}{a_0 n^2} c_\tau s_{\theta_h} \tag{3.3}$$

$$\frac{\mathrm{d}\Delta i_y}{\mathrm{d}\tau} = \frac{k}{a_0 n^2} s_\tau s_{\theta_h} \tag{3.4}$$

If θ_h is assumed to be piecewise-constant, then Eqs. (3.3) and (3.4) can be integrated analytically between $\tau = 0$ and $\tau = \tau_f$ (Fig. 3.1). Let us first analyze the single-switch strategy, and let τ_c' correspond to the switch angular position. Then, if θ_{h_1} is the constant yaw angle between $\tau = 0$ and $\tau = \tau_c'$ and θ_{h_2} is the yaw angle between $\tau = \tau_c'$ and $\tau = \tau_f$, we can write

$$\Delta i_x = g s_{\theta_{h_1}} \int_0^{\tau_c'} c_\tau \,\mathrm{d}\tau + g s_{\theta_{h_2}} \int_{\tau_c'}^{\tau_f} c_\tau \,\mathrm{d}\tau \tag{3.5}$$

$$\Delta i_y = g s_{\theta_{h_1}} \int_0^{\tau_c'} s_\tau \,\mathrm{d}\tau + g s_{\theta_{h_2}} \int_{\tau_c'}^{\tau_f} s_\tau \,\mathrm{d}\tau \tag{3.6}$$

The constant g represents a normalized acceleration with $g = k/(a_0 n^2)$. Carrying out the integration results in

$$\Delta i_x = g s_{\tau_c'} s_{\theta_{h_1}} + g\left(s_{\tau_f} - s_{\tau_c'}\right) s_{\theta_{h_2}} = \Delta i c_\beta \tag{3.7}$$

$$\Delta i_y = -g(c_{\tau_c'} - 1) s_{\theta_{h_1}} - g(c_{\tau_f} - c_{\tau_c'}) s_{\theta_{h_2}} = \Delta i s_\beta \tag{3.8}$$

These two equations can be solved simultaneously to provide the angles θ_{h_1} and θ_{h_2} for given Δi, β, τ_c', τ_f, and g:

$$s_{\theta_{h_1}} = \frac{\Delta i(c_{\beta-\tau_c'} - c_{\beta-\tau_f})}{g(s_{\tau_f-\tau_c'} + s_{\tau_c'} - s_{\tau_f})} \tag{3.9}$$

$$s_{\theta_{h_2}} = \frac{\Delta i(c_{\beta-\tau_c'} - c_\beta)}{g(s_{\tau_f-\tau_c'} + s_{\tau_c'} - s_{\tau_f})} \tag{3.10}$$

Equations (3.9) and (3.10) uniquely define θ_{h_1} and θ_{h_2} because $-\pi/2 \le \theta_{h_1} \le \pi/2$ and $-\pi/2 \le \theta_{h_2} < \pi/2$ such that $|s_{\theta_{h_1}}| \le 1$ and

$|s_{\theta_{h_2}}| \leq 1$ or $s^2_{\theta_{h_1}} \leq 1$ and $s^2_{\theta_{h_2}} \leq 1$; Δi is the magnitude of the change in inclination and is therefore a positive quantity. The constant g is also strictly positive. The largest eclipse arc in LEO never exceeds some 140 deg, so that 220 deg $\leq \tau_f \leq$ 360 deg. This, in turn, requires that $s_{\tau_f} \leq 0$. It now can be shown that the factor in the denominator, namely, $(s_{\tau_f - \tau'_c} + s_{\tau'_c} - s_{\tau_f}) \geq 0$, is positive semidefinite. From

$$s_{\tau_f - \tau'_c} + s_{\tau'_c} - s_{\tau_f} \geq 0 \tag{3.11}$$

it follows that

$$s_{\tau'_c}(1 - c_{\tau_f}) - s_{\tau_f}(1 - c_{\tau'_c}) \geq 0$$

Both $(1 - c_{\tau_f})$ and $(1 - c_{\tau'_c})$ are greater than or equal to 0, and with $s_{\tau_f} \leq 0$, the condition in Eq. (3.11) is equivalent to

$$s_{\tau'_c} \geq \frac{s_{\tau_f}(1 - c_{\tau'_c})}{(1 - c_{\tau_f})} \tag{3.12}$$

where the right-hand side is less than or equal to 0. If $s_{\tau'_c} \geq 0$, then the inequality in Eq. (3.12) is always satisfied. If $s_{\tau'_c} \leq 0$, then Eq. (3.12) can be written as

$$s_{\tau'_c}(1 - c_{\tau_f}) \geq s_{\tau_f}(1 - c_{\tau'_c})$$

where the right- and left-hand sides are less than or equal to 0. Squaring, one obtains

$$s^2_{\tau'_c}(1 - c_{\tau_f})^2 \geq s^2_{\tau_f}(1 - c_{\tau'_c})^2$$

which reduces to

$$c^2_{\tau'_c}(1 - c_{\tau_f}) - s^2_{\tau_f} c_{\tau'_c} + c_{\tau_f}(1 - c_{\tau_f}) \geq 0$$

However, $s^2_{\tau_f} = (1 - c_{\tau_f})(1 + c_{\tau_f})$, and because $(1 - c_{\tau_f}) \geq 0$, we can divide this inequality by $(1 - c_{\tau_f})$ to obtain the following condition:

$$Y = c^2_{\tau'_c} - (1 + c_{\tau_f})c_{\tau'_c} + c_{\tau_f} \geq 0 \tag{3.13}$$

The roots of this quadratic are

$$c_{\tau'_c} = \frac{(1 + c_{\tau_f}) \pm (1 - c_{\tau_f})}{2} = 1, c_{\tau_f}$$

The inequality in Eq. (3.13) is satisfied for $c_{\tau'_c} \leq c_{\tau_f}$, and because, in this discussion, $s_{\tau'_c} \leq 0$ or $\pi \leq \tau'_c \leq \tau_f$, $c_{\tau'_c} \leq c_{\tau_f}$ is always satisfied. In short, for both $s_{\tau'_c} \geq 0$ and $s_{\tau'_c} \leq 0$, the expression in Eq. (3.11) is always greater than or equal to 0. Figure 3.3 shows a plot of Eq. (3.13) or $Y = f(c_{\tau'_c})$ for three values of τ_f,

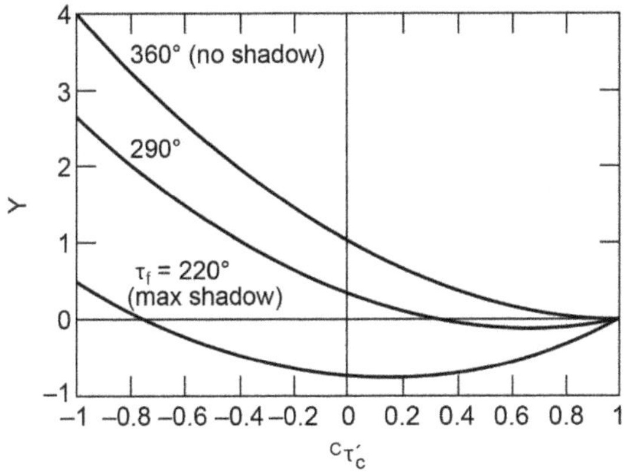

Fig. 3.3 Plot of function *Y* vs $c_{\tau_c'}$ for τ_f = 220, 290, and 360 deg.

namely, 220, 290, and 360 deg; the last of these corresponds to the case of no shadow, whereas the first is for maximum shadow. In Fig. 3.3, the lowest parabola is for $\tau_f =$ 220 deg, and the uppermost parabola is for $\tau_f =$ 360 deg. The y and x intercepts are of equal value.

This discussion thus shows that $s_{\theta_{h_1}}$ has the same sign as the difference $(c_{\beta-\tau_c'} - c_{\beta-\tau_f})$ and $s_{\theta_{h_2}}$ has the same sign as the difference $(c_{\beta-\tau_c'} - c_\beta)$ in Eqs. (3.9) and (3.10), respectively, and because $|s_{\theta_{h_1}}| \leq 1$ and $|s_{\theta_{h_2}}| \leq 1$, the following conditions must be satisfied:

$$\Delta i \leq \frac{g(s_{\tau_f-\tau_c'} + s_{\tau_c'} - s_{\tau_f})}{|c_{\beta-\tau_c'} - c_\beta|} \tag{3.14}$$

$$\Delta i \leq \frac{g(s_{\tau_f-\tau_c'} + s_{\tau_c'} - s_{\tau_f})}{|c_{\beta-\tau_c'} - c_{\beta-\tau_f}|} \tag{3.15}$$

Then, given values for β and τ_f, τ_c' must be such that Δi is maximized. However, because Eqs. (3.14) and (3.15) must also be satisfied, Δi is the minimum of the following two maxima:

$$\Delta i_1 = \frac{g(s_{\tau_f-\tau_c'} + s_{\tau_c'} - s_{\tau_f})}{|c_{\beta-\tau_c'} - c_{\beta-\tau_f}|} \tag{3.16}$$

$$\Delta i_2 = \frac{g(s_{\tau_f-\tau_c'} + s_{\tau_c'} - s_{\tau_f})}{|c_{\beta-\tau_c'} - c_\beta|} \tag{3.17}$$

These two expressions are plotted in Fig. 3.4 as functions of τ_c' for $\tau_f =$ 220 deg and $\beta = 60$ deg. The upper (solid) curve corresponds to Δi_2, which has singularities at $\tau_c' = 0$ and 120 deg in this case because $|c_{\beta-\tau_c'} - c_\beta| = 0$ at those values. The dashed curve for Δi_1 remains below Δi_2 until about $\tau_c' \simeq$ 162 deg, after which its value becomes the greater than that of Δi_2 until $\tau_c' = \tau_f = 220$ deg, where the Δi_1 curve shows a singularity because $|c_{\beta-\tau_c'} - c_{\beta-\tau_f}| = 0$ there. These singularities are not worrisome because Δi

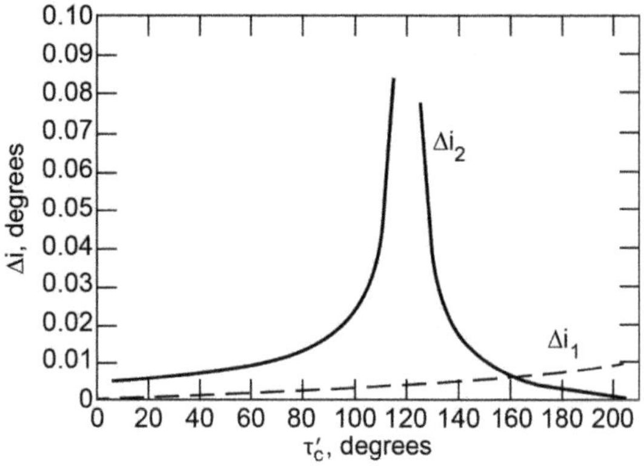

Fig. 3.4 Change in inclination vs τ_c' for τ_f = 220 deg and β = 60 deg.

must be less than the smaller of Δi_1 and Δi_2 for a feasible solution and such solutions always exist. In the range from $\tau_c' = 0$ to $\tau_c' \simeq$ 160 deg, (the crossover point), Δi must be less than or equal to Δi_1 (i.e., $\Delta i \le \Delta i_1$), whereas after the crossover point, $\Delta i \le \Delta i_2$ such that Δi is less than or equal to the smaller of Δi_1 and Δi_2. In other words, the feasible domain is the area in Fig. 3.4, between the x axis and the lowest branch of either curve. Clearly, there exists an optimal value of τ_c' ($\simeq$162 deg in this example) such that Δi is maximized. This maximum corresponds to $\Delta i_1 = \Delta i_2$. The values of the various parameters used in Fig. 3.4 are $a_0 = 7000$ km, $\mu = 398{,}601.3\ \mathrm{km}^3/\mathrm{s}^2$, and $k = 3.5 \times 10^{-7}\ \mathrm{km/s}^2$, corresponding to $g = 4.302544 \times 10^{-5}$. For given values of Δi, τ_f, β, and g, the location of τ_c' is uniquely defined provided that Δi corresponds to the crossover point of Δi_1 and Δi_2, where $\Delta i = \Delta i_1 = \Delta i_2$ and $|\, s_{\theta_{h_1}} | = |\, s_{\theta_{h_2}} | = 1$. To cover both cases, namely, the pairs $(\theta_{h_1} = \pi/2,\ \ \theta_{h_2} = -\pi/2)$ and $(\theta_{h_1} = -\pi/2,\ \ \theta_{h_2} = \pi/2)$, Eqs. (3.7) and (3.8) reduce to

$$\pm g(2s_{\tau_c'} - s_{\tau_f}) = \Delta i c_\beta \tag{3.18}$$

$$\pm g(1 + c_{\tau_f} - 2c_{\tau_c'}) = \Delta i s_\beta \tag{3.19}$$

From which

$$s_{\tau_c'} = \frac{1}{2}\left(\frac{\pm \Delta i c_\beta}{g} + s_{\tau_f}\right) \tag{3.20}$$

$$c_{\tau_c'} = -\frac{1}{2}\left[\frac{\pm \Delta i s_\beta}{g} - (1 - c_{\tau_f})\right] \tag{3.21}$$

such that

$$s_{\tau_c'} = \tan^{-1}\left[\frac{s_{\tau_f} + \left(\pm \dfrac{\Delta i}{g}\right) c_\beta}{(1 + c_{\tau_f}) - \left(\pm \dfrac{\Delta i}{g}\right) s_\beta}\right] \tag{3.22}$$

From Fig. 3.4, $\beta = 60$ deg, $\tau_f = 220$ deg, $g = 4.302544 \times 10^{-5}$, and $\Delta i = 6.1 \times 10^{-3}$ deg such that, with the plus sign selected in Eq. (3.22), $\tau_c' \simeq 162.7$ deg. This is clearly the unique solution because, if we select the minus sign instead, Eq. (3.22) will yield $\tau_c' \simeq 321.6$ deg, which is larger than τ_f and therefore not acceptable. Conversely, if we select, as an example, $\tau_c' \simeq 150$ deg instead, then from Eqs. (3.16) and (3.17), $\Delta i_1 = 0.00009535$ rad and $\Delta i_2 = 0.00017919$ rad, so that, satisfying the inequalities in Eqs. (3.14) and (3.15), we can select any value of Δi provided that $\Delta i \le \min(\Delta i_1, \Delta i_2)$ or $\Delta i \le \Delta i_1 = 0.00009535$ rad. At $\Delta i = \Delta i_1$ exactly, Eqs. (3.9) and (3.10) yield $s_{\theta_{h_1}} = 1.0$ and $s_{\theta_{h_2}} = -0.532089$, or $\theta_{h_1} = \pi/2$ and $\theta_{h_2} = -32.146$ deg. It is then clear that, for a given τ_c' value different from the optimal value, any value of $\Delta i \le \min(\Delta i_1, \Delta i_2)$ can be achieved with a unique combination of θ_{h_1} and θ_{h_2} values with either $|\theta_{h_1}| = \pi/2$ or $|\theta_{h_2}| = \pi/2$ if the maximum feasible value of Δi is selected as described earlier, where $\Delta i = \Delta i_1$ was selected. For lower values of Δi, both $|\theta_{h_1}|$ and $|\theta_{h_2}|$ are less than $\pi/2$, and only at the optimal τ_c' value, where $\Delta i_1 = \Delta i_2 = \Delta i_{\max}$, is it the case that $|\theta_{h_1}| = |\theta_{h_2}| = \pi/2$. If τ_c' is very close to τ_f, it may be preferable to select a smaller value of τ_c' for operational considerations by rotating the orbit a little bit less along the required line of nodes during the revolution. Other considerations also may limit the θ_{h_1} angles to less than a given value, for example, much less than the maximum of $\pm\pi/2$, such that, once again, an appropriate τ_c' value other than the optimal τ_c' would be selected.

Now, returning to the preceding discussion, it has been determined that $s_{\theta_{h_1}}$ has the same sign as $c_{\beta-\tau_c'} - c_{\beta-\tau_f}$ and that $s_{\theta_{h_2}}$ has the same sign as $c_{\beta-\tau_c'} - c_\beta$, such that $s_{\theta_{h_1}}$ and $s_{\theta_{h_2}}$ have opposite signs whenever $c_{\beta-\tau_f}$ and c_β have opposite signs. The functions c_β and $c_{\beta-\tau_f}$ for $\tau_f = 220$ deg have opposite signs for $\tau_f - 3\pi/2 \le \beta \le \pi/2$, $\tau_f - \pi/2 \le \beta \le 3\pi/2$, and $\tau_f + \pi/2 \le \beta \le 2\pi$. This is shown in Fig. 3.5 for three values of τ_f, namely, $\tau_f = 220$ deg (maximum shadow case), $\tau_f = 270$ deg, and $\tau_f = (360 - \varepsilon)$ deg, where ε is small. The heavy line indicates the regions where the value of β is such that $s_{\theta_{h_1}}$ and $s_{\theta_{h_2}}$ are of opposite sign. Where τ_f is close to 2π, as in the last case, the values of β for which $s_{\theta_{h_1}}$ and $s_{\theta_{h_2}}$ are of opposite sign are such that $\tau_f - 3\pi/2 \le \beta \le \pi/2$ and $\tau_f - \pi/2 \le \beta \le 3\pi/2$, corresponding to a very small region along the orbit. This means that, for most orientations of $\boldsymbol{\Delta i}$, θ_h will not change sign if we restrict ourselves to this single-switch theory. In turn, this condition will be achieved at the expense of vanishingly small changes in inclination Δi, because most of the thrust will be wasted to satisfy the rotation along the required line of nodes. This fact then leads us to extend this analysis to include the case in which two switches in the θ_h angle are allowed so that larger changes in relative inclination can be carried out along any given orientation of the line of nodes. Here, we restrict our analysis to yaw angles of 90 and −90 deg only, so that the orbit remains perfectly circular at all times, while maximizing the relative change in inclination along the desired line of nodes.

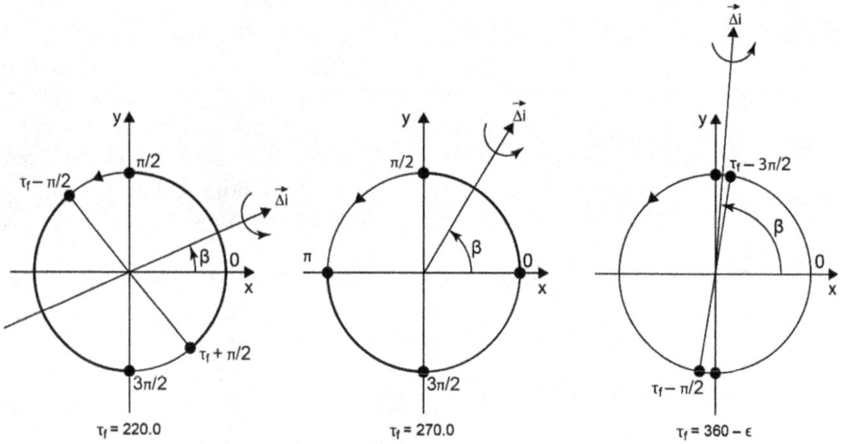

Fig. 3.5 Yaw angle switch regions for τ_f = 220, 270, and nearly 360 deg.

3.3 ALGORITHMS OF THE TWO-SWITCH TRANSFER

Let us assume that $|\theta_{h_i}| = \pi/2$ such that the thrust vector is directed purely normal to the orbit plane. Because we are considering, at most, two switches in the yaw angle, the yaw program is $\pi/2, -\pi/2, \pi/2$ or $-\pi/2, \pi/2, -\pi/2$. From Eqs. (3.3) and (3.4), it follows that

$$\Delta i_x = \pm g\left(\int_0^{\tau'_{c_1}} c_\tau \, \mathrm{d}\tau - \int_{\tau'_{c_1}}^{\tau'_{c_2}} c_\tau \, \mathrm{d}\tau + \int_{\tau'_{c_2}}^{\tau_f} c_\tau \, \mathrm{d}\tau\right) \tag{3.23}$$

$$\Delta i_y = \pm g\left(\int_0^{\tau'_{c_1}} s_\tau \, \mathrm{d}\tau - \int_{\tau'_{c_1}}^{\tau'_{c_2}} s_\tau \, \mathrm{d}\tau + \int_{\tau'_{c_2}}^{\tau_f} s_\tau \, \mathrm{d}\tau\right) \tag{3.24}$$

or

$$\Delta i_x = \pm g\left(2s_{\tau'_{c_1}} - 2s_{\tau'_{c_2}} + s_{\tau_f}\right) \tag{3.25}$$

$$\Delta i_y = \pm(-g)\left(2c_{\tau'_{c_1}} - 1 - 2c_{\tau'_{c_2}} + c_{\tau_f}\right) \tag{3.26}$$

The plus sign is used for the $\pi/2, -\pi/2, \pi/2$ sequence, and the minus sign is used for the $-\pi/2, \pi/2, -\pi/2$ sequence. Because $\Delta i_x = \Delta i c_\beta$ and $\Delta i_y = \Delta i s_\beta$, the expressions in Eqs. (3.25) and (3.26) can be written as

$$\Delta i_x = g\left(2s_{\tau'_{c_1}} - 2s_{\tau'_{c_2}} + s_{\tau_f}\right) = \pm \Delta i c_\beta \tag{3.27}$$

$$\Delta i_y = -g\left(2c_{\tau'_{c_1}} - 2c_{\tau'_{c_2}} + c_{\tau_f} - 1\right) = \pm \Delta i s_\beta \tag{3.28}$$

which can be further cast into the following form:

$$s_{\tau'_{c1}} - s_{\tau'_{c2}} = \pm\frac{\Delta i}{2g}c_\beta - \frac{s_{\tau_f}}{2} \tag{3.29}$$

$$c_{\tau'_{c2}} - c_{\tau'_{c1}} = \pm\frac{\Delta i}{2g}s_\beta - \frac{(1 - c_{\tau_f})}{2} \tag{3.30}$$

Let

$$k_1 = \frac{\Delta i}{2g}s_\beta \tag{3.31}$$

$$k_2 = \frac{(1 - c_{\tau_f})}{2} \tag{3.32}$$

$$k_3 = \frac{\Delta i}{2g}c_\beta \tag{3.33}$$

$$k_4 = \frac{s_{\tau_f}}{2} \tag{3.34}$$

Then

$$s_{\tau'_{c1}} - s_{\tau'_{c2}} = \pm k_3 - k_4 \tag{3.35}$$

$$c_{\tau'_{c1}} - c_{\tau'_{c2}} = \pm(-k_1) + k_2 \tag{3.36}$$

Given the orientation of the line of nodes defined by the angle β and the magnitude of the change in relative inclination Δi, as well as the normalized acceleration g, it is possible to solve for τ'_{c1} and τ'_{c2}, the switch angular positions, from Eqs. (3.35) and (3.36). From Eq. (3.36), we can write

$$\pm\left(1 - s^2_{\tau'_{c2}}\right)^{1/2} = \pm\left(1 - s^2_{\tau'_{c1}}\right)^{1/2} - [\pm(-k_1) + k_2]$$

Squaring then yields

$$\left(1 - s^2_{\tau'_{c2}}\right) = 1 - s^2_{\tau'_{c1}} + [\pm(-k_1) + k_2]^2 - (\pm 2)\left(1 - s^2_{\tau'_{c1}}\right)^{1/2}[\pm(-k_1) + k_2]$$

However, $s_{\tau'_{c2}}$ can be eliminated by means of Eq. (3.35) because

$$s_{\tau'_{c2}} = s_{\tau'_{c1}} - (\pm k_3 - k_4)$$

$$s^2_{\tau'_{c2}} = s^2_{\tau'_{c1}} + (\pm k_3 - k_4)^2 - 2s_{\tau'_{c1}}(\pm k_3 - k_4)$$

Therefore, after carrying out this elimination, we obtain

$$K - 2s_{\tau'_{c_1}}(\pm k_3 - k_4) = \pm 2\left(1 - s^2_{\tau'_{c_1}}\right)^{1/2}[\pm(-k_1) + k_2]$$

which is squared once again to yield the quadratic form

$$As^2_{\tau'_{c_1}} + Bs_{\tau'_{c_1}} + C = 0 \tag{3.37}$$

where

$$K = (\pm k_3 - k_4)^2 + [\pm(-k_1) + k_2]^2 \tag{3.38}$$

$$A = K \tag{3.39}$$

$$B = -K(\pm k_3 - k_4) \tag{3.40}$$

$$C = -[\pm(-k_1) + k_2]^2 + \frac{K^2}{4} \tag{3.41}$$

The solution of Eq. (3.37) is

$$s_{\tau'_{c_1}} = \frac{-B \pm \left(B^2 - 4AC\right)^{1/2}}{2A} \tag{3.42}$$

Now, from Eq. (3.35), we can write

$$\pm\left(1 - c^2_{\tau'_{c_2}}\right)^{1/2} = \pm\left(1 - c^2_{\tau'_{c_1}}\right)^{1/2} - (\pm k_3 - k_4)$$

Eliminating $c_{\tau'_{c_2}}$ this time from Eq. (3.36) and using the same manipulations as were employed to generate Eq. (3.37), a quadratic from in $c_{\tau'_{c_1}}$ can be obtained as

$$Ac^2_{\tau'_{c_1}} + B'c_{\tau'_{c_1}} + C' = 0 \tag{3.43}$$

where

$$A = K \tag{3.44}$$

$$B' = -K[\pm(-k_1) + k_2] \tag{3.45}$$

$$C' = -(\pm k_3 - k_4)^2 + \frac{K^2}{4} \tag{3.46}$$

The solution of Eq. (3.43) is

$$c_{\tau'_{c_1}} = \frac{-B' \pm (B'^2 - 4AC')^{1/2}}{2A} \tag{3.47}$$

Once τ'_{c_1} is computed, τ'_{c_2} can be obtained from Eqs. (3.35) and (3.36) such that

$$s_{\tau'_{c_2}} = s_{\tau'_{c_1}} - (\pm k_3 - k_4) \tag{3.48}$$

$$c_{\tau'_{c_2}} = c_{\tau'_{c_1}} - [\pm(-k_1) + k_2] \tag{3.49}$$

The condition $B^2 - 4AC \geq 0$ must be satisfied such that

$$B^2 - 4AC = K\left\{K(\pm k_3 - k_4)^2 + 4[\pm(-k_1) + k_2]^2 - K^2\right\} \geq 0$$

which reduces to the condition

$$K\left\{4 - [\pm(-k_1) + k_2]^2 - (\pm k_3 - k_4)^2\right\}[\pm(-k_1) + k_2]^2 \geq 0$$

However, $K \geq 0$ and the last bracket are also greater than or equal to 0, such that the preceding condition can be replaced by the simpler condition

$$4 - [\pm(-k_1) + k_2]^2 - (\pm k_3 - k_4)^2 \geq 0 \tag{3.50}$$

If we chose the plus sign, which corresponds to the yaw sequence $\pi/2$, $-\pi/2$, $\pi/2$, the condition in Eq. (3.50) can be written as

$$y' = -x^2 + [s_\beta(1 - c_{\tau_f}) + c_\beta s_{\tau_f}]x + \left[4 - \frac{1}{2}(1 - c_{\tau_f})\right] \geq 0 \tag{3.51}$$

The roots of this quadratic are given by

$$x = \frac{[s_\beta(1 - c_{\tau_f}) + c_\beta s_{\tau_f}] \mp \left\{[s_\beta(1 - c_{\tau_f}) + c_\beta s_{\tau_f}]^2 + 4\left[4 - \frac{1}{2}(1 - c_{\tau_f})\right]\right\}^{1/2}}{2} \tag{3.52}$$

where $x = \Delta i/2g$ is proportional to Δi. Given β, g, and τ_f, we seek the maximum value of Δi that satisfies the condition in Eq. (3.51). The solutions given by Eq. (3.52) always exist because the square-root term is greater than or equal to 0. This is because

$$4\left[4 - \frac{1}{2}(1 - c_{\tau_f})\right] \geq 0$$

If $\left[s_\beta(1 - c_{\tau_f}) + c_\beta s_{\tau_f}\right] < 0$, the plus sign must be chosen in Eq. (3.52) so that $\Delta i > 0$ or $x > 0$. Conversely, if $\left[s_\beta(1 - c_{\tau_f}) + c_\beta s_{\tau_f}\right] > 0$, the plus sign must again be chosen to obtain x or $\Delta i > 0$. Therefore, the range of x is $0 \leq x \leq x_2$, where x_2 is given by

$$x_2 = \frac{[s_\beta(1 - c_{\tau_f}) + c_\beta s_{\tau_f}] + \left\{[s_\beta(1 - c_{\tau_f}) + c_\beta s_{\tau_f}]^2 + 4\left[4 - \frac{1}{2}(1 - c_{\tau_f})\right]\right\}^{1/2}}{2} \tag{3.53}$$

Obviously, we must choose $x = x_2$ because it corresponds to the largest value of Δi (i.e., Δi_{max}) that can be achieved along the required line of nodes, unless a smaller value $\Delta i < \Delta i_{max}$ is needed to fine tune the final inclination, in which case all of the coefficients are defined, leading to the solution of τ'_{c_1} and τ'_{c_2}. If one chooses $x = x_2$, then $y' = 0$, and therefore, $B^2 - 4AC = 0$, which simplifies Eq. (3.42) to

$$s_{\tau'_{c_1}} = -\frac{B}{2A} \tag{3.54}$$

If $\beta = 0$ and $\tau_f = 360$ deg (no eclipse condition), then $x_2 = 2$, and $\Delta i_{max} = 4g$, which is the maximum angle by which the circular orbit can be rotated after one revolution of continuous thrusting. Returning to the condition in Eq. (3.5), if we now select the yaw sequence $-\pi/2,\ \pi/2,\ -\pi/2$, then that condition will reduce to

$$y' = -x^2 - [s_\beta(1 - c_{\tau_f}) + c_\beta s_{\tau_f}]x + \left[4 - \frac{1}{2}(1 - c_{\tau_f})\right] \geq 0 \tag{3.55}$$

whose roots are given by

$$x = \frac{-[s_\beta(1-c_{\tau_f}) + c_\beta s_{\tau_f}] \mp \left\{[s_\beta(1-c_{\tau_f}) + c_\beta s_{\tau_f}]^2 + 4\left[4 - \frac{1}{2}(1-c_{\tau_f})\right]\right\}^{1/2}}{2} \tag{3.56}$$

Once again, if $[s_\beta(1 - c_{\tau_f}) + c_\beta s_{\tau_f}] < 0$, the plus sign must be chosen to obtain $\Delta i > 0$, and conversely, if $[s_\beta(1 - c_{\tau_f}) + c_\beta s_{\tau_f}] > 0$, the plus sign also must be chosen to yield $\Delta i > 0$.

The maximum value Δi_{max} is achieved for $x = x'_2$, where x'_2 is given by

$$x'_2 = \frac{-[s_\beta(1-c_{\tau_f}) + c_\beta s_{\tau_f}] + \left\{[s_\beta(1-c_{\tau_f}) + c_\beta s_{\tau_f}]^2 + 4\left[4 - \frac{1}{2}(1-c_{\tau_f})\right]\right\}^{1/2}}{2} \tag{3.57}$$

and the range of x is now $0 \leq x \leq x'_2$. At $x = x'_2$, $y' = 0$, and $B^2 - 4AC = 0$ as well, such that, as in Eq. (3.54)

$$s_{\tau'_{c_1}} = -\frac{B}{2A} \tag{3.58}$$

The value Δi_{max} is therefore achieved as $\Delta i_{max} = \max(2gx_2, 2gx'_2)$. The same discussion carried out so far for $s_{\tau'_{c_1}}$ also can be applied to $c_{\tau'_{c_1}}$ because, from Eq. (3.47)

$$B'^2 - 4AC' = K\left\{K[\pm(-k_1) + k_2]^2 + 4(\pm k_3 - k_4)^2 - K^2\right\} \geq 0$$

reduces to the condition

$$K\Big\{4-(\pm k_3-k_4)^2-[\pm(-k_1)+k_2]^2\Big\}(\pm k_3-k_4)^2\geq 0$$

which is identical to the condition in Eq. (3.50). The selection of x_2 or x_2' once again will result in $B'^2-4AC'=0$; therefore, the maximum value of Δi will be such that

$$c_{\tau_{c_1}'}=-\frac{B'}{2A} \tag{3.59}$$

3.3.1 ALGORITHM 1

This algorithm uses very simple logic, as follows: Let the sequence $\pi/2$, $-\pi/2$, $\pi/2$ for the yaw angle correspond to $S=1$ and the sequence $-\pi/2$, $\pi/2$, $-\pi/2$ correspond to $S=-1$. If $0\leq\beta<\pi/2$ and $3\pi/2<\beta\leq 2\pi$, then let $S=1$, and if $\pi/2\leq\beta\leq 3\pi/2$, then let $S=-1$. If $S=1$, x_2 is computed from Eq. (3.53), and if $S=-1$, x_2' is computed from Eq. (3.57). The change in inclination is evaluated next from $\Delta i=2gx_2$ or $\Delta i=2gx_2'$ depending on whether $S=1$ or $S=-1$, respectively. Now, the coefficients k_1, k_2, k_3, and k_4 are evaluated from Eqs. (3.31), (3.32), (3.33), and (3.34), respectively, whereas the constants K, A, B, C, B', and C' are evaluated from Eqs. (3.28), (3.39), (3.40), (3.41), (3.45), and (3.46), respectively, with the plus sign chosen if $S=1$ and the minus sign chosen if $S=-1$. Next, the first switch point is evaluated from $s_{\tau_{c_1}'}=-B/2A$ and $c_{\tau_{c_1}'}=-B'/2A$, and the second switch point is evaluated from Eqs. (3.48) and (3.49) with, once again, the plus sign for $S=1$ and the minus sign for $S=-1$. If $\tau_{c_1}'<\tau_{c_2}'<\tau_f$, the solution is well-defined. If $\tau_{c_2}'>\tau_f$, we set $\tau_{c_2}'=\tau_f$ because this is the maximum value that τ_{c_2}' can have. However, τ_{c_1}' now must be modified to carry out the inclination change for the particular β of interest. This modification will result in a change in inclination that is less than the maximum value (i.e., $\Delta i<\Delta i_{\max}$), because clearly, $\Delta i_{\max}$ is not feasible in this case. To evaluate Δi and τ_{c_1}' in this case, we return to Eqs. (3.48) and (3.49), which we write as

$$s_{\tau_{c_1}'}=\frac{s_{\tau_f}}{2}\pm\frac{\Delta i}{2g}c_\beta \tag{3.60}$$

$$c_{\tau_{c_1}'}=\frac{(1+c_{\tau_f})}{2}\pm\left(-\frac{\Delta i}{2g}\right)s_\beta \tag{3.61}$$

because we fixed $\tau_{c_2}'=\tau_f$. This system of equations yields the values of τ_{c_1}' and Δi such that, regardless of whether $S=1$ or $S=-1$, we obtain

$$\tau_{c_1}'=\beta\pm\cos^{-1}\left[\frac{1}{2}(c_\beta+c_{\tau_f-\beta})\right] \tag{3.62}$$

$$\Delta i = 2g\left\{\left(s_{\tau'_{c_1}} - \frac{s_{\tau_f}}{2}\right)^2 + \left[\frac{(1+c_{\tau_f})}{2} - c_{\tau'_{c_1}}\right]^2\right\}^{1/2} \quad (3.63)$$

The final difficulty now is choosing between the plus and minus signs in Eq. (3.62). Let us choose the plus sign first. If $\tau'_{c_1} > \tau_f$, then we choose the minus sign instead. If $\tau'_{c_1} > \tau_f$ still holds, then we return to the plus sign and subtract 2π from the answer because, in that case, τ'_{c_1} would also be larger than 2π. In short, we choose the sign that yields $0 < \tau'_{c_1} < \tau_f$. When $\tau'_{c_2} = \tau_f$, there is only one switch in the angle located at τ'_{c_1}, and the solution is analogous to the one described in Sec. 3.2 with $|\theta_{h_i}| = \pi/2$. In Fig. 3.6, the angles τ'_{c_1} and τ'_{c_2} are given as functions of the eclipse entry angle τ_f, which defines the duration of the eclipse, for $\beta = 0$ deg. The lower curve is obviously for τ'_{c_1} and the upper curve for τ'_{c_2}. From $\tau_f = 220$ deg to about $\tau_f = 250$ deg, $\tau'_{c_2} = \tau_f$, indicating that these transfers require only one switch, given by τ'_{c_1}. For larger values of τ_f, the transfer will make use of two switches, and because $\tau_f = 360$ deg, corresponding to the case of no eclipse, $\tau'_{c_1} = 90$ deg and $\tau'_{c_2} = 270$ deg, thereby recovering the well-known solutions valid in full sunlight. All of the numerical examples studied here use $a_0 = 7000$ km, $\mu = 398{,}601.3$ km^3/s^2, $k = 3.5 \times 10^{-7}$ km/s^2, and $g = 4.302544 \times 10^{-5}$, as in Sec. 3.2.

In Fig. 3.6, the maximum change in inclination Δi is given as a function of τ_f for $\beta = 0$ deg. As expected, the maximum rotation occurs for $\tau_f = 360$ deg with continuous thrusting along the orbit. This maximum is $\Delta i = 9.8607 \times 10^{-3}$ deg. In Fig. 3.7, which corresponds to $\beta = 60$ deg, one can see that single-switch transfers are preponderant for τ_f values in the range from 220 to 320 deg, after which the two-switch solutions become possible. Figure 3.8 is for $\beta = 120$ deg with only two-switch solutions regardless of the length of the eclipse arc. For $\beta = 240$ deg, the solutions are identical to the case where $\beta = 60$ deg, except, of course, $S = -1$ instead of $S = 1$. This is not surprising because $\beta \pm \pi$ and β define the same line of nodes of the initial and final orbits, with the difference being that the rotations are of opposite signs (i.e., clockwise vs counterclockwise). The case of $\beta = 300$ deg is identical to the case of

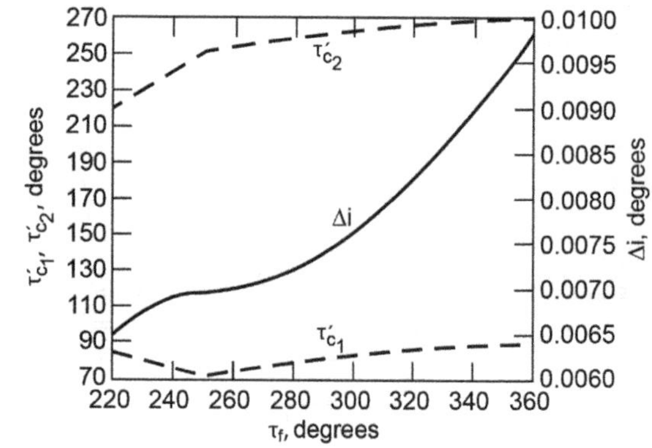

Fig. 3.6 Evolution of τ'_{c_1} and τ'_{c_2} and maximum change in inclination vs τ_f for $\beta = 0$ deg.

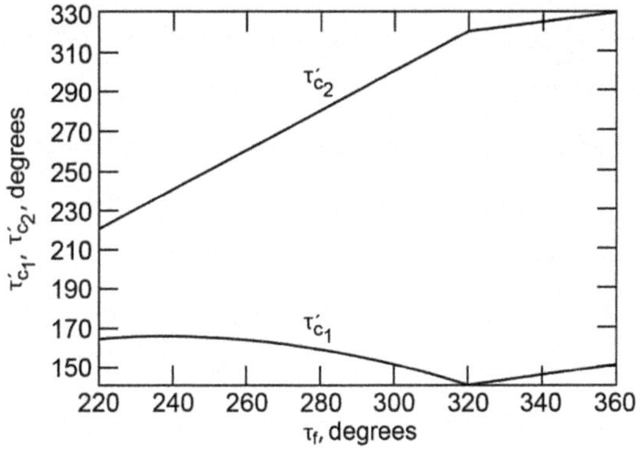

Fig. 3.7 τ'_{c_1} **and** τ'_{c_2} **vs** τ_f **for** $\beta = 60$ **deg.**

$\beta = 120$ deg except, once again, for the sign of S. In Fig. 3.8, Δi is shown for these particular β values as a function of τ_f, with equal maxima at $\tau_f = 2\pi$. So far in this section, we have selected the sequence $\pi/2$, $-\pi/2$, $\pi/2$ or the sequence $-\pi/2$, $\pi/2$, $-\pi/2$ for the yaw angle, corresponding to $S = 1$ or $S = -1$, respectively, on a purely arbitrary basis, thinking that it is very near-optimal to let $S = 1$ for $3\pi/2 < \beta < \pi/2$ and $S = -1$ for $\pi/2 \leq \beta \leq 3\pi/2$.

3.3.2 ALGORITHM 2

This algorithm is somewhat more complex than the preceding one because it computes both solutions, namely, for $S = 1$ and $S = -1$, and compares the Δi values achieved in each case. If one solution does not exist, then Δi is set equal to zero so that the other solution wins out. Furthermore, the two solutions, when they exist simultaneously, are not necessarily of the two-switch type, depending on β, because, in many cases, the second switch τ'_{c_2} becomes equal to τ_f, the eclipse entry point. As an example, for $\beta = 71$ deg, τ'_{c_1} and τ'_{c_2} are evaluated as functions of τ_f for the sequences $S = 1$ and $S = -1$, respectively. For this particular β value, two distinct solutions do indeed exist, but they yield the same Δi value. If β is increased to 80 deg, the two solutions will start to differ in the value of Δi that they yield because now the solution obtained with $S = -1$ provides a slightly superior Δi value. Operational considerations may favor one or the other solution,

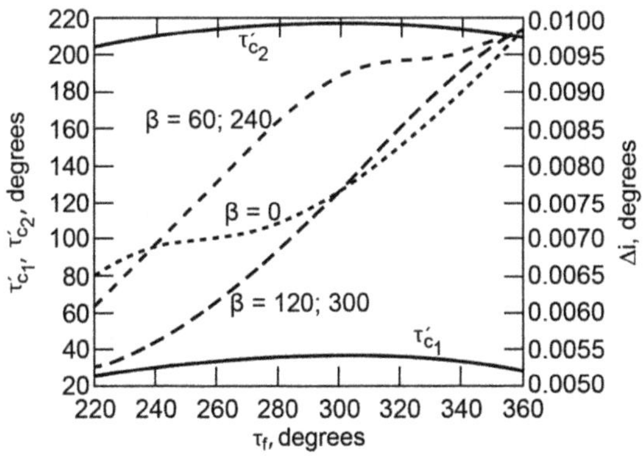

Fig. 3.8 τ'_{c_1} **and** τ'_{c_2} **vs** τ_f **for** $\beta = 120$ **deg, and inclination change vs** τ_f **for** $\beta = 0$, **60, 120, 240, and 300 deg.**

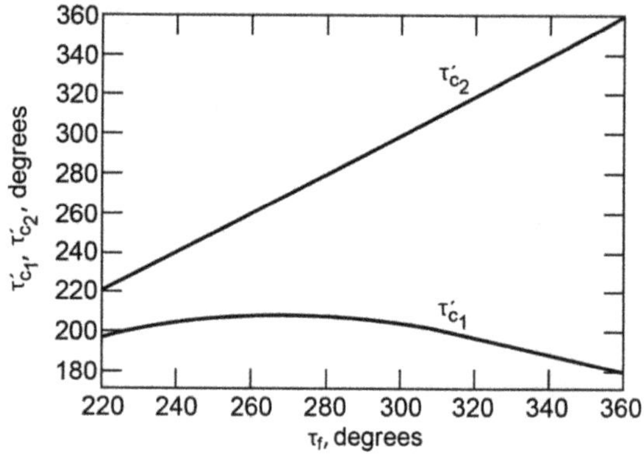

Fig. 3.9 τ'_{c_1} **and** τ'_{c_2} **vs** τ_f **for** $\beta = 89$ **deg and** $S = 1$**.**

depending, for example, on how fast the vehicle can be configured in attitude. At $\beta = 89$ deg, the $S = -1$ solution provides a maximum increase in Δi of about 4% over that of the $S = 1$ solution, showing that these solutions are almost equally good, as is clear from Figs. 3.9 and 3.10. This example of LEO shows that the $S = -1$ region should be extended to cover β values between 74 and 90 deg at the expense of the $S = 1$ solutions. Because of symmetry, the $S = 1$ solutions now will cover the 254 deg $< \beta <$ 270 deg range as well.

Figure 3.11 shows that, for $\tau_f = 220$ deg, $S = 1$ for $0 \leq \beta \leq 73$ deg, $S = -1$ for 74 deg $\leq \beta \leq$ 253 deg, and $S = 1$ for 254 deg $\leq \beta \leq$ 360 deg. As τ_f approaches 360 deg, the $S = 1$ and $S = -1$ regions will tend to correspond to $\beta = 90$ deg and $\beta = 270$ deg, respectively, as they should for this limiting case of no shadow. In short, Fig. 3.11, which is valid for LEO and corresponds to the most severe case of shadowing, shows how to select the yaw sequence for given τ_f and β. For example, for $\tau_f = 360$ deg, if 270 deg $\leq \beta \leq$ 90 deg, the sequence $S = 1$ is selected; otherwise, for all other values of β, it is $S = -1$ that will yield the largest Δi. Figure 3.12 shows how the optimal τ'_{c_1} and τ'_{c_2} values vary as functions of β for the given value of $\tau_f = 220$ deg. Finally, in Fig. 3.13, Δi vs β is plotted for various values of τ_f, namely, 220, 260, and 310 deg. The fluctuations vanish at $\tau_f = 360$ deg, and Δi becomes independent of β, displaying the constant value of 9.86×10^{-3} deg. Finally, for given β, Δi increases as τ_f is increased, as expected.

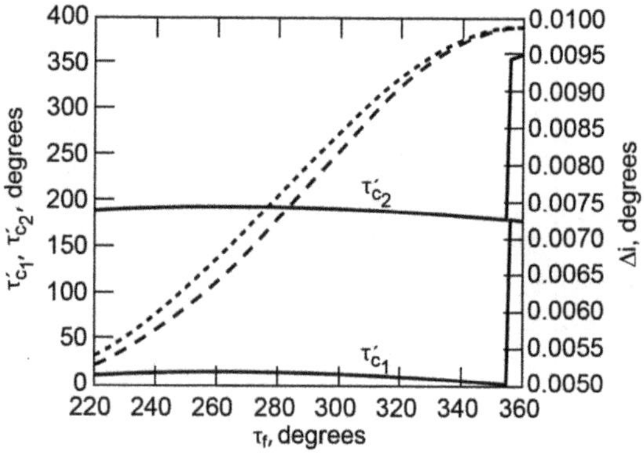

Fig. 3.10 τ'_{c_1} **and** τ'_{c_2} **vs** τ_f **for** $\beta = 89$ **deg and** $S = -1$**, and inclination change vs** τ_f **for** $S = -1$ **and** $S = 1$ **solutions.**

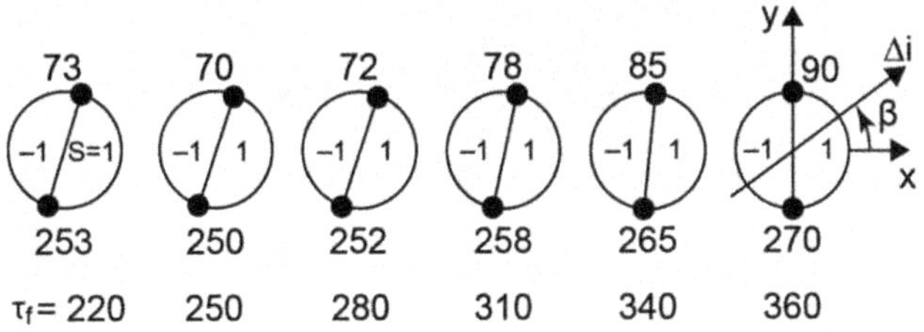

Fig. 3.11 $S = 1$ **and** $S = -1$ **regions of maximum-**Δi **solutions in LEO vs** τ_f.

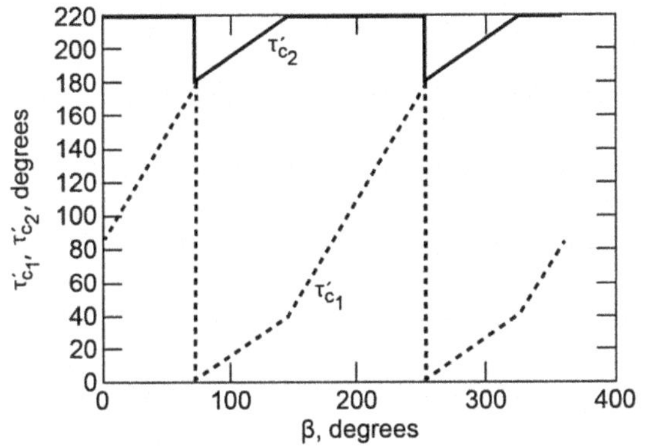

Fig. 3.12 Optimal τ'_{c_1} **and** τ'_{c_2} **vs** β **for** $\tau_f = 220$ **deg.**

3.4 CONCLUSION

In this chapter, the problem of low-thrust inclination control in near-circular orbit $(0 \leq e \leq 10^{-2})$ and in the presence of Earth shadow has been analyzed using simple steering laws consisting of piecewise-constant yaw angle selection. The linearized form of the variation-of-parameters equations is used, providing analytic expressions for the components of the inclination change vector due to out-of-plane thrusting. Two-switch transfer algorithms are analyzed, and their performances are compared. Algorithm 2, based on an optimal two-switch strategy, represents the overall better algorithm, and it provides a robust suboptimal transfer mode amenable to onboard autonomous guidance applications for future EOTVs.

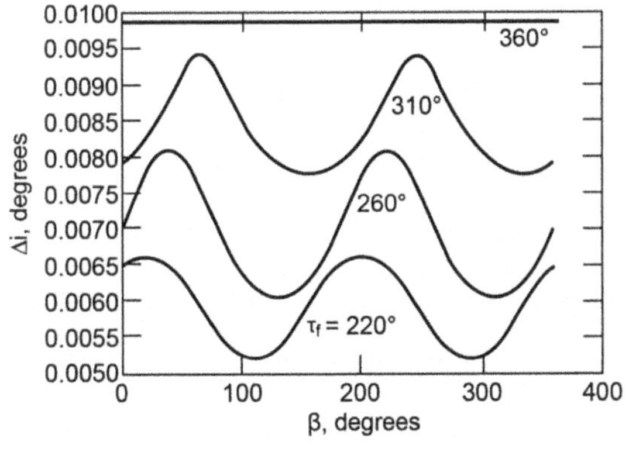

Fig. 3.13 Inclination change vs β **for** $\tau_f = 220, 260, 310,$ **and 360 deg.**

These algorithms are simple to implement and achieve maximum rotation of the orbit plane about the desired line of nodes of the current and final orbits using constant yaw angles, with the thrust turned off during shadowing.

REFERENCES

[1] Edelbaum, T. N., "Propulsion requirements for controllable satellites," *ARS Journal*, Vol. 31, No. 8, Aug. 1961, pp. 1079–1089.

[2] Wiesel, W. E., and Alfano, S., "Optimal many-revolution orbit transfer," AAS/AIAA Astrodynamics Specialist Conference, AAS Paper 83-352, Lake Placid, NY, Aug. 1983.

[3] Cass, J. R., "Discontinuous low thrust orbit transfer," M.S. Thesis, Rept. AFIT/GA/AA/83D-1, School of Engineering, U.S. Air Force Institute of Technology, Wright-Patterson AFB, OH, Dec. 1983.

[4] McCann, J. M., "Optimal launch time for a discontinuous low thrust orbit transfer," M.S. Thesis, Rept. AFIT/GA/AA/88D-7, School of Engineering, U.S. Air Force Institute of Technology, Wright-Patterson AFB, OH, Dec. 1988.

[5] Dickey, M. R., Klucz, R. S., Ennix, K. A., and Matuszak, L. M., "Development of the electric vehicle analyzer," Rept. AL-TR-90-006, Astronautics Laboratory, U.S. Air Force Space Technology Center, Edwards AFB, CA, June 1990.

[6] Kechichian, J. A., "Low-thrust eccentricity-constrained orbit raising," AAS Paper 91-156, AAS/AIAA Spaceflight Mechanics Meeting, Houston, TX, Feb. 1991.

[7] Kechichian, J. A., "Orbit raising with low-thrust tangential acceleration in the presence of Earth shadow," AAS/AIAA Astrodynamics Specialist Conference, AAS Paper 91-513, Durango, CO, Aug. 1991.

[8] Sackett, L. L., Malchow, H. L., and Edelbaum, T. N., "Solar electric geocentric transfer with attitude constraints: analysis," Report R901, Charles Stark Draper Laboratory, Inc., Cambridge, MA, Aug. 1975; also NASA CR-134927.

[9] Sackett, L. L., Malchow, H. L., and Edelbaum, T. N., "Solar electric geocentric transfer with attitude constraints: program manual," Report R902, Charles Stark Draper Laboratory, Inc., Cambridge, MA, Jan. 1975.

[10] Suskin, M. A., and Horsewood, J. L., "Enhancements to SECKSPOT: convergence in the presence of shadowing," AAS/AIAA Astrodynamics Specialist Conference, AAS Paper 93-667, Victoria, BC, Canada, Aug. 1993.

[11] Marec, J.-P., *Optimal Space Trajectories*, Elsevier, Amsterdam, 1979, pp. 159–168.

[12] Kechichian, J. A., "Low-thrust inclination control in the presence of Earth shadow," AAS Paper 91-157, AAS/AIAA Spaceflight Mechanics Meeting, Houston, TX, Feb. 1991.

[13] Kechichian, J. A., "Low-thrust inclination control in presence of Earth shadow," *Journal of Spacecraft and Rockets*, Vol. 35, No. 4, July–Aug. 1998, pp. 526–532.

CHAPTER 4

Orbit Plane Control Strategies for Inclined Geosynchronous Satellite Constellations

NOMENCLATURE

i^* = wedge angle between the premaneuver and postmaneuver orbits
i_M = ecliptic inclination of the lunar orbit
i_m = equatorial inclination of the lunar orbit
i_{max} = maximum orbit inclination
i_{min} = minimum orbit inclination
i_s = equatorial inclination of the Sun's apparent orbit
J_2 = second zonal harmonic of Earth's potential, 1.0827×10^{-3}
n = mean motion of the satellite orbit, rad/s
n_m = mean motion of the lunar orbit, 0.23 rad/day
n_s = mean motion of the Sun's apparent orbit, 0.017203 rad/day
R = perturbing function due to J_2 and luni-solar gravity
R_e = equatorial radius of Earth, 6378.14 km
V = velocity, km/s
ΔV = velocity change, km/s
μ = ratio of the mass of the moon to the combined masses of the Earth and moon, 1/82.3
μ_e = gravitational constant of Earth, 398,601.3 km^3/s^2
Ω_M = ecliptic ascending node of the lunar orbit
Ω_m = right ascension of the ascending node of the lunar orbit

4.1 INTRODUCTION

The dynamics and control of the motion of the orbit planes of an example constellation of five satellites in inclined geosynchronous orbits are analyzed in this chapter. Orbit maintenance strategies that confine each individual orbit inclination within a predefined tolerance deadband are designed using fast analytic orbit prediction approximations, as well as more exacting numerically integrated solutions. The initial orbit inclination of each satellite is chosen optimally within the tolerance band such that each satellite orbit remains confined within the limits of the band for the longest possible time before an inclination adjust maneuver is carried out. The total change in velocity required by each satellite over the constellation lifetime is thus rapidly determined, and the ideal initial nodes of the initially evenly spaced orbit planes that result in the minimum total velocity change accumulated by all five satellites is established. This total velocity change is also shown to be dependent on the ecliptic node of the lunar orbit within its 18.6-year regression cycle. A more elaborate scheme that maintains even spacing between the orbit planes at the equator by controlling each satellite node is also presented analytically, to optimize the initial nodal configuration that minimizes the total accumulated velocity change of all five satellites over the constellation lifetime.

The inclination i and the ascending node Ω of a satellite under geosynchronous conditions are affected by the perturbations due to the triaxiality of the Earth, as well as the gravitational perturbations due to the Sun and moon [1–13]. Single satellite control strategies are described by Billik [6], Balsam [8], Eckstein and Hechler [11], Kamel and Wagner [12], and Slavinskas et al. [14]. Murdoch and Pocha [15], and Walker [16], and in particular, Hubert and Swale [17] address the dynamics and control of a cluster of satellites in geostationary orbit with the cluster residing within well-defined limits in longitude and latitude. This chapter considers the problem of the orbit plane control of a constellation of geosynchronous satellites whose orbits are identically inclined with evenly spaced ascending nodes. It is assumed that each orbit plane contains a single satellite. Given a constellation lifetime, it is desired to maintain the inclination of each satellite orbit within a small tolerance band centered at the nominal inclination and determine the total accumulated change in velocity required by each satellite during its lifetime. The control strategies developed for the geostationary satellites are no longer applicable here because the nominal inclination for the application at hand is not near zero but rather 12 deg. Furthermore, because each satellite orbit starts from a different node and, therefore, experiences a different nodal rate, the ideal even spacing between nodes at the initial time is distorted if left uncontrolled. Using a combination of graphical, analytical, and numerical techniques, the following sections present orbit maintenance strategies that control either the inclination or both the inclination and node of each satellite, such that the tolerance band in inclination is never violated and, in the latter case, the evenly spaced nodal configuration is maintained over the constellation lifetime. Thus, the velocity change requirements for each satellite are rapidly determined. Because these requirements depend on both the initial nodal configuration of the evenly spaced constellation at the initial time and the lunar orbit ecliptic node, they are, in turn, minimized by the optimal selection of the initial nodes for a given start time and by the optimal selection of both the initial nodes and the start time within the 18.6-year regression cycle of the lunar orbit ecliptic node. The analysis in this chapter first appeared as a conference paper [18] and later as a journal article [19].

4.2 GENERAL DISCUSSION

For 24-h near-equatorial circular orbits, an inclination deadband is defined, and the satellite is placed initially at the maximum allowed inclination with an appropriate node Ω such that a maximum period of free drift occurs with both i and Ω varying and with i decreasing during the first half of that period and increasing during the second half until the inclination constraint is violated [9, 20]. An impulsive maneuver is then applied to target certain ideal initial conditions in i and Ω, such that the subsequent motion of the orbit plane complies with the inclination tolerance until the end of the mission. For small deadbands, these ideal initial conditions can be selected in such a manner that the time between two successive maneuvers is maximized. However, this simple orbit maintenance scheme becomes much more complicated if the satellite orbit is inclined at a much larger angle with respect to the equator. Unlike the near-equatorial case, for which the inclination deadband is defined by $0 \leq i \leq i_{max}$, the deadband is now defined by $i_{min} \leq i \leq i_{max}$. The maximum-duration free-drift period mentioned earlier is no longer feasible because i will violate the i_{min} constraint before the completion of the drift. Furthermore, because the satellites are positioned in five different planes that are evenly spaced in Ω, their Ω-drift rates are not identical, resulting in the distortion of the nodal spacing with time. In a first analysis, use is made of the orbit planes of the five satellites oscillating slowly about a dynamically steady invariant plane inclined at roughly 7.5 deg. Furthermore, by employing the $h_1 - h_2$ plane to represent the various orbit plane precessions, with $h_1 = s_i s_\Omega$ and $h_2 = s_i c_\Omega$ and with s_i and c_i representing sin i and cos i, respectively, one can define an inclination deadband in the form of an annulus consisting of the region confined between the two concentric circles centered at $h_1 = h_2 = 0$ with respective radii of $s_{i_{min}}$ and $s_{i_{max}}$. The nonsingular variables h_1 and h_2 effectively replace the pair (i, Ω) because the coupled system of differential equations for i and Ω has a singularity at $i = 0$. Thus, the use of h_1 and h_2 extends the applicability of this analysis to the near-equatorial case without any restrictions. A target locus confined to the tolerance annulus is also defined; it consists of four separate curves depending on the range of Ω, such that the impulsive maneuvers always target the orbit plane orientation parameters to given points on this locus curve. The symmetrical Ω spacing is controlled by adjusting the nodal spacing at the time of the inclination control maneuvers. This chapter shows how these maneuvers are automated and designed to control the constellation drift within well-defined bounds. Numerical integration is also used to generate more accurate requirements for changes in velocity for the entire constellation. We first use approximate graphical techniques consisting of the depiction of the evolution of the projection of the orbital angular momentum vector onto the equatorial plane to obtain preliminary evaluations of the free-drift characteristics of the constellation. The precession of a geosynchronous satellite orbit plane about the equilibrium plane at $i = 7.5$ deg and $\Omega = 0$ deg takes roughly 54 years for a complete 360 deg variation in Ω. The inclination deadband for a nominal 12 deg inclination with a ± 1 deg tolerance consists of the annulus between the two concentric circles centered at $h_1 = h_2 = 0$ with respective radii of $s_{i_{min}}$ and $s_{i_{max}}$ that are labeled "11°" and "13°," respectively, as shown in Fig. 4.1. The invariant plane is

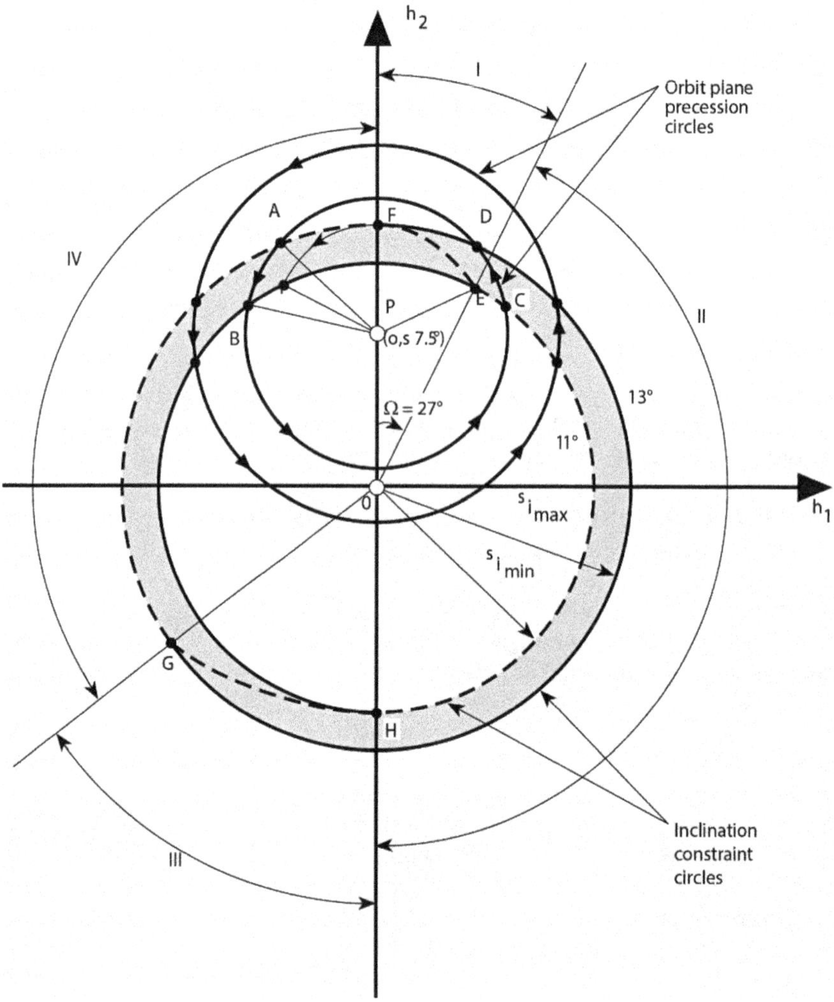

Fig. 4.1 Inclination deadband for minimum and maximum constraints.

represented by the point $P(h_1 = 0, h_2 = s_{7.5^\circ})$ on the h_2 axis, and the orbit-plane precessions shown by concentric circles centered at P. Each point (h_1, h_2) corresponds to a well-defined orbit-plane orientation whose parameters are given by i and Ω. Because of the definitions of h_1 and h_2, a point in the (h_1, h_2) plane has polar coordinates (s_i, Ω) with the angle Ω measured from the h_2 axis and with s_i being its distance from the origin O. Starting from a given (i, Ω) pair, or equivalently an (h_1, h_2) pair, the orbit plane will precess such that the subsequent motion in the $h_1 - h_2$ plane describes, to the zeroth order, a circle centered at the equilibrium point P. Two such precession circles are shown in Fig. 4.1. Only those portions of these precession circles that are within the annulus are acceptable in the sense that they comply with and satisfy the inclination tolerance deadband. For small tolerances on the nominal inclination, say, on the order of ± 1 or 2 deg, let us adopt the dotted line *EFAGHE* as the locus of the initial orbits to target, such that, starting from any point on this locus, it will take the longest possible time for the orbit to violate the inclination deadband constraint.

The portions *EF* and *GH* are circular arcs centered at P and lying entirely within the annulus. The circular arcs *FG* and *HE* are centered at O and lie on the $s_{i_{max}}$ and $s_{i_{min}}$ circles, respectively. The tolerance annulus can therefore be divided into four regions, namely, I, II, III, and IV, as indicated in Fig. 4.1. For a constant precession rate around P and a nominal 10-year constellation lifetime, a nominal 10-year precession will result in an angular motion of nearly 66 deg. As shown in Fig. 4.2, with $i_{min} = 11$ deg and $i_{max} = 13$ deg, if we place the common orbit of the five satellites at point E, the precession arc *EFI* remains within the tolerance annulus for the longest possible time and covers a central angle of 130 deg in roughly 20 years. If we place the orbit plane at G, the precession arc *GHQ* will cover only 68 deg before deadband violation. This strategy is also acceptable because it results in a 10-year precession without any maneuvering. Point E corresponds to an initial value of $\Omega_i = 27$ deg at $i = 11$ deg. Figure 4.2 also shows a larger deadband with $i_{min} = 7$ deg and $i_{max} = 17$ deg or a ± 5 deg tolerance from the nominal $i = 12$ deg inclination. The arc $E'F'I'$ covers a central angle of 263 deg in nearly 39 years, and the arc $G'H'Q'$ covers an angle of 189 deg in 28 years with no deadband violation or any maneuvering requirements. There are many more feasible initial orbits that satisfy the 10-year lifetime requirement with no maneuvering for this larger-deadband example, as expected.

Let us now consider each satellite flying in a different orbit plane with the five planes equally spaced in Ω. In Fig. 4.3, $i = 12$ deg, the nominal value, and the satellite planes are given by $\Omega = 0$, 72, 144, 216, and 288 deg, corresponding to the points J, K, L, M, and N, respectively, with equal spacing of $\Delta\Omega = 72$ deg. The lines labeled 1, 2, 3, 4, and 5 identify the five nodal directions. A 10-year precession of the five satellite orbits will result in the precession arcs JJ', KK', LL', MM', and NN', each covering an equal central angle of 66 deg at P, the equilibrium point. The smallest and largest inclinations achieved are $i = 4.5$ deg and $i = 19.8$ deg by satellites 5 and 3, respectively, measured by the corresponding segments ON'

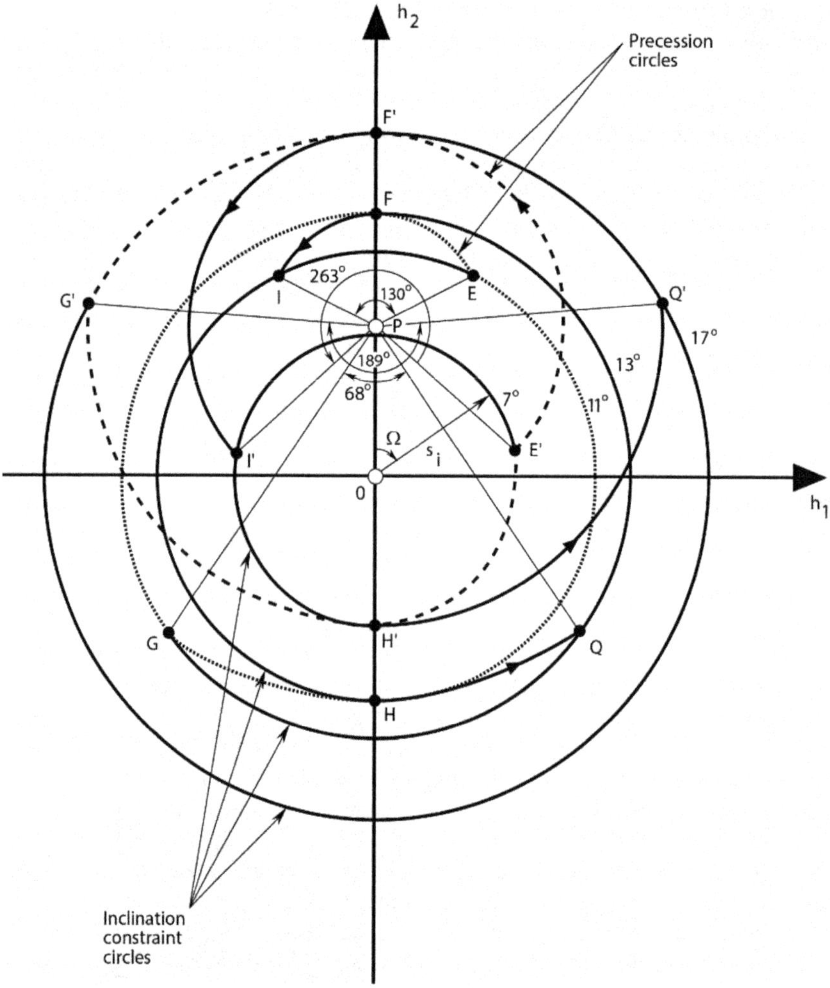

Fig. 4.2 Maximum-duration compliance strategies for given tolerance deadbands with satellites in common orbit plane.

and OL' equal to the sines of the corresponding inclinations. However, the nodes of the orbits are now given by the directions of the lines $1'$, $2'$, $3'$, $4'$, and $5'$ measured at O from the h_2 axis, showing significant distortion with respect to the initial symmetric configuration. The largest nodal separation is between satellites 1 and 5 with $\Delta\Omega = 142$ deg or twice the initial separation. This indicates that a tolerance deadband in the node must also be considered to maintain proper spacing, in addition to the inclination deadband discussed so far.

In Fig. 4.4, the initial nodal configuration is given by $\Omega = 24, 96, 168, 240$, and 312 deg represented by the lines 1, 2, 3, 4, and 5, respectively, such that, at the end of the 10-year drift, the nodes of the satellites' orbits are given by the directions $1'$, $2'$, $3'$, $4'$, and $5'$, respectively. The largest nodal separation is between satellites 4 and 5 with $\Delta\Omega = 132$ deg, or slightly less than for the configuration in Fig. 4.3. Furthermore, point N' is closest to the origin O, with $ON' = s_i$ and $i = 4$ deg representing the largest deviation from the nominal 12 deg. In Fig. 4.5, the initial Ω distribution is given by $\Omega = 48$, 120, 192, 264, and 336 deg with final configuration along $1'$, $2'$, $3'$, $4'$, and $5'$ and the largest $\Delta\Omega$ between satellites 4

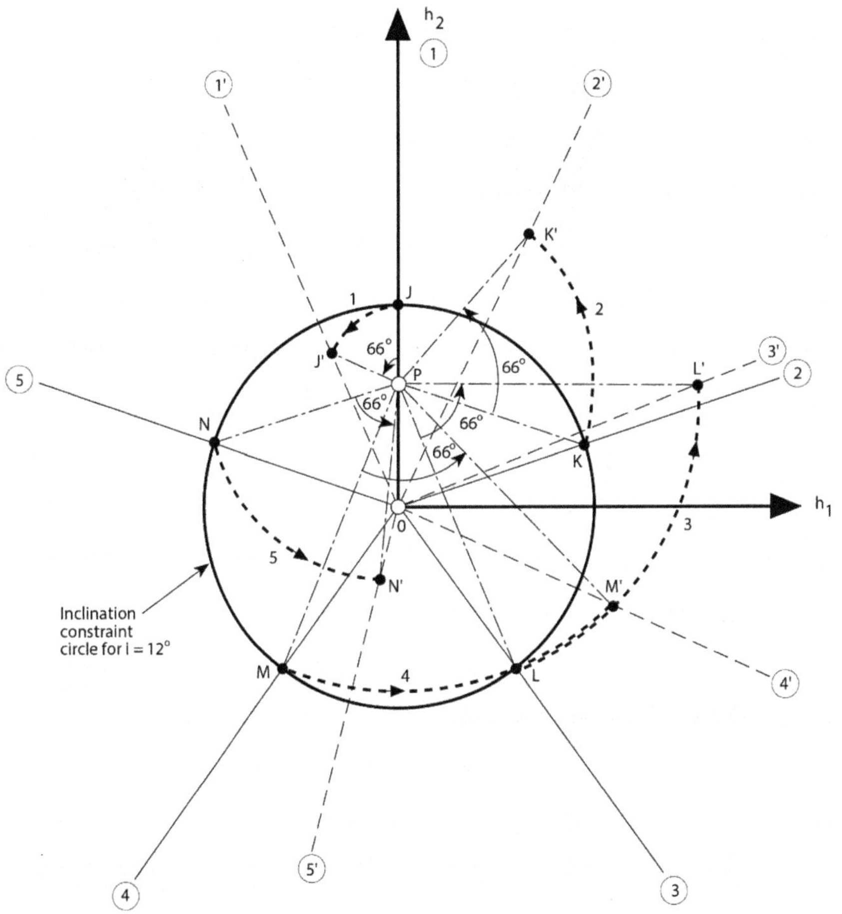

Fig. 4.3 Orbit plane precession of constellation with initial inclination of 12 deg and initial nodes of 0, 72, 144, 216, and 288 deg.

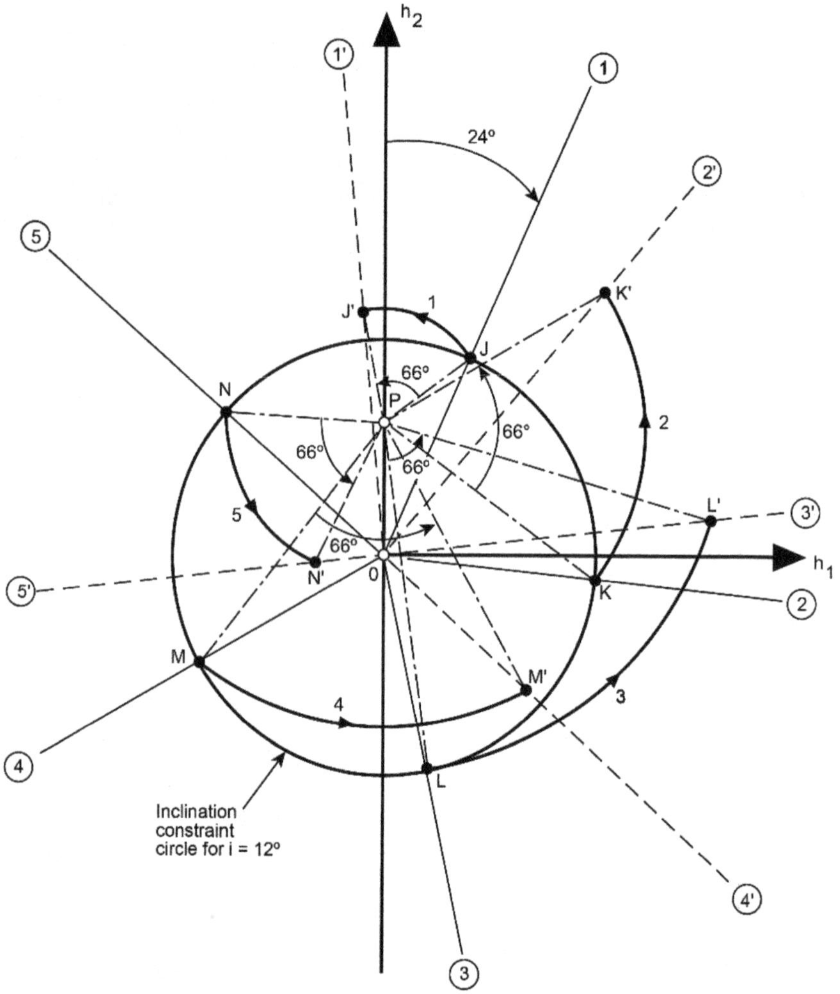

Fig. 4.4 Orbit plane precession of constellation with initial inclination of 12 deg and initial nodes of 24, 96, 168, 240, and 312 deg.

and 5 with $\Delta\Omega = 157$ deg. The largest inclination excursion is achieved by satellite 2 with $OK' = s_i$ and $i = 20.1$ deg. This shows that there exists an optimum initial Ω configuration that will result in the smallest inclination excursion from the 12 deg nominal inclination at the end of the 10-year drift period. However, the distortion in the initial uniform Ω distribution must also be taken into account, and the overall optimum initial configuration must be found to minimize the inclination excursion as well as the Ω distortion.

If we move the initial configuration to $\Omega = 72$, 144, 216, 288, and 360 deg, corresponding to a further rotation of 24 deg from the configuration of Fig. 4.5, then we will recover the initial configuration of Fig. 4.3. This means that the smallest inclination excursion in the 10-year drift period corresponding to the optimum initial configuration can be iterated from these three figures. In fact, the locus of all possible precession end points at the 10-year mark that originate from the common initial nominal inclination at 12 deg and any initial Ω value between 0 and 360 deg is shown in Fig. 4.6. The smallest and largest final inclinations are shown by the points N' and M', respectively, with $ON' = s_{i_{\min}}$ and

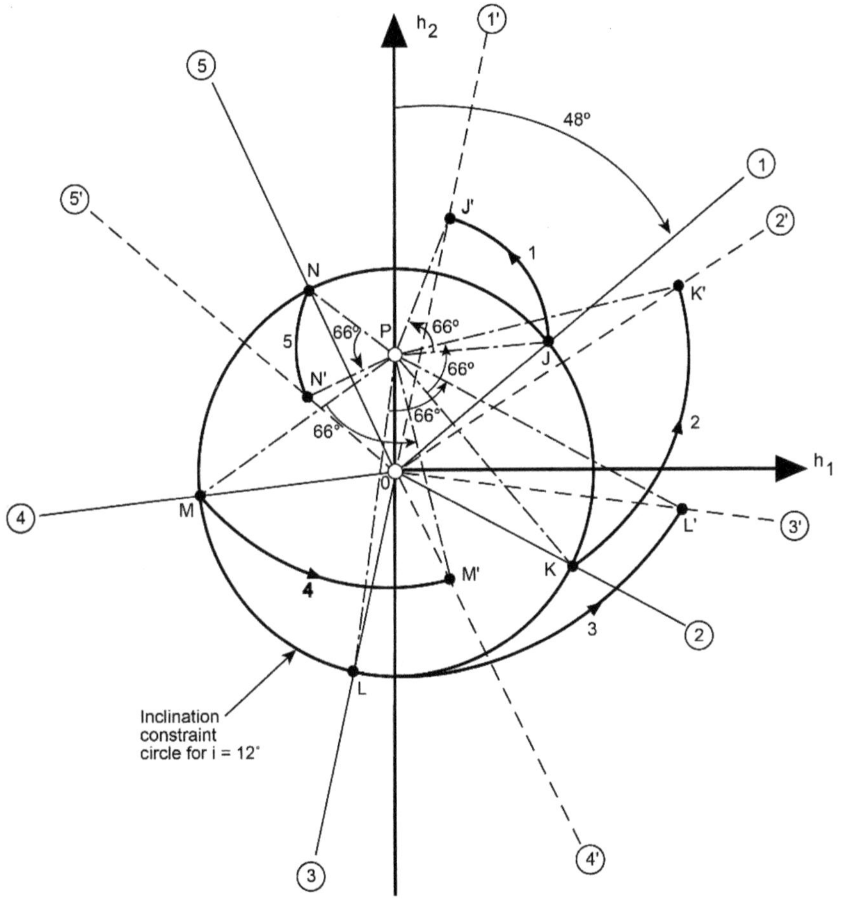

Fig. 4.5 Orbit plane precession of constellation with initial inclination of 12 deg and initial nodes of 48, 120, 192, 264, and 336 deg.

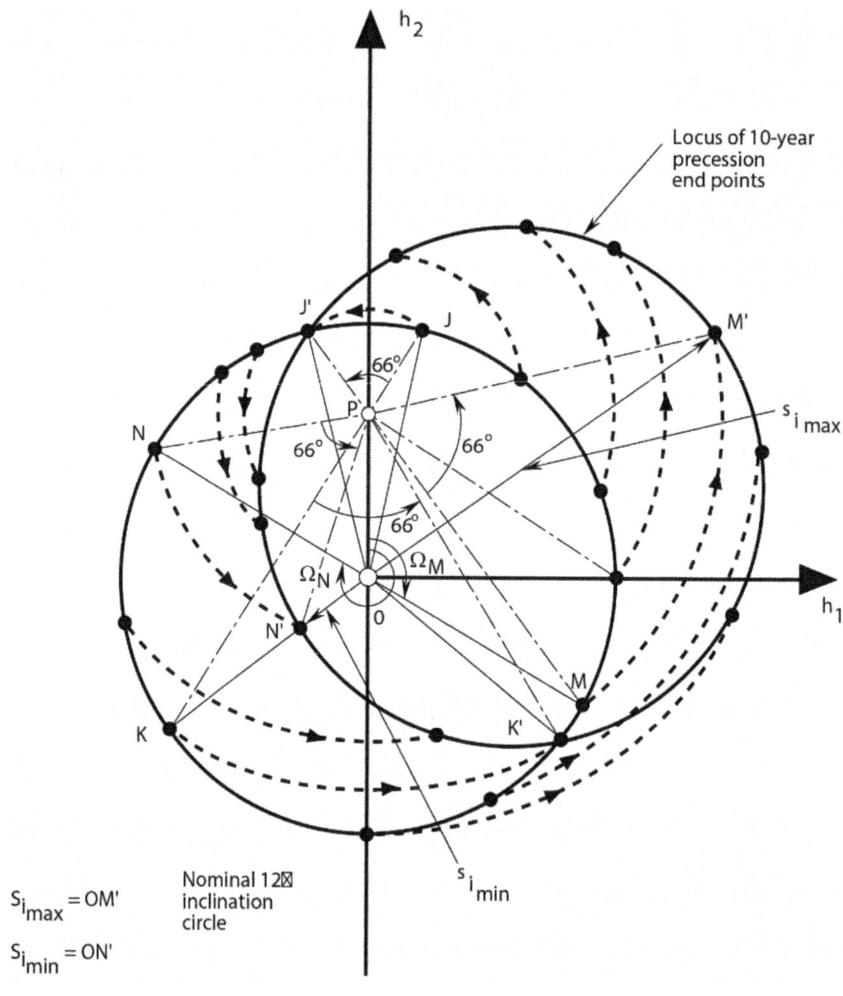

Fig. 4.6 Locus of 10-year precession end points for an initial inclination of 12 deg.

$OM' = s_{i_{max}}$ and with $i_{min} \approx 3.6$ deg and $i_{max} \approx 20.1$ deg, indicating that the largest change in the inclination is on the order of 8.5 deg. Similar loci can be plotted for different values of the satellites' lifetime to determine the largest possible changes in the inclination over that particular lifetime. The two points of intersection between a given locus and the nominal inclination constraint circle are points that result in zero change in the inclination. For example, the precessions JJ' and KK' start and end at the nominal inclination, with small and acceptable deviations during the precessions.

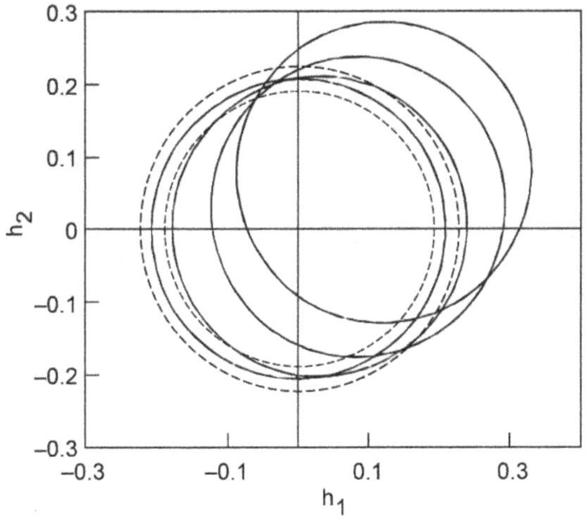

Fig. 4.7 Loci of 2-, 6-, and 10-year precession end points for an initial inclination of 12 deg.

Figure 4.7 shows the 2-, 6-, and 10-year precession end points for the case of 12 deg nominal inclination. These loci are circles that are rotated by the corresponding precession angles about the equilibrium point situated on the h_2 axis. Figure 4.8 shows the 2-, 4-, 6-, 8-, and 10-year mappings of the original inclination constraint circle, whereas Fig. 4.9 shows the $h_1 - h_2$ mappings for the much longer 10-, 20-, 30-, 40-, and 50-year precessions. The case of 54-year precession maps the original nominal constraint circle back onto itself after a complete 360 deg rotation about the same equilibrium point. This is how the satellite orbit plane oscillates slowly about its idealized equilibrium orientation with a period of 54 years. The automation of these approximate zeroth-order analytic precessions was described by Musen et al. [2] and also briefly in Sec. 4.3 of this chapter. Initially, the satellite orbits are evenly spaced in Ω and lying on the dotted line of Fig. 4.1. When one of the satellites reaches the boundary of the annulus, a maneuver must place it back on the dotted curve for another period of free-drift. This maneuver can consist of a pure inclination change at constant Ω. However, the other 4 satellites experience different Ω rates such that they too must be maneuvered back to the dotted curve,

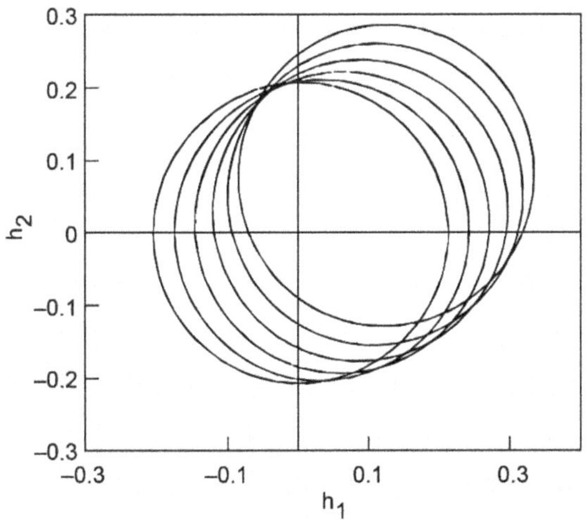

Fig. 4.8 Loci of 2-, 4-, 6-, 8-, and 10-year precession end points for an initial inclination of 12 deg.

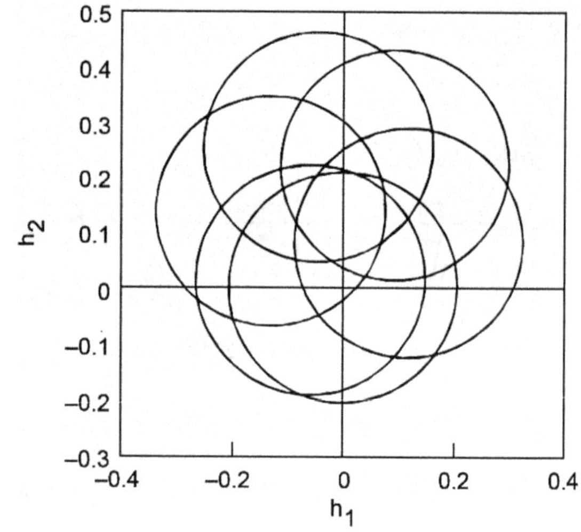

Fig. 4.9 Loci of 10-, 20-, 30-, 40-, and 50-year precession end points for an initial inclination of 12 deg.

but this time, each satellite must also adjust its Ω value in such a way that all five orbits are once again evenly spaced in Ω. This scenario is then repeated until the 10-year satellite lifetime is consumed. The maximum change in velocity needed by the most active satellite can thus be evaluated. Furthermore, the optimal initial configuration can also be determined to minimize this maximum ΔV budget. For larger inclination deadbands, such as in Fig. 4.2, it may be prohibitive to maneuver the vehicles back to the dotted target locus, thereby requiring the implementation of more elaborate targeting strategies to minimize fuel consumption over the mission lifetime.

4.3 INCLINATION CONTROL STRATEGY WITH ANALYTIC ORBIT PREDICTION

Given the pair of inclination constraints $(i_{\min}, i_{\max})$, the coordinates of points E and G are first computed as follows: Point E is the intersection of the $s_{i_{\min}}$ constraint circle and the EF precession arc of radius $PF = r_e = s_{i_{\max}} - h_{2e}$, where $h_{2e} = s_{7.5\text{deg}}$ such that

$$h_1^2 + h_2^2 = s_{i_{\min}}^2 \tag{4.1}$$

$$h_1^2 + (h_2 - h_{2e})^2 = r_e^2 = (s_{i_{\max}} - h_{2e})^2 \tag{4.2}$$

The solution is given by

$$(h_2)_E = s_{i_{\max}} - \frac{(s_{i_{\max}}^2 - s_{i_{\min}}^2)}{2h_{2e}} \tag{4.3}$$

$$(h_1)_E = \left[s_{i_{\min}}^2 - (h_2)_E^2\right]^{1/2} \tag{4.4}$$

The plus sign is chosen in front of the radical in Eq. (4.4) because point E will always be on the right-hand side of the diagram in Fig. 4.1, such that $(h_1)_E > 0$.

The node and inclination are obtained from the general formulas

$$\Omega = \tan^{-1}(h_1/h_2) \tag{4.5}$$

$$i = \sin^{-1}\left[\left(h_1^2 + h_2^2\right)^{1/2}\right] \tag{4.6}$$

Point G is at the intersection of the circle of radius $s_{i_{\max}}$ and precession arc GH of radius $r_g = PG = s_{i_{\min}} + h_{2e}$ such that

$$h_1^2 + h_2^2 = s_{i_{\max}}^2 \tag{4.7}$$

$$h_1^2 + (h_2 - h_{2e})^2 = \left(s_{i_{\min}} + h_{2e}\right)^2 \tag{4.8}$$

The solution is given by

$$(h_2)_G = \frac{\left(s_{i_{\max}}^2 - s_{i_{\min}}^2\right)}{2h_{2e}} - s_{i_{\min}} \tag{4.9}$$

$$(h_1)_G = -\left[s_{i_{\max}}^2 - (h_2)_G^2\right]^{1/2} \tag{4.10}$$

Like point E, point G can be either above or below the h_1 axis; however, $(h_1)_G < 0$ because G is always on the left-hand side of the (h_1, h_2) diagram. Therefore, the minus sign in front of the radical must be used in Eq. (4.10). We now have evaluated the pairs (i_E, Ω_E) and (i_G, Ω_G) with of course $i_E = i_{\min}$ and $i_G = i_{\max}$. The regions I, II, III, and IV are defined by the intervals $0 \le \Omega < \Omega_E$, $\Omega_E \le \Omega < \pi$, $\pi \le \Omega < \Omega_G$, and $\Omega_G \le \Omega < 2\pi$, respectively. At time zero, the orbits are evenly spaced in Ω and must be located on the target locus defined in Fig. 4.1. For region I, the initial orbit must be located on the circular arc FE at the intersection of this arc of equation

$$h_1^2 + (h_2 - h_{2e})^2 = (s_{i_{\max}} - h_{2e})^2$$

and the line $h_1 = (\tan\Omega)h_2$. Solving for h_2 yields

$$h_2 = h_{2e}c_\Omega^2 \pm \left[h_{2e}^2 c_\Omega^4 + c_\Omega^2\left(s_{i_{\max}}^2 - 2h_{2e}s_{i_{\max}}\right)\right]^{1/2} \tag{4.11}$$

If $\Omega > \pi/2$, then the condition $h_2 < 0$ requires the selection of the minus sign in front of the radical. The term $\left(s_{i_{\max}} - 2h_{2e}\right)$ can be either positive or negative depending on the value of $i_{\max}$. If $i_{\max} > 15.132$ deg, then $\left(s_{i_{\max}} - 2h_{2e}\right) > 0$, such that the radical term in Eq. (4.11) is larger than the first term $h_{2e}c_\Omega^2$, which is always positive. In this case, the two intersection points between the FE precession arc centered at P and the straight line given by $h_1 = (\tan\Omega)h_2$ are such that one is above the h_1 axis with $h_2 > 0$ and the other is below the h_1 axis with $h_2 < 0$, with the plus sign selected in front of the radical in the first case and the minus sign in the second case. If $i_{\max} < 15.132$ deg, then $\left(s_{i_{\max}} - 2h_{2e}\right) < 0$, and this time, the intersection points are both above the h_1

axis with $h_2 > 0$. The radical term is now smaller than the leading $h_{2e}c_\Omega^2$ term, with the plus sign corresponding to the higher intersection point. In either case

$$\text{if } \Omega < \pi/2, \quad h_2 = h_{2e}c_\Omega^2 + \left[h_{2e}^2 c_\Omega^4 + c_\Omega^2\left(s_{i_{\max}}^2 - 2h_{2e}s_{i_{\max}}\right)\right]^{1/2} \tag{4.12}$$

and

$$\text{if } \Omega > \pi/2, \quad h_2 = h_{2e}c_\Omega^2 - \left[h_{2e}^2 c_\Omega^4 + c_\Omega^2\left(s_{i_{\max}}^2 - 2h_{2e}s_{i_{\max}}\right)\right]^{1/2} \tag{4.13}$$

Once h_2 is determined, the value of h_1 is obtained from $h_1 = (\tan\Omega)h_2$. For region II, we need $i = i_{\min}$, and therefore, the coordinates of the corresponding initial orbit are given by

$$h_1 = s_{i_{\min}} s_\Omega \tag{4.14}$$

$$h_2 = s_{i_{\min}} c_\Omega \tag{4.15}$$

Similarly, for region IV, we need $i = i_{\max}$, and therefore

$$h_1 = s_{i_{\max}} s_\Omega \tag{4.16}$$

$$h_2 = s_{i_{\max}} c_\Omega \tag{4.17}$$

For region III, given a value of Ω, the (h_1, h_2) coordinates at time zero are given by the intersection of the *GH* precession arc centered at *P* with the equation

$$h_1^2 + (h_2 - h_{2e})^2 = \left(s_{i_{\min}} + h_{2e}\right)^2$$

and the straight line $h_1 = (\tan\Omega)h_2$. The solution is given by

$$h_2 = h_{2e}c_\Omega^2 \pm \left[h_{2e}^2 c_\Omega^4 + c_\Omega^2\left(s_{i_{\min}}^2 + 2h_{2e}s_{i_{\min}}\right)\right]^{1/2} \tag{4.18}$$

Even though $\Omega > \pi$, such that $h_1 < 0$, h_2 can be either positive or negative depending on whether $\Omega > 3\pi/2$. Because $\left(s_{i_{\min}}^2 + 2h_{2e}s_{i_{\min}}\right)$ is always positive, the radical term in Eq. (4.18) is always larger than the positive first term $h_{2e}c_\Omega^2$ such that the two solutions corresponding to the two intersection points always have $h_2 > 0$ for the first point and $h_2 < 0$ for the other point, because these two points can never be on the same side of the h_1 axis with the same sign for their h_2 values. Therefore

$$\text{if } \Omega < 3\pi/2, \quad h_2 = h_{2e}c_\Omega^2 - \left[h_{2e}^2 c_\Omega^4 + c_\Omega^2\left(s_{i_{\min}}^2 + 2h_{2e}s_{i_{\min}}\right)\right]^{1/2} \tag{4.19}$$

and

$$\text{if } \Omega > 3\pi/2, \quad h_2 = h_{2e}c_\Omega^2 + \left[h_{2e}^2 c_\Omega^4 + c_\Omega^2\left(s_{i_{\min}}^2 + 2h_{2e}s_{i_{\min}}\right)\right]^{1/2} \tag{4.20}$$

As before, the corresponding h_1 coordinate is obtained from $h_1 = (\tan\Omega)h_2$, and finally, as for the case of region I, the intermediate value of the inclination

$i_{\min} \le i \le i_{\max}$ is obtained from Eq. (4.21) without ambiguity as

$$i = \sin^{-1}\left[\left(h_1^2 + h_2^2\right)^{1/2}\right] \tag{4.21}$$

For a given initial set values Ω_1, Ω_2, Ω_3, Ω_4, and Ω_5 separated by $360/5 = 72$ deg each, the initial (h_1, h_2) coordinates of all five satellites are computed according to the regions in which they initially lie, and these coordinates are then predicted forward in time using the analytic circular precession approximation discussed earlier. Given a time Δt, say, on the order of a few days, and given the precession angular velocity $\omega = 2\pi/P$ where P is the period of these precessions centered at P with a value of roughly 54 years, the angular motion $\Delta\theta'' = \omega\Delta t$ is first calculated, after which the radius of the particular precession circle is evaluated from the expression

$$r = \left\{h_1^2(i) + [h_2(i) - h_{2e}]^2\right\}^{1/2}$$

Next, an angle ε is computed as described by Musen et al. [2]

$$\varepsilon = \tan^{-1}\left[\frac{h_1(i)}{h_{2e} - h_2(i)}\right] \tag{4.23}$$

the linear distance d is determined from $d = 2r\,\sin(\Delta\theta''/2)$, and finally, the predicted coordinates are obtained as [2]

$$h_1(i+1) = h_1(i) + d\cos\left(\varepsilon + \frac{\Delta\theta''}{2}\right) \tag{4.24}$$

$$h_2(i+1) = h_2(i) + d\sin\left(\varepsilon + \frac{\Delta\theta''}{2}\right) \tag{4.25}$$

with time $i+1$ being equal to $i + \Delta t$. These calculations are repeated every Δt step for each of the satellites in the constellation, and their trajectories in the h_1–h_2 plane are described. At each step, the inclination of each satellite orbit is tested for possible violation of the $i_{\min}$ or $i_{\max}$ constraint. As soon as a violation occurs, inclination of that satellite's orbit is retargeted to the target locus curve, leaving its node unchanged. The change in velocity needed for the inclination adjust maneuver is evaluated from $\Delta V = 2V\sin(i^*/2)$, where V is the geosynchronous orbit velocity $V = \sqrt{\mu_e/a_s}$, $\mu_e = 398{,}601.3\ \mathrm{km^3/s^2}$ is the gravitational constant of Earth, $a_s = 42{,}164$ km is the orbit semimajor axis, and i^* is the wedge angle between the premaneuver and postmaneuver orbits given by the (Ω_1, i_1) and (Ω_1, i_2) pairs, respectively [2]

$$c_{i^*} = s_{i_1}s_{i_2} + c_{i_1}c_{i_2} \tag{4.26}$$

i_2 is, of course, the target inclination that will place that particular orbit on the target locus for another period of maneuver-free precession. Each satellite

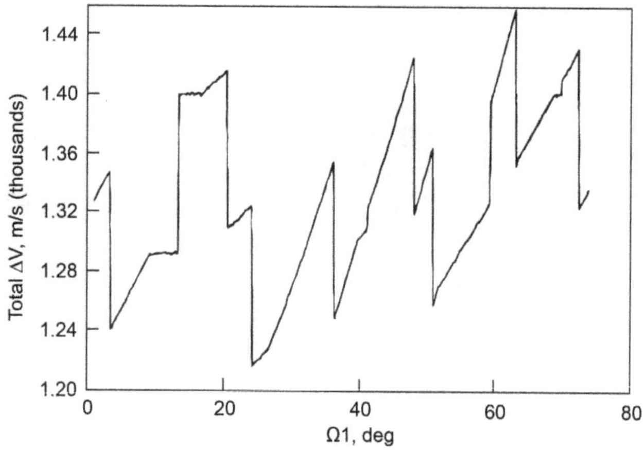

Fig. 4.10 Total ΔV requirements for a five-satellite constellation at nominal 12 deg inclination and ± 1 deg tolerance vs initial node configuration for a 4000-day lifetime.

performs such a maneuver a number of times during a predefined constellation lifetime, say, 10 years, and the total change in velocity needed by each satellite is evaluated by adding the individual small changes as they occur. If we now add the total velocity changes from the individual satellites, then the overall velocity change ΔV_{tot} needed by all five satellites can be calculated. This calculation is repeated for each initial value of Ω_1, and therefore, the initial values of $\Omega 2, \ldots, \Omega_5$, because they are incremented by 72 deg at time zero, and the total quantity ΔV_{tot} is plotted as a function of Ω_1, as in Fig. 4.10, with Ω_1 varying between 0 and 72 deg. The minimum ΔV_{tot} value occurs for $\Omega_1 = 24.3$ deg at 1216 m/s over a period of 4000 days, whereas the maximum ΔV_{tot} value corresponds to an initial $\Omega_1 = 63$ deg at 1456 m/s. Figure 4.11 shows the trajectories of the five satellites for the overall ΔV-minimizing initial configuration given by nodal values of $\Omega_1 = 24.3$ deg, $\Omega_2 = 96.3$ deg, $\Omega_3 = 168.3$ deg, $\Omega_4 = 240.3$ deg, and $\Omega_5 = 312.3$ deg. Figure 4.12 shows the variation of the nodes over the same 4000-day lifetime for an initial configuration given by $\Omega_1 = 0$ deg, $\Omega_2 = 72$ deg, $\Omega_3 = 144$ deg, $\Omega_4 = 216$ deg, and $\Omega_5 = 288$ deg. Because the nodes are left uncontrolled, the even nodal spacing is quickly distorted, as was clear from the earlier graphical discussion.

The h_1–h_2 trajectories in this latter example are shown in Fig. 4.13, with spacecraft 1, 2, 3, 4, and 5 requiring one, three, four, one, and four inclination adjust maneuvers, respectively. Figure 4.14 shows the accumulated velocity change for each satellite, with individual totals of 322.611, 430.421, 215.374, 350.309, and 107.558 m/s, respectively.

These values are also reported in Table 4.1.

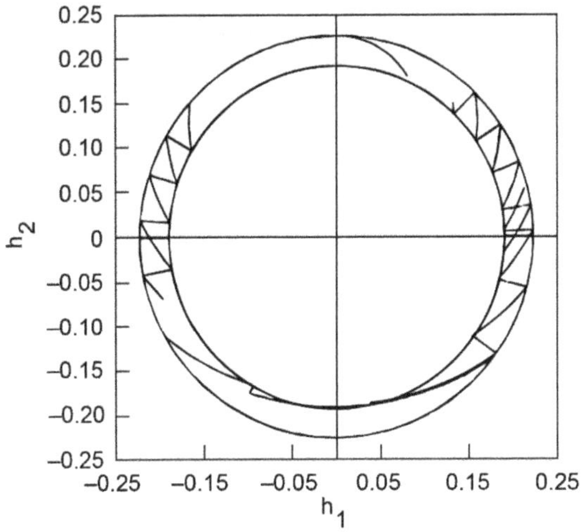

Fig. 4.11 Constellation trajectories for an optimal initial node configuration of $\Omega_1 = 24.3$ deg, $\Omega_2 = 96.3$ deg, $\Omega_3 = 168.3$ deg, $\Omega_4 = 240.3$ deg, and $\Omega_5 = 312.3$ deg over a 4000-day lifetime.

4.4 INCLINATION CONTROL AND NODE ADJUST MANEUVER STRATEGY WITH ANALYTIC ORBIT PREDICTION

In this section, a simple strategy is presented that controls the nodal distortion by adjusting the nodal spacing at each inclination control event. The vehicle carrying out an inclination change maneuver leaves its node unchanged, as before. However, the other four vehicles that do not require any inclination change at this stage adjust their respective nodes such that the even nodal spacing with

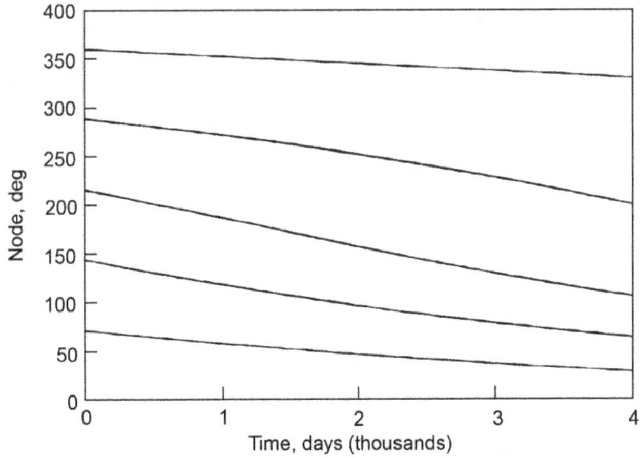

Fig. 4.12 Uncontrolled nodal variation of a five-satellite constellation over a 4000-day lifetime for initial values of $\Omega_1 = 0$ deg, $\Omega_2 = 72$ deg, $\Omega_3 = 144$ deg, $\Omega_4 = 216$ deg, and $\Omega_5 = 288$ deg.

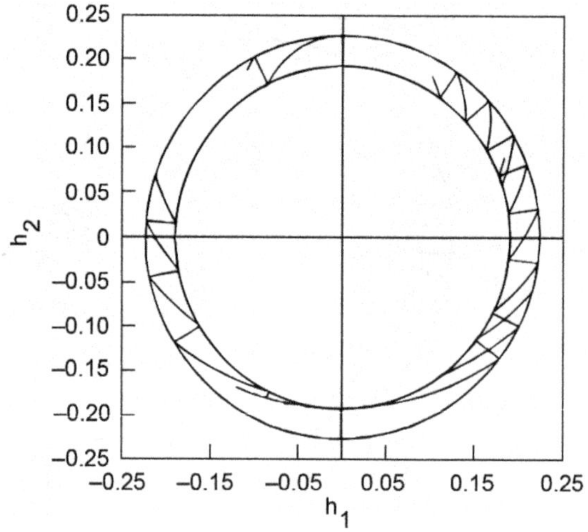

Fig. 4.13 Constellation trajectories over a 4000-day lifetime for initial values of $\Omega_1 = 0$ deg, $\Omega_2 = 72$ deg, $\Omega_3 = 144$ deg, $\Omega_4 = 216$ deg, and $\Omega_5 = 288$ deg and a nominal inclination of 12 deg with a ± 1 deg tolerance.

respect to the vehicle performing the inclination change, is recovered, meaning that all five nodes are once again equally distributed with a spacing of 72 deg. Figure 4.15 shows the evolution of the constellation satellite nodes over a 3650-day period for an initial configuration given by $\Omega_1 = 0$ deg, $\Omega_2 = 72$ deg, $\Omega_3 = 144$ deg, $\Omega_4 = 216$ deg, and $\Omega_5 = 288$ deg, with no excessive distortion. Figure 4.16 depicts the five satellites' polar trajectories, with the nodal maneuvers represented by the short circumferential instantaneous adjustments at constant i whereas the inclination change maneuvers are performed by instantaneous radial adjustments

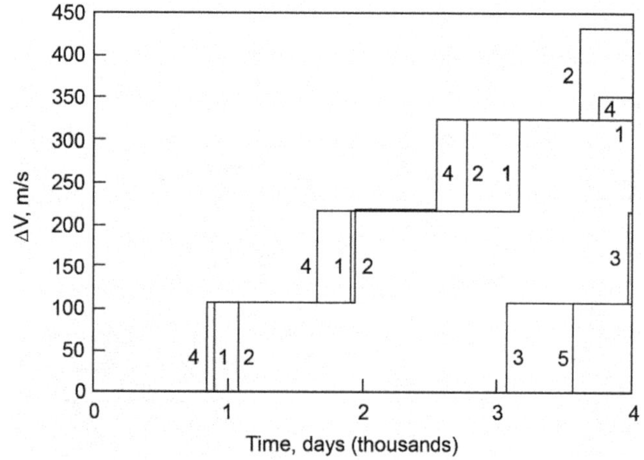

Fig. 4.14 Individual accumulated total ΔV requirements for a five-satellite constellation over a 4000-day lifetime for initial values of $\Omega_1 = 0$ deg, $\Omega_2 = 72$ deg, $\Omega_3 = 144$ deg, $\Omega_4 = 216$ deg, and $\Omega_5 = 288$ deg.

TABLE 4.1 ACCUMULATED INDIVIDUAL SATELLITE VELOCITY CHANGES IN 4000 DAYS WITH INCLINATION CONTROL AND SPECIFIED INITIAL NODES

	Sat. 1 ($\Omega_1 = 0$ deg)	Sat. 2 ($\Omega_2 = 72$ deg)	Sat. 3 ($\Omega_3 = 144$ deg)	Sat. 4 ($\Omega_4 = 216$ deg)	Sat. 5 ($\Omega_5 = 288$ deg)
ΔV (m/s)	322.611	430.421	215.374	350.309	107.558

at constant Ω. The remaining portions of these trajectories consist of circular arcs representing the precessions of the five orbit planes by analytic approximation. The accumulated velocity changes of each spacecraft are depicted in Fig. 4.17, where all satellites are maneuvering simultaneously with total velocity changes at the 3650-day mark of 528.867, 552.639, 436.577, 559.457, and 564.554 m/sec, respectively, for satellites 1, 2, 3, 4, and 5. These values are also reported in Table 4.2.

As an example, in Fig. 4.18, note the inclination time history for satellite 4 starting at $\Omega_4 = 216$ deg. The small discontinuities in the inclination rates are due to the node change carried out impulsively at the times of inclination adjust maneuvers performed by the other vehicles. The combined total velocity change for all five satellites stands at 2642.096 m/s. If we now compute this

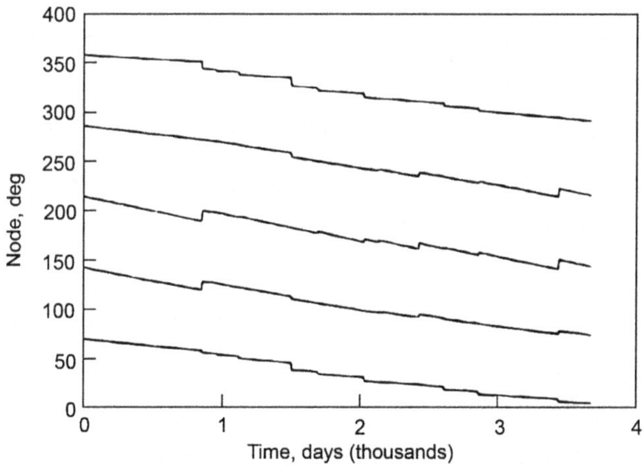

Fig. 4.15 Controlled nodal variation of a five-satellite constellation over a 3650-day lifetime for initial values of $\Omega_1 = 0$ deg, $\Omega_2 = 72$ deg, $\Omega_3 = 144$ deg, $\Omega_4 = 216$ deg, and $\Omega_5 = 288$ deg.

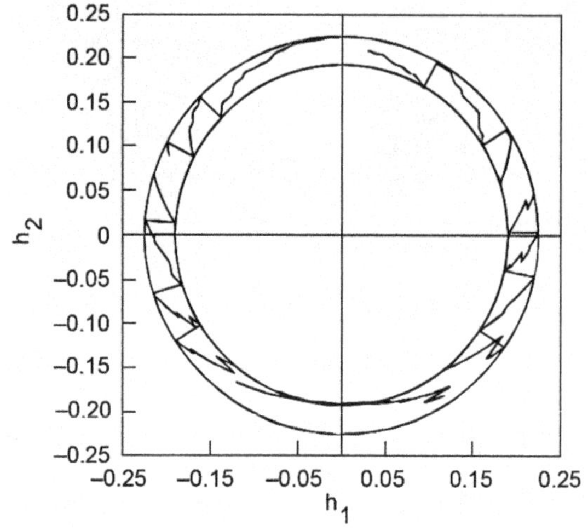

Fig. 4.16 Constellation trajectories over a 3650-day lifetime with both inclination and node control for initial values of $\Omega_1 = 0$ deg, $\Omega_2 = 72$ deg, $\Omega_3 = 144$ deg, $\Omega_4 = 216$ deg, and $\Omega_5 = 288$ deg.

total velocity change for values of Ω_1 spanning the interval (0, 72) deg, we can produce the plot in Fig. 4.19, which shows that the optimal initial configuration that results in the lowest total velocity change over the 3650-day lifetime of the constellation is given by $\Omega_1 = 33.1$ deg, $\Omega_2 = 105.1$ deg, $\Omega_3 = 177.1$ deg, $\Omega_4 = 249.1$ deg, and $\Omega_5 = 321.1$ deg, with a corresponding ΔV_{tot} value of 2364.709 m/s and individual satellite totals of 435.897, 489.508, 506.379, 369.913, and 563.011 m/s, respectively. These

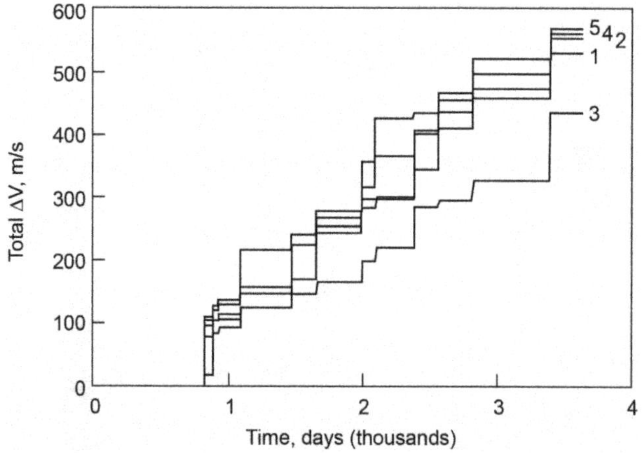

Fig. 4.17 Individual accumulated total velocity change requirements for a five-satellite constellation over a 3650-day lifetime with inclination and node control for initial values of $\Omega_1 = 0$ deg, $\Omega_2 = 72$ deg, $\Omega_3 = 144$ deg, $\Omega_4 = 216$ deg, and $\Omega_5 = 288$ deg.

TABLE 4.2 ACCUMULATED INDIVIDUAL SATELLITE VELOCITY CHANGES IN 3650 DAYS WITH INCLINATION AND NODE CONTROL AND SPECIFIED INITIAL NODES

	Sat. 1	Sat. 2	Sat. 3	Sat. 4	Sat. 5
Initial node	($\Omega_1 = 0$ deg)	($\Omega_2 = 72$ deg)	($\Omega_3 = 144$ deg)	($\Omega_4 = 216$ deg)	($\Omega_5 = 288$ deg)
ΔV (m/s)	528.867	552.639	436.577	559.457	564.554
Initial node	$\Omega_1 = 33.1$	$\Omega_2 = 105.1$	$\Omega_3 = 177.1$	$\Omega_4 = 249.1$	$\Omega_5 = 321.1$
ΔV (m/s)	435.897	489.508	506.379	369.913	563.011

values are included in Table 4.2. Figure 4.20 shows all five corresponding trajectories in the $h_1 - h_2$ plane, whereas Fig. 4.21 shows how the individual satellites accumulate their velocity change budgets as they maneuver over their lifetimes. The two strategies discussed so far made use of the analytic approximation in describing the orbit plane precessions. Section 4.5 revisits the strategy of Sec. 4.3 by replacing the circular precessions, which are epoch-independent, by more exacting representations of those precessions, which, in this case, are simulated by numerically integrating the differential equations for i and Ω, thus properly accounting for the epoch dependency of the orbit plane motion.

4.5 INCLINATION CONTROL STRATEGY WITH NUMERICAL ORBIT PROPAGATION

For a circular orbit, the orbit plane orientation parameters obey the following differential equations

$$\dot{i} = -\frac{1}{na^2 s_i}\frac{\partial R}{\partial \Omega} \tag{4.27}$$

$$\dot{\Omega} = \frac{1}{na^2 s_i}\frac{\partial R}{\partial i} \tag{4.28}$$

where R is the perturbing function due to J_2 and the luni-solar gravity given by

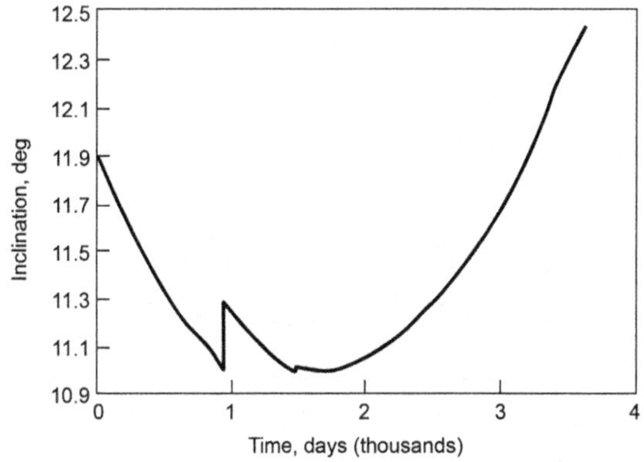

Fig. 4.18 Inclination time history of satellite 4 with an initial node of $\Omega_4 = 216$ deg over a 3650-day lifetime.

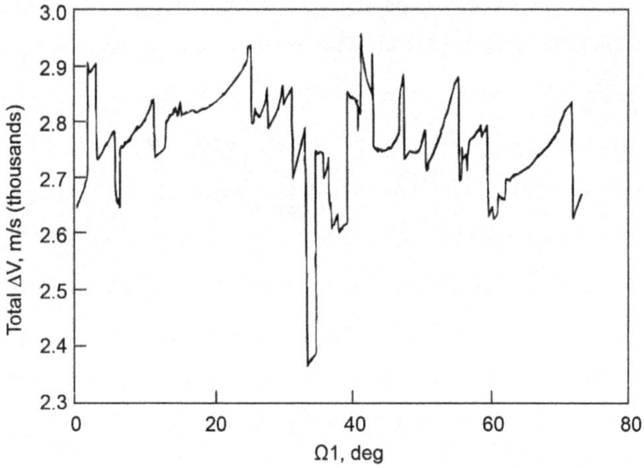

Fig. 4.19 Total ΔV requirements for a five-satellite constellation over a 3650-day lifetime with inclination and node control vs initial node configuration.

Kamel and Tibbitts [9] as

$$R = \frac{1}{4}\mu n_m^2 a^2 \left[\left(1 - \frac{3}{2}s_i^2\right)\left(1 - \frac{3}{2}s_{i_m}^2\right) + \frac{3}{4}s_{2i}s_{2i_m}c_{\Omega-\Omega_m} + \frac{3}{4}s_i^2 s_{i_m}^2 c_{2(\Omega-\Omega_m)} \right]$$
$$+ \frac{1}{4}n_s^2 a^2 \left[\left(1 - \frac{3}{2}s_i^2\right)\left(1 - \frac{3}{2}s_{i_s}^2\right) + \frac{3}{4}s_{2i}s_{2i_s}c_{\Omega} + \frac{3}{4}s_i^2 s_{i_s}^2 c_{2\Omega} \right]$$
$$+ \frac{3}{2}n^2 J_2 R_e^2 \left(\frac{1}{3} - \frac{1}{2}s_i^2\right) \qquad (4.29)$$

In these equations, n and a are the satellite orbit

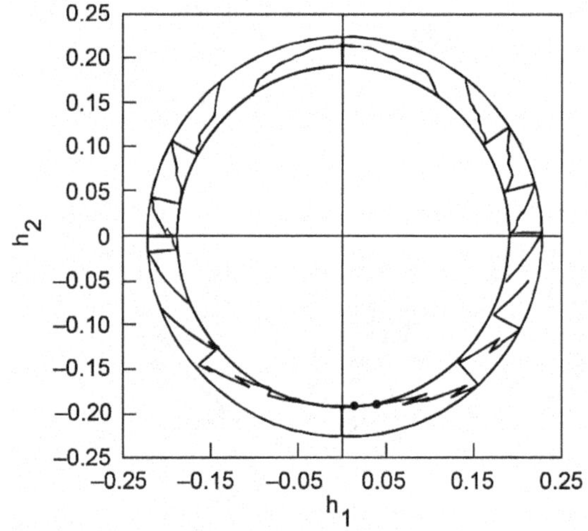

Fig. 4.20 Constellation trajectories over a 3650-day lifetime with inclination and node control for a ΔV-minimizing initial configuration of $\Omega_1 = 33.1$ deg, $\Omega_2 = 105.1$ deg, $\Omega_3 = 177.1$ deg, $\Omega_4 = 249.1$ deg, and $\Omega_5 = 321.1$ deg.

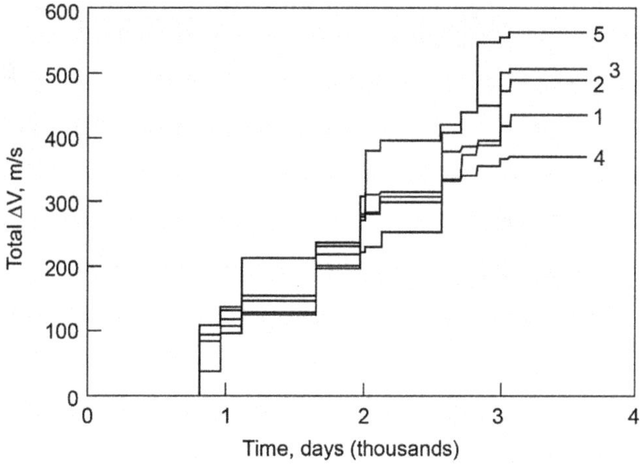

Fig. 4.21 Individual accumulated total ΔV requirements over a 3650-day lifetime with inclination and node control for the overall ΔV-minimizing initial configuration of $\Omega_1 = 33.1$ deg, $\Omega_2 = 105.1$ deg, $\Omega_3 = 177.1$ deg, $\Omega_4 = 249.1$ deg, and $\Omega_5 = 321.1$ deg.

mean motion and semimajor axis, respectively; $\mu = 1/82.3$ is the ratio of the mass of the moon to the combined mass of the moon and Earth; $n_m = 0.23$ rad/day is the lunar orbit mean motion; $n_s = 0.017203$ rad/day is the apparent solar orbit mean motion; i_m and Ω_m are the equatorial inclination and the right ascension of the ascending node of the moon's orbit with respect to the equator, respectively; i_s is the Sun's apparent orbit equatorial inclination; and J_2 and R_e are the second zonal harmonic of Earth's potential and the equatorial radius of Earth, respectively. The values of i_m and Ω_m vary according to the expressions

$$c_{i_m} = c_{i_M} c_{i_s} - s_{i_s} s_{i_M} c_{\Omega_M} \tag{4.30}$$

$$s_{\Omega_m} = s_{i_M} s_{\Omega_M} / s_{i_m} \tag{4.31}$$

where $i_M = 5.145$ deg is the ecliptic inclination of the moon's orbit and $\Omega_M = 259.183$ deg $- 0.05295t$ in degrees is its ascending node, which regresses linearly in time because of the solar gravity. Here, t is measured in Julian days from the epoch of 1 January 1900 at 12 hrs, as described by Kamel and Tibbitts [9]. The regression period of Ω_M is roughly equal to 18.6 years such that the equation for Ω_M in the sentence after Eq. (4.31) is also adopted with the epoch of 1 January 1993 at 12 hrs, adding roughly five regression cycles to the original epoch. This rough calculation is done only for illustration purposes to show the effect of epoch selection on the optimal values of Ω_1–Ω_5 at the start of the simulation and, thereby, on the total ΔV requirements. The Astronomical Almanac lists Ω_M as $\Omega_M = 279.854$ deg $- 0.05295d$, where d is the interval in days from 1 January 1992 at 0 hrs. Now, the use of the nonsingular variables $h_1 = s_i s_\Omega$

and $h_2 = s_i c_\Omega$ by Kamel and Tibbitts [9] led to the nonsingular set:

$$\dot{h}_1 = \frac{c_i}{na^2}\frac{\partial R}{\partial h_2} \tag{4.32}$$

$$\dot{h}_2 = \frac{-c_i}{na^2}\frac{\partial R}{\partial h_1} \tag{4.33}$$

For $i < \pi/2$, we have $c_i = \left(1 - h_1^2 - h_2^2\right)^{1/2}$ because, for all $0 \le i \le \pi$, $s_i = \left(h_1^2 + h_2^2\right)^{1/2}$, and $c_{2i} = \left(1 - 2h_1^2 - 2h_2^2\right)$. Restricting ourselves to $c_i > 0$, or $i < \pi/2$, we also have $s_i c_i = \left(h_1^2 + h_2^2\right)^{1/2}\left(1 - h_1^2 - h_2^2\right)^{1/2}$. From $s_\Omega = h_1/s_i$ and $c_\Omega = h_2/s_i$, we have $s_{2\Omega} = 2h_1h_2/s_i^2$ and $c_{2\Omega} = \left(h_2^2 - h_1^2\right)/s_i^2$ as well as

$$c_{\Omega-\Omega_m} = \frac{h_2}{s_i}c_{\Omega_m} + \frac{h_1}{s_i}s_{\Omega_m} \tag{4.34}$$

$$s_{\Omega-\Omega_m} = \frac{h_1}{s_i}c_{\Omega_m} - \frac{h_2}{s_i}s_{\Omega_m} \tag{4.35}$$

$$c_{2(\Omega-\Omega_m)} = \frac{\left(h_2^2 - h_1^2\right)}{s_i^2}c_{2\Omega_m} + \frac{2h_1h_2}{s_i^2}s_{2\Omega_m} \tag{4.36}$$

$$s_{2(\Omega-\Omega_m)} = \frac{2h_1h_2}{s_i^2}c_{2\Omega_m} - \frac{\left(h_2^2 - h_1^2\right)}{s_i^2}s_{2\Omega_m} \tag{4.37}$$

Equations (4.32) and (4.33) can be written directly in terms of h_1 and h_2 by making the following manipulations

$$\frac{\partial R}{\partial h_1} = \frac{\partial R}{\partial i}\frac{\partial i}{\partial h_1} + \frac{\partial R}{\partial \Omega}\frac{\partial \Omega}{\partial h_1} = \frac{\partial R}{\partial i}\frac{h_1}{s_i c_i} + \frac{\partial R}{\partial \Omega}\frac{h_2}{s_i^2}$$

and

$$\frac{\partial R}{\partial h_2} = \frac{\partial R}{\partial i}\frac{\partial i}{\partial h_2} + \frac{\partial R}{\partial \Omega}\frac{\partial \Omega}{\partial h_2} = \frac{\partial R}{\partial i}\frac{h_2}{s_i c_i} - \frac{\partial R}{\partial \Omega}\frac{h_1}{s_i^2}$$

We have

$$\begin{aligned}\frac{\partial R}{\partial i} &= \frac{1}{4}\mu n_m^2 a^2\left[-3s_ic_i\left(1 - \frac{3}{2}s_{i_m}^2\right) + \frac{3}{2}c_{2i}s_{2i_m}c_{\Omega-\Omega_m} + \frac{3}{2}s_ic_is_{i_m}^2c_{2(\Omega-\Omega_m)}\right]\\ &\quad + \frac{1}{4}n_s^2a^2\left[-3s_ic_i\left(1 - \frac{3}{2}s_{i_s}^2\right) + \frac{3}{2}c_{2i}s_{2i_s}c_\Omega + \frac{3}{2}s_ic_is_{i_s}^2c_{2\Omega}\right] - \frac{3}{2}n^2J_2R_e^2s_ic_i\end{aligned} \tag{4.38}$$

$$\begin{aligned}\frac{\partial R}{\partial \Omega} &= \frac{1}{4}\mu n_m^2 a^2\left[-\frac{3}{4}s_{2i}s_{2i_m}s_{\Omega-\Omega_m} - \frac{3}{2}s_i^2s_{i_m}^2s_{2(\Omega-\Omega_m)}\right]\\ &\quad + \frac{1}{4}n_s^2a^2\left(-\frac{3}{4}s_{2i}s_{2i_s}s_\Omega - \frac{3}{2}s_i^2s_{i_s}^2s_{2\Omega}\right)\end{aligned} \tag{4.39}$$

After making the proper substitutions, the final form of the differential equations of motion is obtained:

$$
\begin{aligned}
\frac{dh_1}{dt} = {} & \frac{\mu}{4n} n_m^2 \left\{ -3h_2 \left(1 - \frac{3}{2} s_{i_m}^2\right) \left(1 - h_1^2 - h_2^2\right)^{1/2} \right. \\
& + \frac{3}{2} s_{2i_m} \left[c_{\Omega_m} \left(1 - h_1^2 - h_2^2\right) - h_2 \left(h_2 c_{\Omega_m} + h_1 s_{\Omega_m}\right) \right] \\
& \left. + \frac{3}{2} s_{i_m}^2 \left(1 - h_1^2 - h_2^2\right)^{1/2} \left(h_2 c_{2\Omega_m} + h_1 s_{2\Omega_m}\right) \right\} \\
& + \frac{n_s^2}{4n} \left[-3h_2 \left(1 - \frac{3}{2} s_{i_s}^2\right) \left(1 - h_1^2 - h_2^2\right)^{1/2} \right. \\
& \left. + \frac{3}{2} s_{2i_s} \left(1 - h_1^2 - 2h_2^2\right) + \frac{3}{2} s_{i_s}^2 h_2 \left(1 - h_1^2 - h_2^2\right)^{1/2} \right] \\
& - \frac{3}{2a^2} n J_2 R_e^2 h_2 \left(1 - h_1^2 - h_2^2\right)^{1/2}
\end{aligned}
\tag{4.40}
$$

$$
\begin{aligned}
\frac{dh_2}{dt} = {} & -\frac{\mu}{4n} n_m^2 \left\{ -3h_1 \left(1 - \frac{3}{2} s_{i_m}^2\right) \left(1 - h_1^2 - h_2^2\right)^{1/2} \right. \\
& + \frac{3}{2} s_{2i_m} \left[s_{\Omega_m} \left(1 - h_1^2 - h_2^2\right) - h_1 \left(h_2 c_{\Omega_m} + h_1 s_{\Omega_m}\right) \right] \\
& \left. + \frac{3}{2} s_{i_m}^2 \left(1 - h_1^2 - h_2^2\right)^{1/2} \left(h_2 s_{2\Omega_m} - h_1 c_{2\Omega_m}\right) \right\} \\
& - \frac{n_s^2}{4n} \left[-3h_1 \left(1 - \frac{3}{2} s_{i_s}^2\right) \left(1 - h_1^2 - h_2^2\right)^{1/2} \right. \\
& \left. - \frac{3}{2} s_{2i_s} h_1 h_2 - \frac{3}{2} s_{i_s}^2 h_1 \left(1 - h_1^2 - h_2^2\right)^{1/2} \right] \\
& + \frac{3}{2} \frac{n}{a^2} h_1 J_2 R_e^2 \left(1 - h_1^2 - h_2^2\right)^{1/2}
\end{aligned}
\tag{4.41}
$$

with $i_s = 23.445$ deg, $J_2 = 1.0827 \times 10^{-3}$, and $R_e = 6378.14$ km.

We can now revisit the strategy on inclination control with analytic orbit prediction (i.e., the strategy in Sec. 4.3) and replace the circular precession arcs by the more exacting precessions obtained by numerically integrating Eqs. (4.40) and (4.41). As before, the maneuvers are triggered as soon as one of the two inclination constraint circles is violated. The values of Ω_M and therefore of i_m and Ω_m are varied during the integration according to Eqs. (4.30) and (4.31). Using the expression $\Omega_M = 259.183$ deg - $0.05295t$ and the initial configuration with $\Omega_1 = 0$ deg, $\Omega_2 = 72$ deg, $\Omega_3 = 144$ deg, $\Omega_4 = 216$ deg, and $\Omega_5 = 288$ deg, we obtain for the 3650-day lifetime the plot in Fig. 4.22, which shows lesser

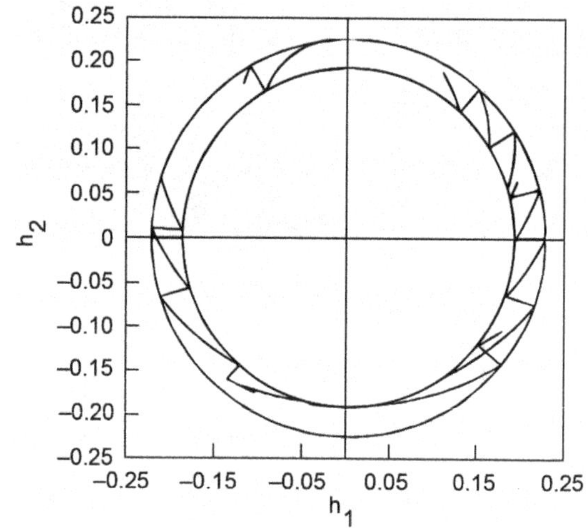

Fig. 4.22 Numerically integrated constellation trajectories over a 3650-day lifetime with inclination control for initial values of $\Omega_1 = 0$ deg, $\Omega_2 = 72$ deg, $\Omega_3 = 144$ deg, $\Omega_4 = 216$ deg, and $\Omega_5 = 288$ deg and a lunar orbit given by $\Omega_M = 259.183$ deg − $0.05295t$.

maneuvering requirements than the equivalent plot in Fig. 4.13 that was generated for a slightly longer lifetime of 4000 days. Figures 4.23–4.26 show how these trajectories are affected by the initial epoch, meaning the lunar orbit ecliptic node and, therefore, the subsequent variations in i_m and Ω_m affecting the perturbations due to the moon. These figures were generated with the following expressions for Ω_M

$$\Omega_M = 259.183 \text{ deg} - 0.05295t$$
$$\Omega_M = 162.549 \text{ deg} - 0.05295t$$
$$\Omega_M = 65.916 \text{ deg} - 0.05295t$$
$$\Omega_M = 329.283 \text{ deg} - 0.05295t$$

corresponding, respectively, to the original epoch and epochs of +5, +10, and +15 years from that epoch. The three subsequent Ω_M values were obtained by successively subtracting an angle of 96.633 deg, corresponding to a 5-year

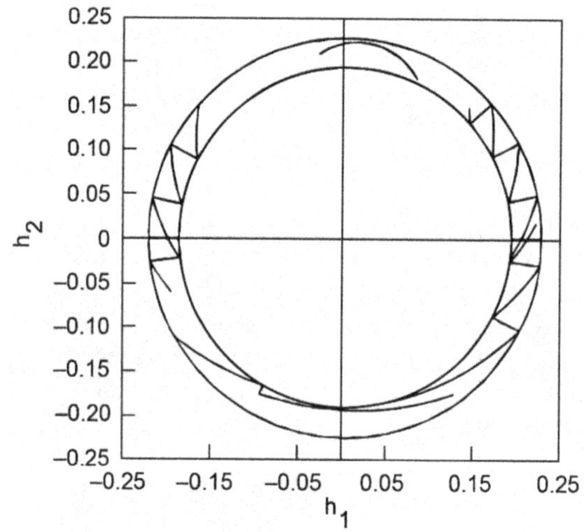

Fig. 4.23 Numerically integrated constellation trajectories over a 3650-day lifetime with inclination control for an initial ΔV-minimizing configuration of $\Omega_1 = 24.3$ deg, $\Omega_2 = 96.3$ deg, $\Omega_3 = 168.3$ deg, $\Omega_4 = 240.3$ deg, and $\Omega_5 = 312.3$ deg and a lunar orbit given by $\Omega_M = 259.183$ deg − $0.05295t$.

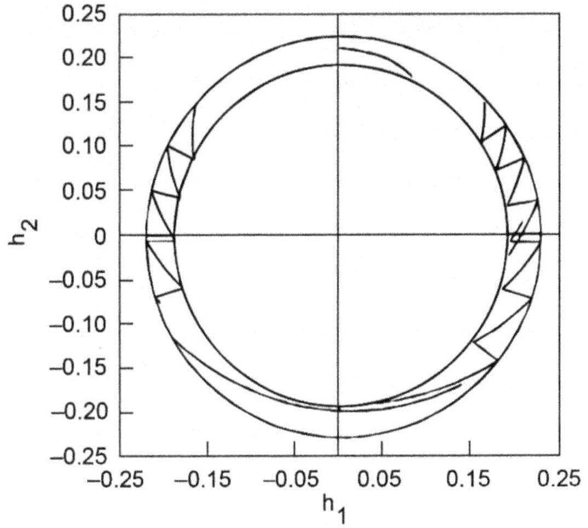

Fig. 4.24 Numerically integrated constellation trajectories over a 3650-day lifetime with inclination control for initial values of $\Omega_1 = 24.3$ deg, $\Omega_2 = 96.3$ deg, $\Omega_3 = 168.3$ deg, $\Omega_4 = 240.3$ deg, and $\Omega_5 = 312.3$ deg and a lunar orbit given by $\Omega_M = 162.549$ deg − $0.05295t$.

lunar orbit ecliptic nodal regression.

These four epoch-dependent exact trajectories are for the initial configuration of Fig. 4.11, given by $\Omega_1 = 24.3$ deg, $\Omega_2 = 96.3$ deg, $\Omega_3 = 168.3$ deg, $\Omega_4 = 240.3$ deg, and $\Omega_5 = 312.3$ deg, which was obtained analytically and which corresponds to the overall analytic minimum-ΔV solution depicted in Fig. 4.10 for the slightly shorter 3650-day lifetime. Note how the trajectory of spacecraft 1 is affected by the Ω_M epoch requiring several maneuvers as it repeatedly hits the outer inclination constraint. The final four figures in this section, namely, Figs. 4.27–4.30 show the cumulative total five-satellite ΔV requirements for the four epochs used as a function of the initial even nodal configuration. Appreciable variability can be observed in both the ΔV

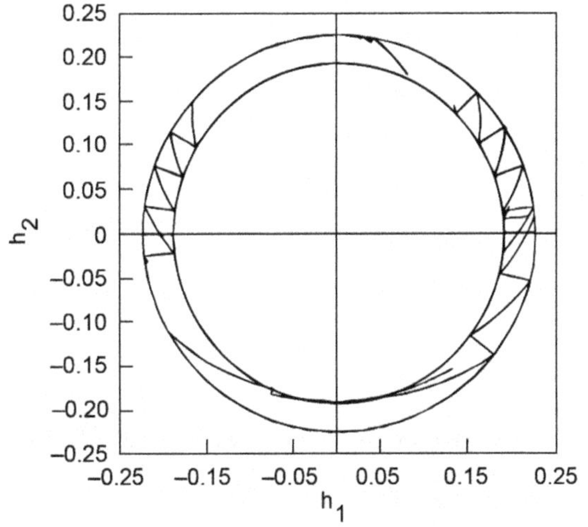

Fig. 4.25 Numerically integrated constellation trajectories over a 3650-day lifetime with inclination control for initial values of $\Omega_1 = 24.3$ deg, $\Omega_2 = 96.3$ deg, $\Omega_3 = 168.3$ deg, $\Omega_4 = 240.3$ deg, and $\Omega_5 = 312.3$ deg and a lunar orbit given by $\Omega_M = 65.916$ deg − $0.05295t$.

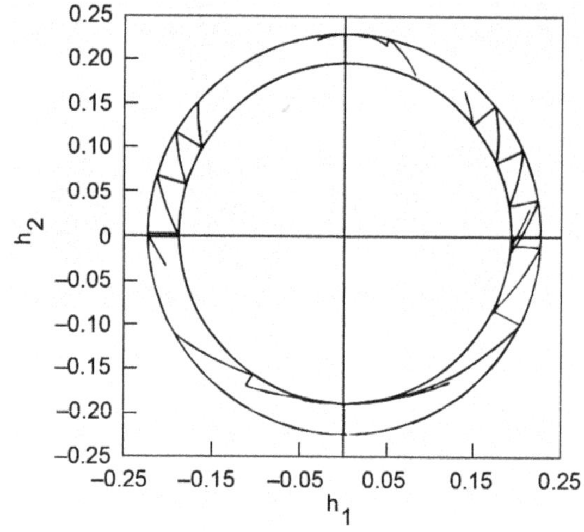

Fig. 4.26 Numerically integrated constellation trajectories over a 3650-day lifetime with inclination control for initial values of $\Omega_1 = 24.3$ deg, $\Omega_2 = 96.3$ deg, $\Omega_3 = 168.3$ deg, $\Omega_4 = 240.3$ deg, and $\Omega_5 = 312.3$ deg and a lunar orbit given by $\Omega_M = 329.283$ deg – $0.05295t$.

requirements and the initial constellation geometry that leads to the minimum overall velocity change. Each of these four plots required close to a 9-hour run on a 486/66 MHz desktop computer using a 0.1 deg step to cover the range, $0\,\text{deg} \leq \Omega_1 \leq 72\,\text{deg}$, as opposed to only a few minutes for the analytic method used in Sec. 4.3 and 4.4. If a series of the exact plots were generated at 1-year epoch intervals spanning the entire 18.6-year regression cycle, then the optimum initial configuration and the

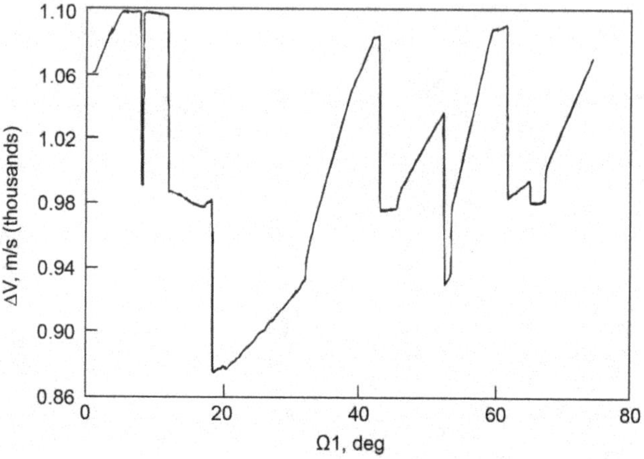

Fig. 4.27 Total ΔV requirements for a five-satellite constellation with inclination control over a 3650-day lifetime vs initial node configuration for a lunar orbit given by $\Omega_M = 259.183$ deg – $0.05295t$.

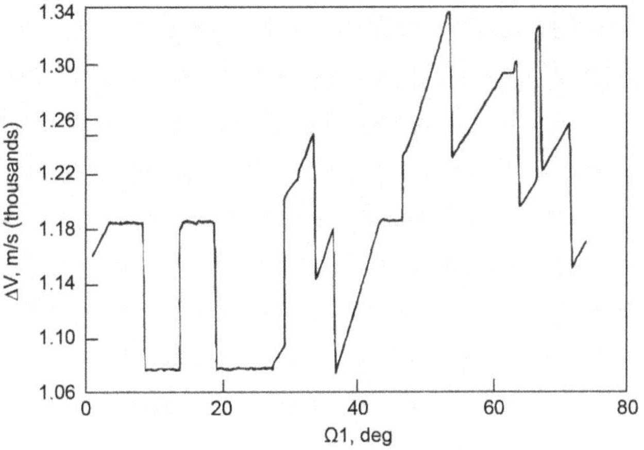

Fig. 4.28 Total ΔV requirements for five-satellite constellation with inclination control over a 3650-day lifetime vs initial node configuration for a lunar orbit given by $\Omega_M = 162.549$ deg − 0.05295t at the initial epoch.

optimum epoch within this cycle could be selected to provide the overall minimum-ΔV requirements. The same exacting simulations could also be carried out for the strategy of Sec. 4.4, which adjusts the nodes periodically to establish the relevant minimum-ΔV requirements.

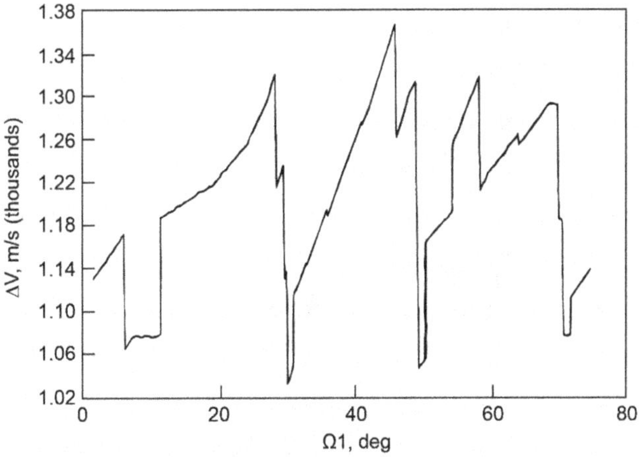

Fig. 4.29 Total ΔV requirements for five-satellite constellation with inclination control over a 3650-day lifetime vs initial node configuration for a lunar orbit given by $\Omega_M = 65.916$ deg − 0.05295t at the initial epoch.

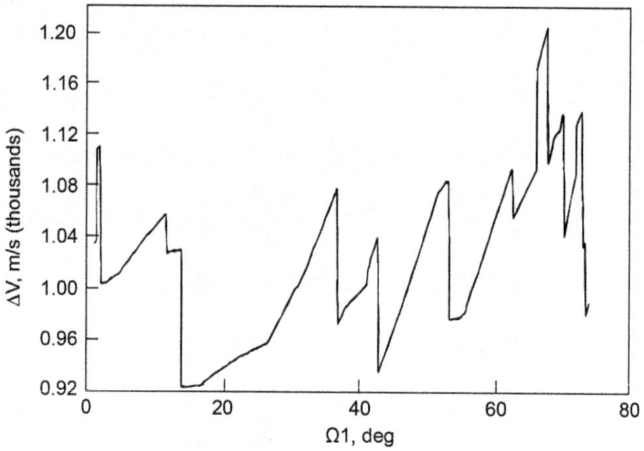

Fig. 4.30 Total ΔV requirements for five-satellite constellation with inclination control over a 3650-day lifetime vs initial node configuration for a lunar orbit given by $\Omega_M = 329.283$ deg − $0.05295t$ at the initial epoch.

4.6 CONCLUSION

This chapter has presented analytic and numerical simulations using a pair of simple suboptimal orbit maintenance strategies to determine the optimal initial orbit plane configuration of a constellation of five satellites in inclined geosynchronous orbits that results in the minimum change in velocity accumulated by the five satellites over the constellation lifetime. These orbit plane control strategies perform either intermittent inclination adjust maneuvers that confine each satellite orbit within a user-defined inclination tolerance deadband centered at the common nominal inclination at all times or both inclination and node adjust maneuvers that, in addition, keep the initially evenly spaced nodal configuration from being distorted. In both cases, the optimal initial nodal values and inclinations are determined to minimize the total overall velocity change accumulated by all five satellites. The orientation of the orbit plane of the moon within its 18.6-year regression cycle is also shown to have a significant effect on the total ΔV requirements of the constellation over its fixed lifetime, resulting in an optimal launch time for the constellation. The suboptimal orbit plane maintenance strategies developed in this chapter are of a general nature, meaning that they are not restricted to a given number of satellites, their common nominal inclination, or a prespecified lifetime. These solutions approach the previous true minimum-ΔV solution; the minimum-ΔV solution itself requires the consideration of the complete state-constrained cooperative differential game, which is quite difficult to analyze. Further savings in fuel are possible by using electric engines and designing the appropriate maneuvering strategies to replace the impulsive strategies employed here.

REFERENCES

[1] Musen, P., "On the long-period lunisolar effect in the motion of the artificial satellite," *Journal of Geophysical Research*, Vol. 66, No. 6, June 1961, pp. 1659–1665.

[2] Musen, P., Bailie, A., and Upton, E., "Development of the lunar and solar perturbations in the motion of an artificial satellite," NASA TN D-494, Jan. 1961.

[3] Musen, P., "On the long-period lunar and solar effects on the motion of an artificial satellite, 2, *Journal of Geophysical Research*, Vol. 66, No. 9, Sept. 1961, pp. 2797–2805; also NASA SP-54.

[4] Frick, R. H., and Garber, T. B., "Perturbations of a synchronous satellite," Report R-399-NASA, Rand Corp., Santa Monica, CA, May 1962.

[5] Allen, R. R., and Cook, G. E., "The long-period motion of the plane of a distant circular orbit," *Proceedings of the Royal Society*, Vol. 280, No. 1380, July 1964, pp. 97–109.

[6] Billik, B. H., "Cross-track sustaining requirements for a 24-hr satellite," *Journal of Spacecraft and Rockets*, Vol. 4, March 1967, No. 3, pp. 297–301.

[7] Frick, R. H., "Orbital regression of synchronous satellites due to the combined gravitational effects of the sun, the moon, and the oblate earth," Report R-454-NASA, Rand Corp., Santa Monica, CA, Aug. 1967.

[8] Balsam, R. E., and Anzel, B. M., "A simplified approach for correction of perturbations on a stationary orbit," *Journal of Spacecraft and Rockets*, Vol. 6, No. 7, July 1969, pp. 805–811.

[9] Kamel, A., and Tibbitts, R., "Some useful results on initial node locations for near-equatorial circular satellite orbits," *Celestial Mechanics*, Vol. 8, Aug. 1973, pp. 45–73.

[10] Kamel, A. A., "Synchronous satellite ephemeris due to earth's triaxiality and luni-solar effects," *AAS/AIAA Astrodynamics Conference, AIAA Paper 78-1441*, Palo Alto, CA, Aug. 1978.

[11] Eckstein, M. C., Leibold, A., and Hechler, F., "Optimal autonomous stationkeeping of geostationary satellites," *AAS/AIAA Astrodynamics Specialist Conference, AAS Paper 81-206*, Lake Tahoe, NV, Aug. 1981.

[12] Kamel, A. A., and Wagner, C. A., "On the orbital eccentricity control of synchronous satellites," *Journal of the Astronautical Sciences*, Vol. 30, No. 1, Jan. 1982, pp. 61–73.

[13] Kamel, A. A., "Geosynchronous satellite perturbations due to Earth's triaxiality and luni-solar effects," *Journal of Guidance, Control, and Dynamics*, Vol. 5, No. 2, March–April 1982, pp. 189–193.

[14] Slavinskas, D. D., Johnson, G. K., and Benden, W. J., "Efficient inclination control for geostationary satellites," AIAA Paper 85-0216, AIAA 23rd Aerospace Sciences Meeting, Reno, NV, Jan. 1985.

[15] Murdoch, J., and Pocha, J. J., "The orbit dynamics of satellite clusters," 33rd International Astronautical Congress, IAF Paper 82-54, Paris, France, Sept. 1982.

[16] Walker, J. G., "The geometry of satellite clusters," Rept. RAE-TR-81084, Royal Aircraft Establishment, Farnborough, England, U.K., 1981

[17] Hubert, S., and Swale, J., "Stationkeeping of a constellation of geostationary communication satellites," *AIAA/AAS Astrodynamics Conference, AIAA Paper 84-2024*, Seattle, WA, Aug. 1984.

[18] Kechichian, J. A., "Orbit plane control strategies for inclined geosynchronous satellite constellation," AAS/AIAA Astrodynamics Specialist Conference, AAS Paper 95-342, Halifax, Nova Scotia, Canada, Aug. 1995.

[19] Kechichian, J. A., "Orbit plane control strategies for inclined geosynchronous satellite constellation," *Journal of Spacecraft and Rockets*, Vol. 35, No. 1, Jan. 1998, pp. 46–54.

[20] Kechichian, J. A., "Optimal steering for north-south stationkeeping of geostationary spacecraft," *Journal of Guidance, Control, and Dynamics*, Vol. 20, No. 3, May–June 1997, pp. 435–444.

CHAPTER 5

Optimal Steering for North–South Stationkeeping of Geostationary Spacecraft

NOMENCLATURE

a_0, a = initial and current orbit semimajor axes, respectively, m
c = exhaust velocity, $I_{sp}g_r$
c_i, s_i = cos i, sin i
E_{jet} = power of the jet
g_r = acceleration due to gravity, m/s^2
H = Hamiltonian
$\boldsymbol{h}$ = angular momentum vector
I_{sp} = specific impulse, s
i^* = wedge angle
i_M = moon's inclination with respect to the ecliptic
i_m = equatorial inclination of the lunar orbit
i_{max} = maximum orbit inclination
i_s = equatorial inclination of the Sun's apparent orbit
J_2 = second zonal harmonic of Earth's potential, 1.0827×10^{-3}
m = spacecraft mass, kg
n = orbit mean motion, rad/s
n_m = mean motion of the lunar orbit, 0.23 rad/day
n_s = mean motion of the Sun's apparent orbit, 0.017203 rad/day
P = thrust power, W
R = perturbing function due to Earth's triaxiality, the moon, and the Sun
R_e = mean equatorial radius of the Earth, 6378.14 km
T = $(t - t_0)$, s or thrust magnitude, N depending on contest
t = time, s or Julian days, depending on context
V = maneuvering velocity
W = precession velocity
$\boldsymbol{\delta h}$ = change in the angular momentum vector
μ = ratio of the mass of the moon to the combined masses of the Earth and moon, 1/82.3
ϕ = heading angle
Ω_M = ecliptic ascending node of the lunar orbit

Ω_m = right ascension of the ascending node of the moon's orbit with respect to Earth's equator
ω = precession rate

5.1 INTRODUCTION

The problem of north–south stationkeeping of geostationary spacecraft using electric thrusters is analyzed in this chapter. Pure yawing with short-duration low-thrust arcs applied infrequently is assumed, and the dynamics are cast in continuous form to obtain an analytic steering law in inclination–node (i, Ω) space that brings the spacecraft back to the ideal initial orbit orientation for the initiation of an optimal free-drift period that satisfies the inclination constraint for the longest possible duration. This problem is posed as a minimum-time navigation problem between two (i, Ω) pairs and is similar to the Zermelo problem of navigating a ship in strong variable currents. The simple linear steering law thus obtained is easy to use and fuel-optimal compared to other suboptimal strategies for travel between two given (i, Ω) pairs. Control strategies for the north–south stationkeeping of geostationary spacecraft with chemical propulsion have been thoroughly documented in the literature [1–7]. There exists an ideal drift in inclination–right ascension of the ascending node $(i–\Omega)$ space that results in the satisfaction of the inclination deadband constraint for the longest possible duration. This inclination deadband is defined by $0 \leq i \leq i_{max}$, where i_{max} is the maximum allowable inclination. Once the deadband is consumed, an impulsive maneuver will target certain optimized initial conditions in i and Ω to continue the satisfaction of the deadband constraint. Considerable fuel can be saved if the low-specific-impulse chemical rockets are replaced by high-specific-impulse electric engines to execute the same stationkeeping maneuvers, thereby extending the operational life of these satellites. However, these maneuvers cannot be carried out in an essentially instantaneous manner, but must be implemented in small incremental steps spanning several weeks or more, depending on the level of the thrust acceleration and the frequency of the incremental maneuvering. The inherent long durations of these low-thrust maneuvers must be factored into the design of the maneuver strategy inasmuch as the strategy must account for the natural drift that occurs before the completion of the maneuver sequence.

This chapter employs, as an example for illustration purposes, the ideal drift cycle in (i, Ω) as the fundamental cycle to repeat and casts the dynamics in a continuous fashion to convert the problem into a navigation problem of minimum-time travel between given (i, Ω) pairs, not unlike the well-known Zermelo problem in optimal control theory. The control is now a steering angle that defines the direction of the change in the (i, Ω) region to be achieved by the small incremental change in velocity at that particular moment in the overall sequence. It is assumed that these small velocity changes are large enough to overcome the natural drift in the opposite direction such that the ideal initial conditions are recovered in time. We begin in Sec. 5.2 with a description of the idealized (i, Ω) drift dynamics due to Kamel and Tibbits [7], which will be used in the design of the optimal steering strategy detailed in the subsequent sections. Section 5.3 shows how the relevant linearized variation-of-parameters equations are used to calculate the changes in the orbit plane orientation due to small thrust arcs. In Sec. 5.4, a suboptimal strategy that does not take into account the natural drift in *his* and Ω between successive thrust arcs is presented. This strategy consists of applying the thrust arc at the current common line of nodes of the current orbit and the final target orbit. The more efficient strategy of Sec. 5.5 creates new intermediate target orbits and applies the thrust arc at the common line of nodes of the current orbit and the intermediate target orbit, thereby taking into account the natural drift in i and Ω between maneuvers into the transfer design. This chapter is based on a previously published conference paper [8] and journal article [9].

5.2 GENERAL ANALYSIS OF NORTH–SOUTH DRIFT: IMPULSIVE MANEUVERING

The equations of motion governing the evolution of the inclination i and ascending node Ω of a near-circular satellite orbit are given by [7]

$$\dot{i} = -\frac{1}{na^2 s_i}\frac{\partial R}{\partial \Omega} \tag{5.1}$$

$$\dot{\Omega} = \frac{1}{na^2 s_i}\frac{\partial R}{\partial i} \tag{5.2}$$

These equations are valid for any value of the semimajor axis a and inclination i but are restricted to small eccentricity e. Here, n is the orbit mean motion, and R is the perturbing function due to Earth's triaxiality, the moon, and the Sun

$$\begin{aligned} R = {} & \frac{1}{4}\mu n_m^2 a^2\left[\left(1-\frac{3}{2}s_i^2\right)\left(1-\frac{3}{2}s_{i_m}^2\right)+\frac{3}{4}s_{2i}s_{2i_m}c_{\Omega-\Omega_m}\right. \\ & \left.+\frac{3}{4}s_i^2 s_{i_m}^2 c_{2(\Omega-\Omega_m)}\right]+\frac{1}{4}n_s^2 a^2\left[\left(1-\frac{3}{2}s_i^2\right)\left(1-\frac{3}{2}s_{i_s}^2\right)\right. \\ & \left.+\frac{3}{4}s_{2i}s_{2i_s}c_{\Omega}+\frac{3}{4}s_i^2 s_{i_s}^2 c_{2\Omega}\right]+\frac{3}{2}n^2 J_2 R_e^2\left(\frac{1}{3}-\frac{1}{2}s_i^2\right) \end{aligned} \tag{5.3}$$

where $\mu = 1/82.3$ is the ratio of the mass of the moon to the combined masses of the Earth and moon, $n_m = 0.23$ rad/day is the moon's orbit mean motion; $n_s = 0.017203$ rad/day is the mean motion of the Sun's apparent orbit; i_m is the equatorial inclination of the lunar orbit; i_s is the equatorial inclination of the Sun's apparent orbit; Ω_m is the right ascension of the ascending node of the moon's orbit with respect to Earth's equator; J_2 is the second zonal harmonic of Earth's potential; and R_e is the mean equatorial radius of the Earth. Now, the moon's inclination with respect to the ecliptic i_M remains constant at 5.145 deg, but its ascending node Ω_M varies as a linear function of time according to $\Omega_M = 259.183\ \text{deg} - 0.05295t$, where t is the time in Julian days measured from 12 hrs on 1 January 1900. The regression rate of 0.05295 deg/day is such that it takes 18.6 years for Ω_M to regress by 360 deg. This regression of the node is due to the perturbation of the moon's orbit caused by solar gravity. The effect of this ecliptic nodal regression is to induce a variation of both the equatorial inclination i_m and ascending node Ω_m of the lunar orbit such that $i_s - i_M \le i_m \le i_s + i_M$ or $18.3\ \text{deg} \le i_m \le 28.59\ \text{deg}$ and $-13.0\ \text{deg} \le \Omega_m \le 13.0\ \text{deg}$. This can be shown from simple spherical trigonometry, such that, from Fig. 5.1 and from the law of cosines and the law of sines, respectively, with $i_s = 23.445$ deg

$$c_{i_m} = c_{i_M} c_{i_s} - s_{i_s} s_{i_M} c_{\Omega_M} \tag{5.4}$$

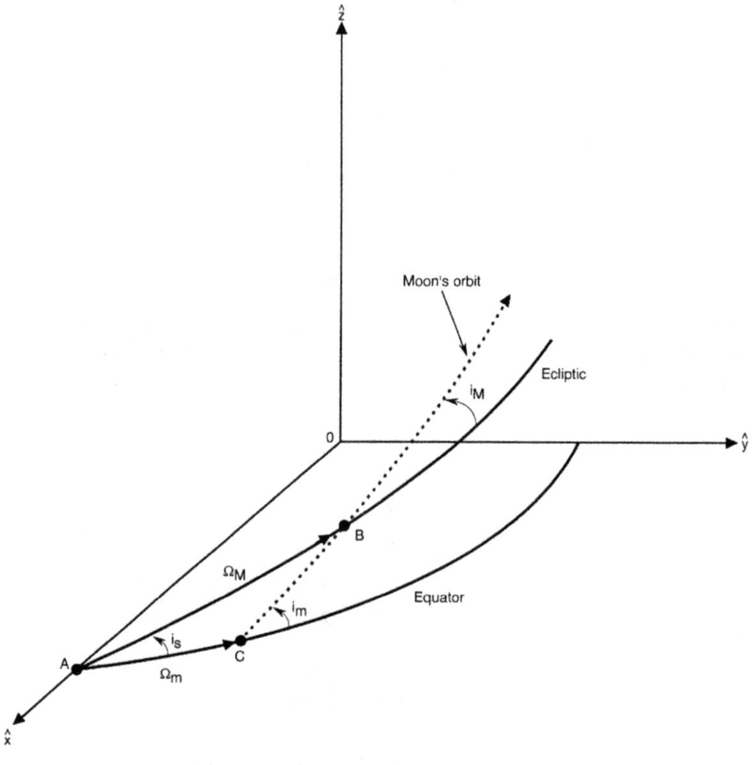

Fig. 5.1 Sun and Moon orbit geometry.

$$s_{\Omega_m} = s_{i_M} s_{\Omega_M} / s_{i_m} \tag{5.5}$$

Both i_M and i_s are fixed, but Ω_M varies according to the linear law provided in the preceding paragraph such that i_m varies according to Eq. (5.4) and Ω_m varies according to Eq. (5.5). The moon's orbit inclination with respect to the equator and its equatorial ascending node vary with the same period as Ω_M (18.6 years). If the variables i and Ω are changed to $h_1 = s_i s_\Omega$ and $h_2 = s_i c_\Omega$, respectively, then Eqs. (5.1) and (5.2), which are singular at $i = 0$, will be transformed to the set

$$\dot{h}_1 = \frac{c_i}{na^2} \frac{\partial R}{\partial h_2} \tag{5.6}$$

$$\dot{h}_2 = \frac{-c_i}{na^2} \frac{\partial R}{\partial h_1} \tag{5.7}$$

This can be seen from

$$\frac{\partial R}{\partial h_1} = \frac{\partial R}{\partial i} \frac{\partial i}{\partial h_1} + \frac{\partial R}{\partial \Omega} \frac{\partial \Omega}{\partial h_1}$$

$$\frac{\partial R}{\partial h_2} = \frac{\partial R}{\partial i} \frac{\partial i}{\partial h_2} + \frac{\partial R}{\partial \Omega} \frac{\partial \Omega}{\partial h_2}$$

In addition, because $s_i = \left(h_1^2 + h_2^2\right)^{1/2}$ and $\tan \Omega = h_1/h_2$, we have $\partial i/\partial h_1 = s_\Omega / c_i$, $\partial i/\partial h_2 = c_\Omega / c_i$, $\partial \Omega / \partial h_1 = c_\Omega / s_i$, and $\partial \Omega / \partial h_2 = -s_\Omega / s_i$ such that the two preceding expressions yield $\partial R/\partial i$ and $\partial R/\partial \Omega$ as

$$\frac{\partial R}{\partial i} = c_i \left(s_\Omega \frac{\partial R}{\partial h_1} + c_\Omega \frac{\partial R}{\partial h_2} \right)$$

$$\frac{\partial R}{\partial \Omega} = s_i \left(c_\Omega \frac{\partial R}{\partial h_1} - s_\Omega \frac{\partial R}{\partial h_2} \right)$$

We can now use these expressions in Eqs. (5.1) and (5.2) with $\dot{h}_1 = c_i s_\Omega \dot{i} + s_i c_\Omega \dot{\Omega}$ to obtain Eq. (5.6), and similarly with $\dot{h}_2 = c_i c_\Omega \dot{i} - s_i s_\Omega \dot{\Omega}$ to obtain Eq. (5.7).

For near-equatorial orbits or small i, $c_i \cong 1$ such that the preceding system can now be cast in the canonical form by means of a quadratic Hamiltonian H in the nonsingular variables h_1 and h_2, after a change of variable from time t to Ω_M [7]

$$\frac{dh_1}{d\Omega_M} = \frac{\partial H}{\partial h_2}$$

$$\frac{dh_2}{d\Omega_M} = -\frac{\partial H}{\partial h_1}$$

with

$$H = A_{11}h_1^2 + A_{12}h_1h_2 + A_{22}h_2^2 + B_1h_1 + B_2h_2$$

The coefficients appearing in the Hamiltonian are functions of J_2, i_m, Ω_m, i_s, μ, n_s, and n_m, as well as the mean motion n and semimajor axis a of the synchronous orbit itself. Because i_m and Ω_m are given in terms of certain trigonometric functions of Ω_M, Kamel and Tibbitts [7] showed that the A_{ij} and B_i coefficients can be expanded in terms of certain Fourier trigonometric series, involving constant and varying parts such that $A_{11} = \overline{A}_{11} + \tilde{A}_{11}$, $A_{12} = \tilde{A}_{12}$, $A_{22} = \overline{A}_{22} + \tilde{A}_{22}$, $B_1 = \tilde{B}_1$, and $B_2 = \overline{B}_2 + \tilde{B}_2$, where the barred elements are the constant parts and the elements with the tilde are the varying parts. The varying parts involve a fundamental harmonic in terms of s_{Ω_M} or c_{Ω_M} and a higher harmonic in terms of $s_{2\Omega_M}$ or $c_{2\Omega_M}$. The terms depending on Ω_M have smaller coefficients than the constant terms such that the differential equations in h_1 and h_2 essentially form a linear system with slowly varying coefficients, which is thus amenable to perturbation analysis based on Lie transforms. If the varying parts in A_{ij} and B_i are neglected, then the differential equations in h_1 and h_2 reduce to a linear system with constant coefficients that can therefore be solved in closed form. This simplification amounts to the effective averaging out of the slowly varying fluctuations in h_1 and h_2 with a period of 18.6 years. The main part of the Hamiltonian whose coefficients are constant is given by

$$H_0 = \overline{A}_{11}h_1^2 + \overline{A}_{22}h_2^2 + \overline{B}_2h_2$$

The differential system involving only H_0 is therefore given by

$$\frac{dh_1}{d\Omega_M} = \frac{\partial H_0}{\partial h_2} = 2\overline{A}_{22}h_2 + \overline{B}_2 \tag{5.8}$$

$$\frac{dh_2}{d\Omega_M} = -\frac{\partial H_0}{\partial h_1} = -2\overline{A}_{11}h_1 \tag{5.9}$$

The equilibrium solution is obtained from $dh_1/d\Omega_M = 0$ and $dh_2/d\Omega_M = 0$, or using the subscript e for equilibrium, $h_{1e} = 0$ and $h_{2e} = -\overline{B}_2/\left(2\overline{A}_{22}\right)$. In view of the definitions $h_1 = s_i s_\Omega$ and $h_2 = s_i c_\Omega$, we obtain the conditions: $\Omega_e = 0$ and $i_e = \sin^{-1}(h_{2e})$. This corresponds to an inertially fixed orbit plane, also called the invariant plane, with its line of nodes along the intersection of the ecliptic and equatorial planes. The equilibrium inclination i_e is a function of the semimajor axis, and for the synchronous altitude, we obtain $a/R_e = 6.61072$ and $i_e = 7.5$ deg. If our inclination tolerance is larger than 7.5 deg, then the optimal strategy consists of placing the satellite in an orbit inclined at exactly 7.5 deg such that it will never require any north–south stationkeeping maneuver. If the inclination constraint were less than 7.5 deg, however, then these maneuvers would be necessary. In this case, optimal maneuvering strategies must be devised such that, for example, the constraint is satisfied for the longest time possible between two such maneuvers. Now, the motion of the unitized angular momentum vector with components h_1 and h_2, that is, the motion of the orbit plane itself described by the Eulerian angles Ω and i, consists of a small oscillation about the equilibrium solution whose period is also a function of the orbit semimajor axis [7]. For the synchronous altitude, it is roughly equal to 54 years. Kamel and Tibbitts [7] showed that the two first-order equations [i.e., Eqs. (5.8) and (5.9)] can be reduced to the harmonic-oscillator type

$$\frac{d^2 h_1}{d\Omega_M^2} + \omega^2 h_1 = 0 \tag{5.10}$$

with $\omega^2 = 4\overline{A}_{22}\overline{A}_{11}$ and, therefore, with the period $P = 2\pi/\omega$. The general solution for the motion is written in terms of two arbitrary constants x and X such that $h_1 = (\delta_1 X)^{1/2} \sin(\omega\Omega_M + x)$ and $h_2 = h_{2e} + 2(X/\delta_1)^{1/2} \cos(\omega\Omega_M + x)$, with $\delta_1 = 2\left(\overline{A}_{22}/\overline{A}_{11}\right)^{1/2}$. For some values of $X > 0$, these equations describe an anticlockwise ellipse with its center at (0, h_{2e}) in the (h_1, h_2) domain, with a semiminor axis of $(\delta_1 X)^{1/2}$ along h_1, and with a semimajor axis of $2(X/\delta_1)^{1/2}$ along the h_2 axis. This ellipse represents the precession cycle of the orbital angular momentum about its equilibrium position with a period of roughly 54 years. Because δ_1 is nearly equal to 2, to zeroth order this ellipse is essentially approximated by a circle of radius $(2X)^{1/2}$ centered at (0, h_{2e}). The polar coordinates (s_i, Ω) are used to represent a point (h_1, h_2) on the precession circle, and if the inclination must remain below a given maximum $i \leq i_{\max}$, then this inclination constraint can be represented by the interior of a circle centered at the origin with radius $s_{i_{\max}}$ (Fig. 5.2). The precession time taken from an initial point (h_{10}, h_{20}) at time $t = t_0$ to a point (h_1, h_2) at time t is proportional to the subtended angle θ, as shown in Fig. 5.2. Then, the linear relationship between θ and $T = (t - t_0)$ is simply given by $T = \theta P/(2\pi)$.

The inclination constraint cannot be satisfied forever, but if the initial values of Ω_0 and i_0 are chosen in an optimal way, then the time between maneuvers can be maximized. Fuel-minimizing strategies can also be devised if higher maneuvering frequencies are tolerated. The former strategy is achieved by choosing $i_0 = i_{\max}$ and $\Omega_0 = 3\pi/2 + \theta_{\max}/2$, with $\theta_{\max} = 2\sin^{-1}\left(s_{i_0}/h_{2e}\right)$ and the total elapsed time between t_0 and $t_0 + T$ is given by $T_{\max} = \theta_{\max} P/(2\pi)$. The arc between t_0 and $t_0 + T_{\max}$ (Fig. 5.2) remains inside the inclination constraint circle for the longest possible time, and at time $t_0 + T_{\max}$, an impulsive maneuver must recover the initial node Ω_0 without changing the inclination, which is now the same as i_0. Figure 5.3 shows the relative geometry of the initial and final orbits at the maneuver time with nodes Ω_1 (premaneuver) and Ω_2 (postmaneuver), respectively, and identical values $i_1 = i_2 = i_{\max}$. The two orbits intersect at point A, where the north–south stationkeeping velocity change is applied. The postmaneuver node Ω_2 is, of course, equal to the target value of Ω_0. The post-velocity-change

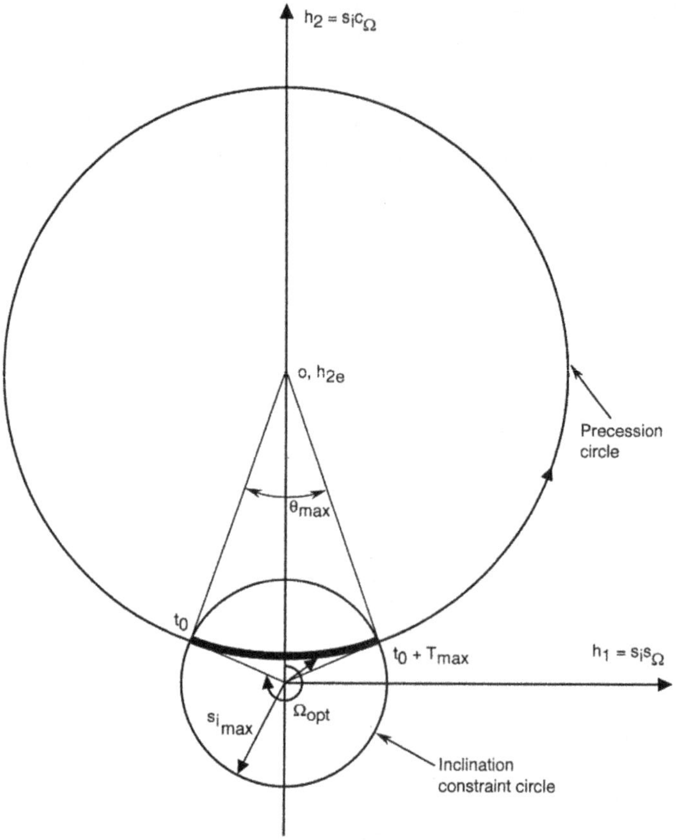

Fig. 5.2 Optimal drift within the inclination constraint circle [1].

orbit will then correspond to the initial orbit that starts the precession cycle until i_{max} is about to be violated again. The inclination decreases from i_{max} to a minimum without going through zero, for one-half of the precession time T_{max}, and then it increases to i_{max} at the end of the cycle with its node at Ω_1. This nodal rotation at constant inclination i_{max} is equivalent to an "inclination change." The geometry of this maneuver is shown in Fig. 5.4, with the premaneuver and postmaneuver orbits defined by the pairs (i_1, Ω_1) and (i_2, Ω_2), respectively, and, once again, $i_1 = i_2 = i_{max}$. The angle i^* is called the wedge angle, and it represents the inclination of one orbit relative to the other. An impulsive change in velocity applied at the intersection of these two orbits rotating the orbit by the angle i^* will, therefore, achieve the condition (i_{max}, Ω_2) that starts the precession cycle.

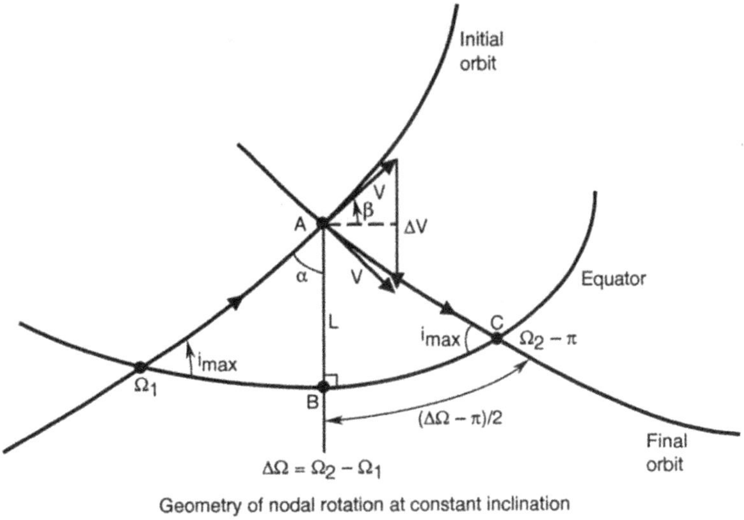

Fig. 5.3 Premaneuver and postmaneuver orbit geometry [1].

From spherical trigonometry, the wedge angle i^* is computed from $c_{i^*} = c_{\Omega_1-\Omega_2} s_{i_1} s_{i_2} + c_{i_1} c_{i_2}$, and with $i_1 = i_2 = i_{\max}$, we obtain $c_{i^*} = c_{\Omega_1-\Omega_2} s^2_{i_{\max}} + c^2_{i_{\max}}$. The velocity change is obtained from $\Delta V = 2V \sin(i^*/2)$.

5.3 MECHANICS OF LOW-THRUST MANEUVERING

We next analyze the mechanics of low-thrust inclination control for the case of a near-circular orbit. The out-of-plane motion is described by the following two equations linearized about a reference circular orbit

$$\frac{\mathrm{d}\Delta i_x}{\mathrm{d}\tau} = \frac{a_0^2}{\mu_e} c_{\theta'} f_h$$
$$\frac{\mathrm{d}\Delta i_y}{\mathrm{d}\tau} = \frac{a_0^2}{\mu_e} s_{\theta'} f_h$$

where Δi_x and Δi_y are the components of the change in the inclination $\boldsymbol{\Delta i}$ along the inertial $\hat{\boldsymbol{x}}$ and $\hat{\boldsymbol{y}}$ directions, as shown in Fig. 5.5. The $\hat{\boldsymbol{x}}-\hat{\boldsymbol{y}}$ plane is the plane of the initial reference circular orbit of semimajor axis a_0 and mean motion n. The angular position of the spacecraft at time t measured from the $\hat{\boldsymbol{x}}$ axis is given by the angle θ', and the rates are with respect to the dimensionless time $\tau = nt$. Here, μ_e is the gravitational constant of Earth. The instantaneous acceleration vector $\boldsymbol{f}$ has components f_r, f_θ, and f_h in the rotating Euler–Hill frame, and the thrust vector $\boldsymbol{T}$ is completely defined by the yaw and pitch angles θ_h and θ_t (Fig. 5.5).

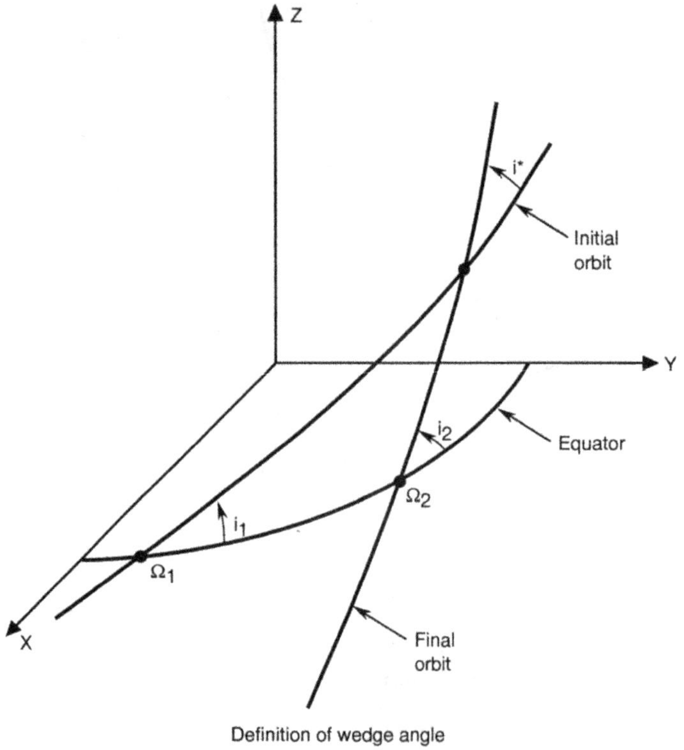

Fig. 5.4 Wedge angle at relative ascending node.

The overall change in inclination is given by the sum of its components along the $\hat{x}$ and $\hat{y}$ directions $\boldsymbol{\Delta i} = \Delta i_x \hat{\boldsymbol{x}} + \Delta i_y \hat{\boldsymbol{y}}$ with $\Delta i_x = \Delta i c_{\theta'}$, $\Delta i_y = \Delta i s_{\theta'}$, and Δi obtained from the variation-of-parameters equation $\mathrm{d}i/\mathrm{d}t = (rf_h/h)c_\theta$, where θ is now the angular position measured from the ascending node. When the spacecraft is initially on the reference circular orbit, we have $\theta = 0$, $r = a_0$, and $h = na_0^2$ such that, for a small impulse $\boldsymbol{\Delta V}$ applied along the direction $\boldsymbol{\beta} = (\beta_r, \beta_\theta, \beta_h)$, we have $\Delta i = (\Delta V/na_0)\beta_h$ because $\Delta tf_h = \Delta V\beta_h$. Therefore, $\Delta i_x = (\Delta V/na_0)\beta_h c_{\theta'}$, $\Delta i_y = (\Delta V/na_0)\beta_h s_{\theta'}$, and $\boldsymbol{\Delta i} = (\Delta i/na_0)\beta_h\hat{\boldsymbol{r}}$. The dimensionless rates in terms of f_h, $\mathrm{d}\Delta i_x/\mathrm{d}\tau$, and $\mathrm{d}\Delta i_y/\mathrm{d}\tau$ are now readily obtained. For near-circular orbits, $\theta' \cong nt = \tau$, $c_{\theta'} \cong c_\tau$, and $s_{\theta'} \cong s_\tau$ with T and m representing the thrust magnitude and spacecraft mass, respectively, the linearized equations reduce to the form

$$\frac{\mathrm{d}\Delta i_x}{\mathrm{d}\tau} = \frac{k}{a_0 n^2} c_\tau s_{\theta_h} \tag{5.11}$$

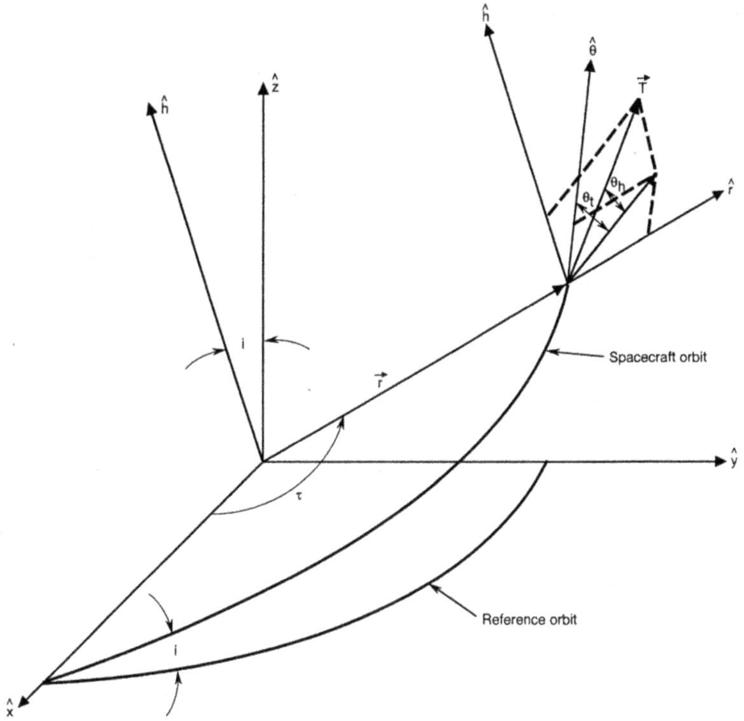

Fig. 5.5 Orbital frame and thrust geometry.

$$\frac{d\Delta i_y}{d\tau} = \frac{k}{a_0 n^2} s_\tau s_{\theta_h} \tag{5.12}$$

where $k = T/m$. Now, if only pure yawing is used (i.e., $\theta_h = \pm\pi/2$), then with $g = k/(a_0 n^2)$ and after a full revolution of continuous thrust with $\theta_h = \pi/2$ for one-half revolution and $\theta_h = -\pi/2$ for the other half, the maximum rotation of the initial orbit plane will be given by

$$\Delta i_{\max} = 4g \tag{5.13}$$

Because $k = T/m = 2P/(mc)$, where P is the thrust power, $c = I_{\text{sp}}\, g_r$ is the exhaust velocity, I_{sp} is the specific impulse, and g_r is the acceleration due to gravity, Eq. (5.13) can be reduced to

$$\Delta i_{\max} = \frac{8P}{m I_{\text{sp}} g_r a_0 n^2}$$

which is given in radians, when P is in watts, m is in kilograms, I_{sp} is in seconds, g_r is in meters per second squared, a_0 is in meters, and n is in radians per second. The power of the jet is the time rate of expenditure of the kinetic energy of the ejected matter

$$P = \dot{E}_{\text{jet}} = \frac{1}{2}\dot{m}c^2 = \frac{1}{2}Tc$$

The power transmitted to the vehicle is $P_{\text{veh}} = TV$, where V is the vehicle velocity. Let the initial orbit semimajor axis be $a_0 = 42{,}241$ km with the corresponding mean motion $n = \left(\mu_e/a_0^3\right)^{1/2} = 7.272205 \times 10^{-5}$ rad/s or 360 deg/day for the two-body geostationary orbit. This value of a_0 is chosen for illustrative purposes, as Earth's inertial rate is actually equal to 360.985 deg/day compared to the 360 deg/day value used here. Let us suppose that the spacecraft mass is $m = 1134$ kg and that it uses a resistojet with a specific impulse of $I_{\text{sp}} = 295$ s and a jet power of 1 kW such that the vehicle is imparted an acceleration of $k = f = T/m = 2P/(mc) = 6.096119 \times 10^{-4}$ m/s^2. The continuous application of this acceleration during one revolution or 24 h will result in a change in velocity of $\Delta V{=}ft = 52.670$ m/s with an equivalent change in maximum orbit inclination given by $\Delta i_{\max}{=}0.62544$ deg. If we now consider a 1-h thrust arc spread around the line of nodes of current and target orbits, then the angular travel will be 360/24 = 15 deg for geosynchronous Earth orbit (GEO). The relative rotation of the orbit plane corresponding to this burn will be obtained from the equations

$$\Delta i_x = k/(a_0 n^2)\int_{-7.5}^{7.5} c_\tau \, d\tau = 7.1240179 \times 10^{-4} \text{ rad}$$

$$\Delta i_y = k/\left(a_0 n^2\right)\int_{-7.5}^{7.5} s_\tau \, d\tau = 0$$

such that $\Delta i = (\Delta i_x^2 + \Delta i_y^2)^{1/2} = 0.040817$ deg. For an inclination deadband of 0.3 deg, with $i_0 = 0.3$ deg and $h_{2e} = s_{7.5\text{deg}} = 0.130526$, we have $\theta_{\max} = 4.5979$ deg, $\Omega_0 = 272.298$ deg, and $T_{\max} = 251.912$ days with a period P of 54 years. The change in ascending node Ω or $\Delta\Omega = \Omega_{\text{init}} - \Omega_{\text{final}}$ is obtained from $\Delta\Omega = -2(90-2.298) = -2\Omega_1 = -175.404$ deg, which is the same as $\Delta\Omega = \Omega_1 - \Omega_2$.

Now, the wedge angle is calculated with $i_1 = i_2 = 0.3$ deg and $\Omega_1 - \Omega_2 = -175.404$ deg such that $i^* = 0.5995$ deg. Because 1 h of pure out-of-plane thrusting rotates the orbit by 0.0408 deg, it will take 0.5995 deg/(0.0408 deg/h) = 14.693 h of total thrust for 1 h each at the line of nodes of the current and final orbits. We can thrust either 1 h per revolution at, say, the ascending common node or 2 h per revolution with 1 h each at the ascending and descending common nodes. The total inclination change can, therefore, be achieved in 14.693 revolutions or days or 7.346 revolutions or days, respectively. This requires a total velocity change of 32.246 m/s. The impulsive change in velocity can be computed from the equation $\Delta V/V = 2i_{\max}\ |\sin(\Delta\Omega/2)| = 0.0104635$. For $i_{\max} = 0.3$ deg, using the relation $V = \left(\mu_e/a_0\right)^{1/2}$, we obtain $\Delta V = 32.142$ m/s, which is only slightly smaller than the 32.246 m/s value obtained for the low-thrust solution. This shows that the 1-h thrust arcs at GEO are effectively as efficient as the impulsive solution with almost no loss of velocity change. However, the continuous 24-h acceleration that rotates the orbit by 0.62544 deg requiring 52.670 m/s of total velocity change can be achieved much more efficiently by applying a single impulse with a velocity change of $\Delta V = 2V\sin(\delta/2)$, with $\delta = 0.62544$ deg and $V = (\mu_e/a_0)^{1/2} = 3.071902$ km/s, such that $\Delta V = 33.532$ m/s, for a savings of 19.138 m/s. This shows that it is not economical to thrust continuously for a complete revolution to achieve an inclination change. Now, the impulsive velocity change to cause an inclination change of 0.5995 deg, which is the wedge angle i^*, can also be computed from the equation $\Delta V = 2V\ \sin(i^*/2) = 32.142$ m/s. If we let $i_{\max} = 0.1$ deg for a tighter tolerance, then $\theta_{\max} = 1.532304$ deg, $\Omega_0 = 270.766152$ deg, $T_{\max} = 83.89$ days, and $\Delta\Omega = -178.467695$ deg, requiring an impulse of $\Delta V = 10.731$ m/s. The wedge angle (Fig. 5.4) is now $i^* = 0.199982$ deg. Because the 1-h low-thrust accelerations produce a relative rotation of the orbit plane of 0.040817 deg, requiring a quasi-impulsive velocity change of $\Delta V = 2.187$ m/s, the total change in velocity of 10.731 m/s for this tighter deadband can be achieved in less than five acceleration cycles requiring three to five revolutions at GEO to achieve. This time frame is still negligible when compared to the drift period of 83 days, and therefore, these low-thrust maneuvers can be accomplished quickly without worrying too much about the drift during the maneuvering period. This will not be the case for lower thrust accelerations. In that case, the orbit rotation to repeat the drift cycle will require a much longer period of time, and therefore, appropriate strategies must be designed to account for the drift accumulated during the longer maneuvering period such that the optimal initial conditions are achieved at the end of the maneuver sequence.

5.4 SUBOPTIMAL STRATEGY

Given the initial (i_1, Ω_1) and final (i_2, Ω_2) orbits, an inclination change through an angle i' at their common line of nodes is required. Because of the lower acceleration level of the low-thrust maneuver, a smaller rotation i'' is achieved instead around point B (Fig. 5.6). The inclination i_j and node Ω_j can be obtained from spherical trigonometry. If we wait in this orbit for some given duration, then i_j and Ω_j will drift to new values at the time of the next thrusting cycle, such that the thrust must now be applied around the new common line of nodes of the present orbit and the target orbit. This suboptimal strategy does not take into account the drift in the maneuver design and is therefore less efficient in recovering the target conditions. From the triangle ABC and the law of sines and cosines, as well as the relation $\Delta\Omega = \Omega_2 - \Omega_1$, we can write

$$s_{AB} = s_{i_2} s_{\Delta\Omega} / s_{i'} \tag{5.14}$$

$$c_{AB} = \frac{\left(-c_{i_2} + c_{i_1} c_{i'}\right)}{s_{i_1} s_{i'}} \tag{5.15}$$

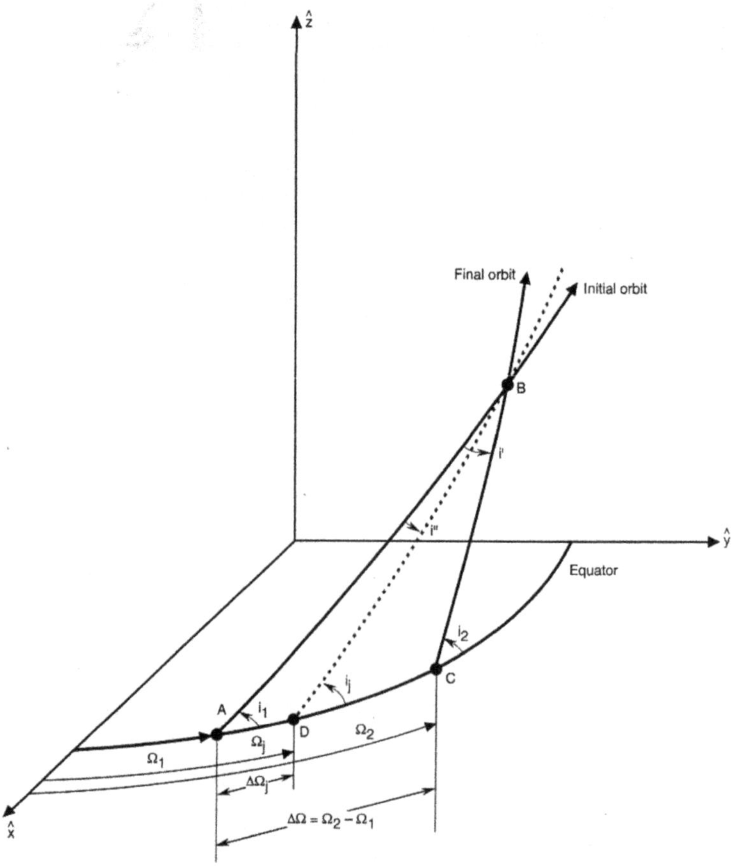

Fig. 5.6 Orbit plane rotation at the relative node.

From the triangle *ABD*, we have

$$-c_{i_j} = -c_{i_1} c_{i''} + s_{i_1} s_{i''} c_{AB}$$

which, upon substitution of Eq. (5.15), gives

$$c_{i_j} = c_{i_1} c_{i''} - \frac{s_{i''}}{s_{i'}} \left(c_{i_1} c_{i'} - c_{i_2} \right) \tag{5.16}$$

This equation determines i_j without ambiguity because $0 \leq i_j \leq \pi$. In a similar manner, we have

$$s_{\Delta\Omega_j} = s_{i''} s_{AB} / s_{i_j} \tag{5.17}$$

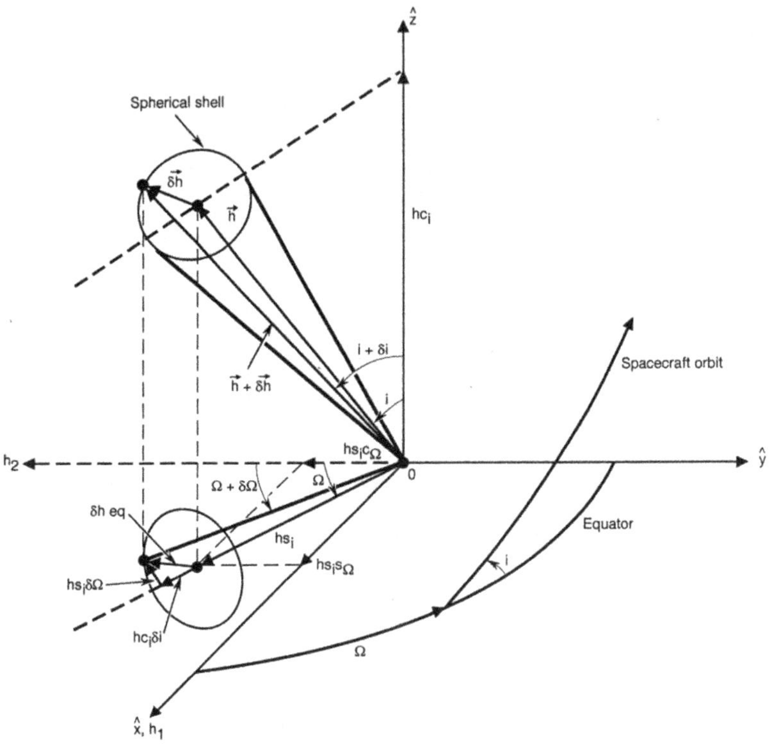

Fig. 5.7 Capability ellipse due to a small thrust arc.

$$c_{i''} = c_{i_1} c_{i_j} + s_{i_1} s_{i_j} c_{\Delta\Omega_j} \tag{5.18}$$

Using Eqs. (5.14) and (5.17), an expression for $\Delta\Omega_j = \Omega_j - \Omega_1$ is obtained as

$$\tan \Delta\Omega_j = \frac{s_{i_1} s_{i''} s_{AB}}{c_{i''} - c_{i_1} c_{i_j}}$$

which, upon substitution with Eq. (5.14), yields

$$\tan \Delta\Omega_j = \frac{s_{i_1} s_{i''} s_{i_2} s_{\Delta\Omega}}{s_{i'} \left(c_{i''} - c_{i_1} c_{i_j} \right)} \tag{5.19}$$

where c_{i_j} is given by Eq. (5.16). Therefore, we can solve for the angles i_j and Ω_j and apply another small low-thrust maneuver at the new line of nodes of this orbit with the unchanged target orbit by replacing Ω_1 by Ω_j and i_1 by i_j and repeating the same calculations until the point (i_2, Ω_2) is finally reached. For the earlier example with i_{max}=0.1 deg, we have $i_1 = i_2 = 0.1$ deg, $i' = 0.199982$ deg, $\delta(i) = i'' = 0.040817$ deg, $i_j = 0.059189$ deg, $\Omega_2 = 270.766152$ deg, $\Omega_1 = 90$ deg $-$ 0.766152 deg $=$ 89.233848 deg, $\Delta\Omega = -2(90$ deg $-$ 0.766152 deg$) =$ -178.467695 deg or, identically, $\Delta\Omega = \Omega_2 - \Omega_1 = 181.532304$ deg and $\Omega_j = \Omega_1 + \Delta\Omega_j = 88.705515$ deg. Also, $h_1 = s_{i_j} s_{\Omega_j} = 1.032783 \times 10^{-3}$, $h_2 = s_{i_j} c_{\Omega_j}$ $= 2.333766 \times 10^{-5}$, $h_1(0) = s_{i_1} s_{\Omega_1} = 1.745172 \times 10^{-3}$, and $h_2(0) = s_{i_1} c_{\Omega_1} =$ 2.333761×10^{-5}, such that $\Delta h_{eq} = \left\{[h_1 - h_1(0)]^2 + [h_2 - h_2(0)]^2\right\}^{1/2} =$ 7.123891×10^{-4} is the change achieved in the h_1–h_2 plane by the 1-h out-of-plane low-thrust maneuver. Similar Δh_{eq} variations will slowly result in achieving the target parameters i_2 and Ω_2 for the initiation of the natural drift cycle. In this strategy, we can ignore the small drift that occurs between the application of two such subsequent maneuvers because i_j and Ω_j will then drift and, therefore, be updated by orbit determination or analytically before the next maneuver. Figure 5.7 shows the angular momentum vector $\boldsymbol{h}$ and its change $\boldsymbol{\delta h}$ due to the low-thrust maneuver, such that $\boldsymbol{h} + \boldsymbol{\delta h}$ is now the postmaneuver angular momentum vector. Because the thrust is applied normal to the orbit plane, the length of $\boldsymbol{h}$ remains unchanged, whereas its direction is changed. Therefore, a thrust arc of duration Δt will place $\boldsymbol{h}$ along the circular edge of the spherical shell shown in Fig. 5.7, for all possible locations of this thrust arc along the spacecraft orbit. The projection of $\boldsymbol{h}$ onto the equatorial plane, which defines the $\hat{\boldsymbol{h}}_1$ and $\hat{\boldsymbol{h}}_2$ axes, is hs_i inclined at an angle Ω from the $\hat{\boldsymbol{h}}_2$ axis or the $-\hat{\boldsymbol{y}}$ inertial axis, and the projection of the spherical shell is, in general, an ellipse centered at the tip of the hs_i vector. The angular momentum vector can be written as

$$\boldsymbol{h} = hs_i s_\Omega \hat{\boldsymbol{x}} - hs_i c_\Omega \hat{\boldsymbol{y}} + hc_i \hat{\boldsymbol{z}}$$

and, for a small rotation δi

$$\boldsymbol{h} + \delta\boldsymbol{h} = hs_{i+\delta i} s_\Omega \hat{\boldsymbol{x}} - hs_{i+\delta i} c_\Omega \hat{\boldsymbol{y}} + hc_{i+\delta i} \hat{\boldsymbol{z}}$$

which leads, after expansion, to

$$\delta\boldsymbol{h} = hc_i s_\Omega \delta i \hat{\boldsymbol{x}} - hc_i c_\Omega \delta i \hat{\boldsymbol{y}} - hs_i \delta i \hat{\boldsymbol{z}}$$

with magnitude $|\delta\boldsymbol{h}| = h\delta i$. The magnitude of the projection of $\delta\boldsymbol{h}$ onto the equatorial plane is given by

$$|\delta\boldsymbol{h}_{eq}| = \left(h^2 c_i^2 \delta i^2\right)^{1/2} = hc_i \delta i \tag{5.20}$$

Similarly, for a small change $\delta\Omega$, the projection of $\delta\boldsymbol{h}$ onto the h_1–h_2 plane has magnitude

$$|\delta\boldsymbol{h}_{eq}| = hs_i \delta\Omega \tag{5.21}$$

For the more general case of simultaneous changes δi and $\delta\Omega$, we can write

$$(\boldsymbol{h} + \delta\boldsymbol{h})_{eq} = hs_{i+\delta i} s_{\Omega+\delta\Omega} \hat{\boldsymbol{x}} - hs_{i+\delta i} c_{\Omega+\delta\Omega} \hat{\boldsymbol{y}}$$

leading to the magnitude

$$|\delta \boldsymbol{h}_{\text{eq}}| = h\left(c_i^2 \delta i^2 + s_i^2 \delta \Omega^2\right)^{1/2} \tag{5.22}$$

Now, from $h_1 = s_i s_\Omega$ and $h_2 = s_i c_\Omega$, we have

$$\delta h_1 = c_i s_\Omega \delta i + s_i c_\Omega \delta \Omega$$
$$\delta h_2 = c_i c_\Omega \delta i - s_i s_\Omega \delta \Omega$$

and therefore

$$\begin{aligned} \delta h_{\text{eq}} &= \left(\delta h_1^2 + \delta h_2^2\right)^{1/2} \\ \delta h_{\text{eq}} &= \left(c_i^2 \delta i^2 + s_i^2 \delta \Omega^2\right)^{1/2} \end{aligned} \tag{5.23}$$

Equation (5.23) shows that $\delta \boldsymbol{h}_{\text{eq}}$ has a component $hc_i\delta$ along hs_i for constant Ω and a component $hs_i\delta\Omega$ along the orthogonal direction for constant δi. The magnitude h is taken equal to 1 for convenience because it is invariant for out-of-plane thrusting. This orthogonal direction is, of course, the direction along the tangent to the inclination constraint circle of radius s_i. For small inclinations, $\boldsymbol{h}$ is essentially along the pole or inertial $\hat{z}$ direction, and the projection of the circular edge of the spherical shell is essentially a circle. When i approaches 90 deg, this projection becomes an elongated ellipse, and in the limit as s_i tends to unity, it becomes a straight line. However, the same rotation δi will be achieved regardless of the value of i because the size of $\delta \boldsymbol{i}$ or $\delta \boldsymbol{h}$ is independent of i, but its projection onto the h_1–h_2 plane is $\delta i c_i$, which tends to zero, reducing the semiminor axis of the projected ellipse to zero as well. For example, the values $\delta i = i_1 - i_j = 0.0408107$ deg and $\delta\Omega = -0.528332$ deg correspond to a value of $\delta h_{\text{eq}} = 7.124629 \times 10^{-4}$ from Eq. (5.23), which is close to the $\Delta h_{\text{eq}} = 7.123891 \times 10^{-4}$ obtained earlier. These two numbers are also comparable to the relative inclination change $\Delta i = 0.040817$ deg obtained earlier. In fact, Δi is essentially equal to Δh_{eq}, and the small difference from δh_{eq} is due to the truncation of the higher-order terms in computing δh_1 and δh_2 and, therefore, δh_{eq}. It is perhaps better to compute Δh_{eq} for a more precise evaluation of the change in the distance between two (h_1, h_2) pairs or, equivalently, two (i, Ω) pairs.

5.5 OPTIMAL (i, Ω) STEERING

This section develops an optimal maneuvering strategy that takes into account the natural drift of the spacecraft orbit between two successive thrusting cycles. Instead of targeting the final conditions (i_2, Ω_2), as in Sec. 5.4, intermediate target parameters for each thrusting cycle are generated, resulting in the overall minimum accumulated thrust arcs. For the same wait period between two thrust arcs as in the suboptimal strategy, this solution will, indeed, provide the minimum-time transfer, which is also fuel-minimizing for this particular transfer with imposed target parameters.

Let us now represent this quantity δh_{eq} by a velocity V, called the maneuvering velocity, as in Fig. 5.8. The natural drift, being in an anticlockwise direction, is represented by concentric circles centered at the equilibrium point $(0, h_{2e})$ with linear velocity W, called the precession velocity, whose components along the h_1 and h_2 axes are given by u and v, respectively. The problem is to return from any given (h_1, h_2) pair to any desired (h_1, h_2) target and, for the example at hand, to the ideal (i_0, Ω_0) initial conditions, in minimum time such that a period of free drift occurs without any maneuvering required. This problem is now cast as a navigation problem that consists of finding the time history of the steering angle ϕ that leads to reaching the point (i_0, Ω_0) in minimum time, thereby using the minimum amount of fuel. This problem is identical to the Zermelo problem of navigating a ship in strong currents [10]. Given the ship's constant velocity V and heading angle ϕ and the current's u and v velocity components along general rectangular coordinates x and y, which, in our case, are the h_1 and h_2 coordinates, respectively, we have [10]

$$\dot{x} = Vc_\phi + u \tag{5.24}$$

$$\dot{y} = Vs_\phi + v \tag{5.25}$$

with the Hamiltonian

$$H = 1 + \lambda_x(Vc_\phi + u) + \lambda_y(Vs_\phi + v) \tag{5.26}$$

for a minimum-time solution. The Euler-Lagrange and optimality conditions can be written as

$$\dot{\lambda}_x = -\partial H/\partial x = -\lambda_x\ \partial u/\partial x - \lambda_y\ \partial v/\partial x \tag{5.27}$$

$$\dot{\lambda}_y = -\partial H/\partial y = -\lambda_x\ \partial u/\partial y - \lambda_y\ \partial v/\partial y \tag{5.28}$$

$$\partial H/\partial \phi = 0 = V(-\lambda_x\ s_\phi + \lambda_y\ c_\phi) \tag{5.29}$$

Equation (5.29) yields the optimal steering angle ϕ with $\tan\phi = \lambda_y/\lambda_x$, and because we are minimizing transit time, the transversality condition is given by $H_f = 0$. The system is autonomous, so that H is constant and, therefore, equal to zero throughout. If we substitute $\lambda_x \tan\phi$ for λ_y into Eq. (5.26) with $H = 0$, then we obtain

$$\lambda_x = -c_\phi/(V + uc_\phi + vs_\phi) \tag{5.30}$$

$$\lambda_y = -s_\phi/(V + uc_\phi + vs_\phi) \tag{5.31}$$

Substituting these two adjoints into Eq. (5.27) with

$$\dot{\lambda}_x = [s_\phi(V + uc_\phi + vs_\phi)\dot{\phi} + c_\phi(vc_\phi - us_\phi)\dot{\phi}]/(V + uc_\phi + vs_\phi)^2$$

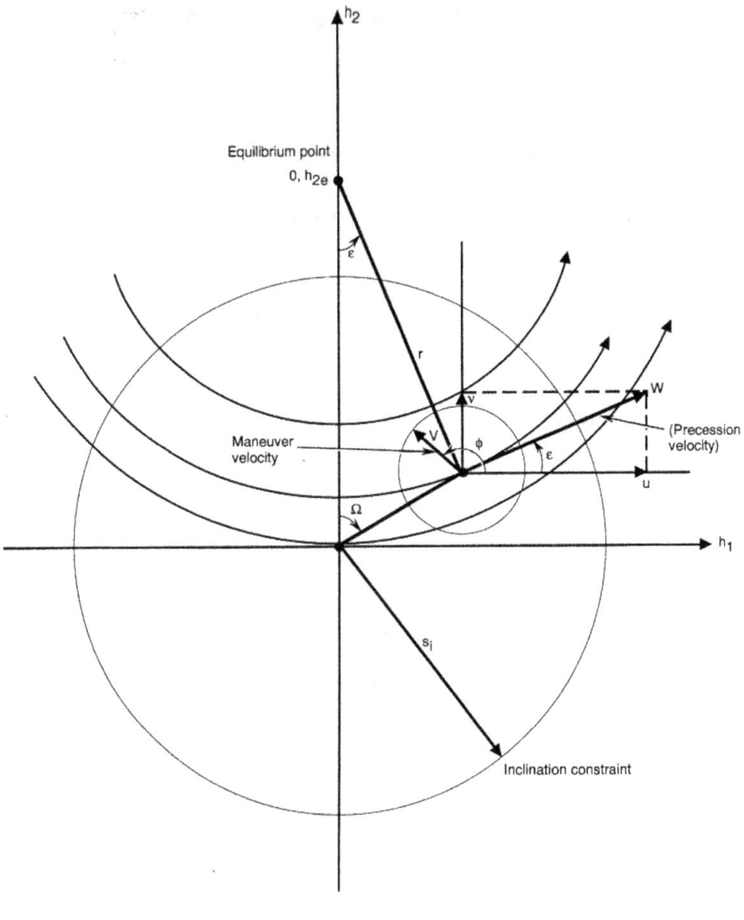

Fig. 5.8 Low-thrust steering geometry.

results in

$$(Vs_{\phi} + v)\dot{\phi} = (V + uc_{\phi} + vs_{\phi})\left(c_{\phi}\,\frac{\partial u}{\partial x} + s_{\phi}\,\frac{\partial v}{\partial x}\right) \tag{5.32}$$

Similarly, from Eq. (5.28), we obtain

$$-(Vc_{\phi} + u)\dot{\phi} = (V + uc_{\phi} + vs_{\phi})\left(c_{\phi}\frac{\partial u}{\partial y} + s_{\phi}\frac{\partial v}{\partial y}\right) \tag{5.33}$$

If we next multiply Eq. (5.32) by s_ϕ and Eq. (5.33) by $-c_\phi$ and add, we obtain

$$\dot{\phi} = s_\phi c_\phi \left(\frac{\partial u}{\partial x} - \frac{\partial v}{\partial y} \right) + s_\phi^2 \frac{\partial v}{\partial x} - c_\phi^2 \frac{\partial u}{\partial y} \tag{5.34}$$

The simultaneous integration of Eqs. (5.24), (5.25), and (5.34) will yield the desired optimal trajectory. If u and v are not constant, $\dot{\phi} \neq 0$ and, therefore, ϕ is not constant, indicating a variable heading strategy. From Fig. 5.8, we have $u = Wc_\varepsilon = \omega r c_\varepsilon$; $v = Ws_\varepsilon = \omega r s_\varepsilon$; and $\omega = \dot{\varepsilon}$, the constant angular velocity of r or the precession rate. Because $rs_\varepsilon = h_1$ and $rc_\varepsilon = h_{2e} - h_2$, we have

$$u = \omega(h_{2e} - h_2) \tag{5.35}$$

$$v = \omega h_1 \tag{5.36}$$

The dynamic equations can now be written as

$$\dot{h}_1 = Vc_\phi + \omega(h_{2e} - h_2) \tag{5.37}$$

$$\dot{h}_2 = Vs_\phi + \omega h_1 \tag{5.38}$$

with a constant Hamiltonian equal to zero

$$H = 1 + \lambda_{h_1}[Vc_\phi + \omega(h_{2e} - h_2)] + \lambda_{h_2}(Vs_\phi + \omega h_1) = 0 \tag{5.39}$$

The Euler–Lagrange equations for the adjoints and the optimality condition are given by

$$\dot{\lambda}_{h_1} = -\lambda_{h_2}\omega \tag{5.40}$$

$$\dot{\lambda}_{h_2} = \lambda_{h_1}\omega \tag{5.41}$$

$$\partial H/\partial\phi = 0 = V(-\lambda_{h_1}s_\phi + \lambda_{h_2}c_\phi) \tag{5.42}$$

As before, $\tan\phi = \lambda_{h_2}/\lambda_{h_1}$, and because $\partial u/\partial h_1 = 0$, $\partial v/\partial h_2 = 0$, $\partial v/\partial h_1 = \omega$, and $\partial u/\partial h_2 = -\omega$, Eq. (5.34) yields $\dot{\phi} = \omega = \text{const}$, such that the optimal steering law is now the simple linear law

$$\phi = \omega t + \phi_0 \tag{5.43}$$

From Eqs. (5.40) and (5.41), we have $\ddot{\lambda}_{h_2} = \dot{\lambda}_{h_1}\omega = -\lambda_{h_2}\omega^2$, and therefore, λ_{h_2} is governed by $\ddot{\lambda}_{h_2} + \lambda_{h_2}\omega^2 = 0$, which is the equation of a linear oscillator with solution $\lambda_{h_2} = K_1 \sin(\omega t + K_2)$, which also results from $\lambda_{h_1} = \dot{\lambda}_{h_2}/\omega$, $\lambda_{h_1} = K_1 \cos(\omega t + K_2)$. Substituting for λ_{h1} and λ_{h2} in $\tan\phi = \lambda_{h_2}/\lambda_{h_1}$ gives $\phi = \omega t + K_2$. Therefore, K_2 is also the value of the control angle ϕ at time zero, ϕ_0. Let us now use the dimensionless variable $\tau' = \omega t$ and convert Eqs. (5.37) and (5.38) to the form

$$\frac{dh_1}{d\tau\prime} = \frac{V}{\omega}\cos(\tau' + \phi_0) + (h_{2e} - h_2) \tag{5.44}$$

$$\frac{dh_2}{d\tau'} = \frac{V}{\omega}\sin(\tau' + \phi_0) + h_1 \tag{5.45}$$

where the derivatives are now with respect to τ'. From Eqs. (5.44) and (5.45), we have

$$\frac{d^2h_2}{d\tau'^2} = \frac{V}{\omega}\cos(\tau' + \phi_0) + \frac{dh_1}{d\tau'} = \frac{2V}{\omega}\cos(\tau' + \phi_0) + (h_{2e} - h_2)$$

and, therefore, the forced-oscillator type of equation

$$\frac{d^2h_2}{d\tau'^2} + h_2 = h_{2e} + \frac{2V}{\omega}\cos(\tau' + \phi_0) \tag{5.46}$$

whose solution is given by

$$h_2 = C_1 \sin(\tau' + C_2) + \int_0^{\tau'} h_{2e} \sin(\tau' - s)ds + \frac{2V}{\omega}\int_0^{\tau'} \cos(s + \phi_0)\sin(\tau' - s)ds$$

which reduces to

$$h_2 = C_1 \sin(\tau' + C_2) + h_{2e}(1 - c_{\tau'}) + \frac{V}{\omega}\left[\tau' \sin(\tau' + \phi_0) - s_{\phi_0}\, s_{\tau'}\right] \tag{5.47}$$

At $\tau' = 0$

$$h_2 = h_{20} = C_1 \sin C_2 \tag{5.48}$$

If we use the solution for h_2 in Eq. (5.44) and solve for h_1 in a manner similar to that used in solving for h_2, we obtain

$$\begin{aligned} h_1 = h_{10} + h_{2e}\, s_{\tau'} + C_1 \cos(\tau' + C_2) - \frac{V}{\omega} s_{\phi_0} c_{\tau'} \\ + \frac{V}{\omega}\tau' \cos(\tau' + \phi_0) - C_1 \cos C_2 + \frac{V}{\omega} s_{\phi_0} \end{aligned} \tag{5.49}$$

From Eqs. (5.45) and (5.47), we have

$$\begin{aligned} \frac{dh_2}{d\tau'} = \frac{V}{\omega}\sin(\tau' + \phi_0) + h_1 = C_1 \cos(\tau' + C_2) + h_{2e}\, s_{\tau'} \\ + \frac{V}{\omega}\left[\sin(\tau' + \phi_0) + \tau' \cos(\tau' + \phi_0) - s_{\phi_0}\, c_{\tau'}\right] \end{aligned}$$

and at $\tau' = 0$

$$\frac{V}{\omega} s_{\phi_0} + h_{10} = C_1 \cos C_2 \tag{5.50}$$

Equations (5.48) and (5.50) yield

$$\tan C_2 = h_{20} \Big/ \left(\frac{V}{\omega} s_{\phi_0} + h_{10} \right) \tag{5.51}$$

and

$$C_1 = \left[h_{20}^2 + \left(\frac{V}{\omega} s_{\phi_0} + h_{10} \right)^2 \right]^{1/2} \tag{5.52}$$

The initial angle ϕ_0 and final time τ'_f are obtained from the boundary conditions at the final time τ'_f, namely, $h_1(\tau'_f) = h_{1f}$ and $h_2(\tau'_f) = h_{2f}$. From Eqs. (5.47) and (5.49), we obtain

$$f_1 = C_1 \sin(\tau'_f + C_2) + h_{2e}(1 - c_{\tau'_f}) + \frac{V}{\omega} \left[\tau'_f \sin(\tau'_f + \phi_0) - s_{\phi_0} s_{\tau'_f} \right] - h_{2f} = 0 \tag{5.53}$$

$$f_2 = h_{2e} s_{\tau'_f} + C_1 \cos\left(\tau'_f + C_2 \right) - \frac{V}{\omega} s_{\phi_0} c_{\tau'_f} + \frac{V}{\omega} \tau'_f \cos\left(\tau'_f + \phi_0 \right) - h_{1f} = 0 \tag{5.54}$$

A Newton–Raphson scheme is used to solve for ϕ_0 and τ'_f from the equations

$$\begin{pmatrix} \dfrac{\partial f_1}{\partial \phi_0} & \dfrac{\partial f_1}{\partial \tau'_f} \\ \dfrac{\partial f_2}{\partial \phi_0} & \dfrac{\partial f_2}{\partial \tau'_f} \end{pmatrix} \begin{pmatrix} \delta\phi_0 \\ \delta\tau'_f \end{pmatrix} = - \begin{pmatrix} f_1 \\ f_2 \end{pmatrix} \tag{5.55}$$

$$\begin{pmatrix} \phi_0 \\ \tau'_f \end{pmatrix}_{\text{new}} = \begin{pmatrix} \phi_0 \\ \tau'_f \end{pmatrix}_{\text{old}} + \begin{pmatrix} \delta\phi_0 \\ \delta\tau'_f \end{pmatrix} \tag{5.56}$$

The Jacobian partials are given in analytical form by

$$\begin{aligned} \frac{\partial f_1}{\partial \phi_0} &= \sin\left(\tau'_f + C_2 \right) \frac{\partial C_1}{\partial \phi_0} + C_1 \cos\left(\tau'_f + C_2 \right) \frac{\partial C_2}{\partial \phi_0} \\ &\quad + \frac{V}{\omega} \left[\tau'_f \cos\left(\tau'_f + \phi_0 \right) - c_{\phi_0} s_{\tau'_f} \right] \end{aligned} \tag{5.57}$$

$$\begin{aligned} \frac{\partial f_1}{\partial \tau'_f} &= C_1 \cos\left(\tau'_f + C_2 \right) + h_{2e} s_{\tau'_f} \\ &\quad + \frac{V}{\omega} \left[\sin\left(\tau'_f + \phi_0 \right) + \tau'_f \cos\left(\tau'_f + \phi_0 \right) - s_{\phi_0} c_{\tau'_f} \right] \end{aligned} \tag{5.58}$$

$$\frac{\partial f_2}{\partial \phi_0} = \cos\left(\tau_f' + C_2\right)\frac{\partial C_1}{\partial \phi_0} - C_1 \sin\left(\tau_f' + C_2\right)\frac{\partial C_2}{\partial \phi_0} - \frac{V}{\omega}\left[\tau_f' \sin\left(\tau_f' + \phi_0\right) + c_{\phi_0}\, c_{\tau_f'}\right] \tag{5.59}$$

$$\frac{\partial f_2}{\partial \tau_f'} = h_{2e}\, c_{\tau_f'} - C_1 \sin\left(\tau_f' + C_2\right) - \frac{V}{\omega}\left[\tau_f' \sin\left(\tau_f' + \phi_0\right) - \cos\left(\tau_f' + \phi_0\right) - s_{\phi_0}\, s_{\tau_f'}\right] \tag{5.60}$$

The partials $\partial C_1/\partial \phi_0$ and $\partial C_2/\partial \phi_0$ appearing in the preceding expressions are readily obtained from Eqs. (5.51) and (5.52), after making use of the relation $\partial \tan C_2/\partial \phi_0 = (1 + \tan^2 C_2)(\partial C_2/\partial \phi_0)$. The result is

$$\frac{\partial C_1}{\partial \phi_0} = \frac{V}{\omega C_1} c_{\phi_0}\left(\frac{V}{\omega} s_{\phi_0} + h_{10}\right) = \frac{V c_{\phi_0}\, h_{20}}{\omega C_1 \tan C_2} \tag{5.61}$$

$$\frac{\partial C_2}{\partial \phi_0} = \frac{-\, V c_{\phi_0}\, h_{20}}{\omega C_1^2} \tag{5.62}$$

A few iterations are needed to solve for ϕ_0 and τ_f', which, in turn, completely define the optimal trajectory $[h_1 = f(\tau'), h_2 = f(\tau')]$ that starts from the given point (h_{10}, h_{20}) and achieves the final desired point (h_{1f}, h_{2f}) in the minimum amount of time.

5.6 RESULTS

Let us use $\Delta h = 7.123891 \times 10^{-4}$, which corresponds to our 1-h low-thrust acceleration. If this maneuver is applied, for example, once per week or time interval Δt, then our velocity is $V = \Delta h/\Delta t = 1.017 \times 10^{-4}$/day. The precession rate ω is obtained from $\omega = 2\pi/(54 \times 365.25) = 3.1856 \times 10^{-4}$ rad/day. For $i_{max} = 0.3$ deg, we have $\Omega_1 = 87.701$ deg, $\Omega_2 = 272.299$ deg, $h_{10} = 0.52317 \times 10^{-2}$, $h_{20} = 0.21003 \times 10^{-3}$, $h_{1f} = -0.52317 \times 10^{-2}$, and $h_{2f} = 0.21003 \times 10^{-3}$. The solution is found with $\phi_0 = 178.414$ deg and $\tau_f' = 0.553569 \times 10^{-1}$, corresponding to a total time of $t_f = \tau_f'/\omega = 173.77068$ days. The free-drift time $T_{max} = \theta_{max}/\omega$ with $\theta_{max}/2 = \tan^{-1}[h_{1f}/(h_{2e} - h_{2f})] = -2.298$deg is $T_{max} = 251.91244$ days. Figure 5.9 shows both the optimal free-drift and low-thrust return trajectories with a combined cycle time of 425.68312 days. Figure 5.10 shows the optimal free-drift and maneuvering trajectories for a 4-deg inclination deadband. Here, $\Omega_1 = 57.695$ deg, $\Omega_2 = 302.305$ deg, $h_{10} = 0.58959 \times 10^{-1}$, $h_{20} = 0.37279 \times 10^{-1}$, $h_{1f} = -0.58959 \times 10^{-1}$, and $h_{2f} = 0.37279 \times 10^{-1}$. The solution for the linear ϕ steering program is given by $\phi_0 = 165.585$ deg with $\tau_f' = 0.503157$, corresponding to

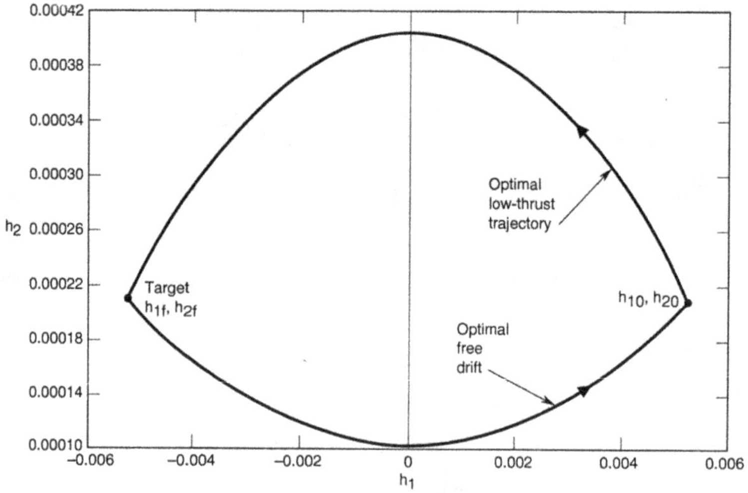

Fig. 5.9 Optimal low-thrust return trajectory for a 0.3-deg inclination deadband.

$t_f = 1579.45845$ days or 4.32432 years. The free-drift time is $T_{max} = 3539.81281$ days or 9.69147 years for a total cycle time of 5119.27126 days. Figure 5.10 also shows a smaller cycle consisting of returning to the zero-inclination target or the origin of the (h_1, h_2) frame. The free-drift portion of this smaller cycle consists

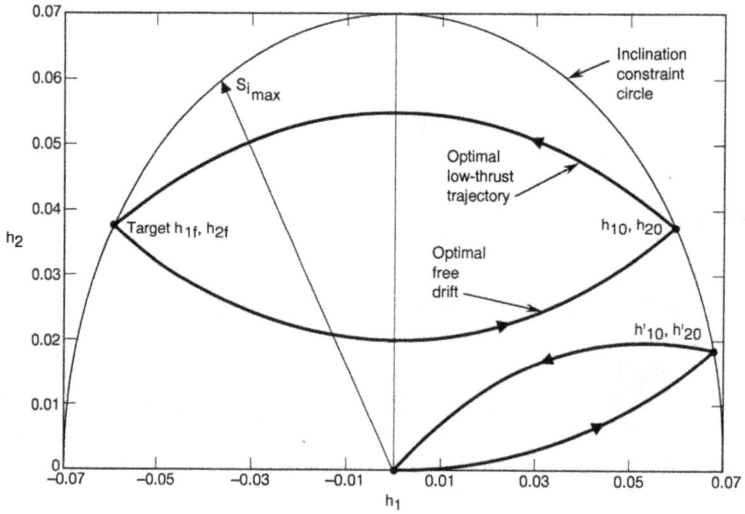

Fig. 5.10 Optimal low-thrust return trajectory for 4-deg inclination deadband.

of one-half of the total arc that passes through the origin at midtime. This is the so called zero-inclination (ZI) strategy, which is not optimal because the deadband is consumed much faster than is the case with the optimal drift [7]. The total free-drift time for this suboptimal drift is obtained from $T_1 = \theta_{ZI}P/(2\pi)$, with $\theta_{ZI} = 4\sin^{-1}[s_{i_0}/(2h_{2e})]$ [7]. The coordinates of the point of intersection with the constraint circle h'_{10} and h'_{20} are obtained from the simultaneous solution of the following two equations:

$$h'^2_{10} + h'^2_{20} = s^2_{i_{max}} \tag{5.63}$$

$$(h'_{20} - h_{2e})^2 = h^2_{2e} - h'^2_{10} \tag{5.64}$$

Equation (5.64) is the equation of the precession circle that passes through the origin and, therefore, has radius h_{2e}. This leads to

$$h'_{10} = \frac{s_{i_{max}}}{2h_{2e}}\left(4h^2_{2_e} - s^2_{i_{max}}\right)^{1/2} \tag{5.65}$$

$$h'_{20} = \frac{s^2_{i_{max}}}{2h_{2e}} \tag{5.66}$$

We have $h'_{10} = 0.67219 \times 10^{-1}$ and $h'_{20} = 0.18639 \times 10^{-1}$, with $\phi_0 = 185.364$ deg and $\tau'_f = 0.353740$, corresponding to $t_f = 1110.42294$ days. The free-drift time is given by $T_1/2 = 1698.24424$ days, yielding a combined cycle time of 2808.66718 days. This shows that, in the free-drift trip-time-maximizing optimal cycle case, the thrust period is 30.853% of the total cycle time, whereas it is at a much higher proportion of 39.535% for the ZI suboptimal cycle. The optimal cycle shows similar gains over other suboptimal strategies whose target coordinates are below the optimal free-drift curve. However, these gains are important for large deadbands on the order of a few degrees, such as in this example. They become vanishingly small for small tolerances, say, on the order of a fraction of a degree. For example, for the case of $i_{max} = 0.3$ deg, we have for the ZI target strategy $h'_{10} = 0.52349 \times 10^{-2}$ and $h'_{20} = 0.10501 \times 10^{-3}$, with $\phi_0 = 180.354$ deg and $\tau'_f = 0.027737$, corresponding to $t_f = 87.070701$ days. The free-drift time is given by $T_1/2 = 125.93087$ days, for a combined cycle time of 213.00157 days. This shows that, for the ZI case, the thrust period is 40.878% of the total cycle time, whereas it is only slightly better at 40.821% for the free-drift trip-time-maximizing cycle. If the mission lifetime is very large, so that several such cycles are required, and if the maximum free-drift time is not desired, then it is more fuel-efficient to use smaller-size cycles by targeting each time to a point on the constraint circle lying above the target that corresponds to the maximum free-drift time. For the case of $i_{max} = 4$deg, let $h''_{10} = 0.3 \times 10^{-1}$, $h''_{20} = 0.629759 \times 10^{-1}$, $h''_{1f} = -0.3 \times 10^{-1}$, and $h''_{2f} = 0.629759 \times 10^{-1}$, with a free-drift period of 2623.95476 days and a trajectory lying above the optimal free-drift trajectory of Fig. 5.10. The low-thrust return from (h''_{10}, h''_{20}) to (h''_{1f}, h''_{2f}) is given by $\phi_0 = 173.222$ deg and $\tau'_f = 0.236569$, corresponding to

$t_f = 742.61420$ days. The thrust period for this smaller cycle is now 22.058% of the total cycle time, as opposed to the 30.853% value for the long free-drift-maximizing cycle in Fig. 5.10. This represents a relative decrease of some 28%, which can be further improved for smaller cycles near the top of the constraint circle. In the limit, it is most fuel-efficient to remain at the top of the constraint circle and use the appropriate level of thrusting to counter the effect of the precession or, in practice, to journey through infinitesimal cycles. This is because W has the smallest value there, or in Zermelo's sense, the current is the weakest. This observation is also true for the impulsive case, which, for the two examples with $i_{\max} = 4$deg, we need $\Delta\Omega \simeq 115$deg and $\Delta V \simeq 362$ m/s for the long cycle compared to $\Delta\Omega \simeq 51$ deg and $\Delta V \simeq 184$ m/s for the shorter cycle, or a decrease from 1.02×10^{-1} to 7.01×10^{-2} m·s^{-1}·day^{-1} with a savings of some 31%. The fuel-minimizing solution would also consist of journeying through infinitesimal cycles near the top of the constraint circle, as for the low-thrust case. If we now assume that the short thrust arcs are quasi-instantaneous compared to the period of drift between such thrust arcs, then we can discretize the continuous trajectories by using the optimal control law $\phi = \tau' + \phi_0$ from the continuous

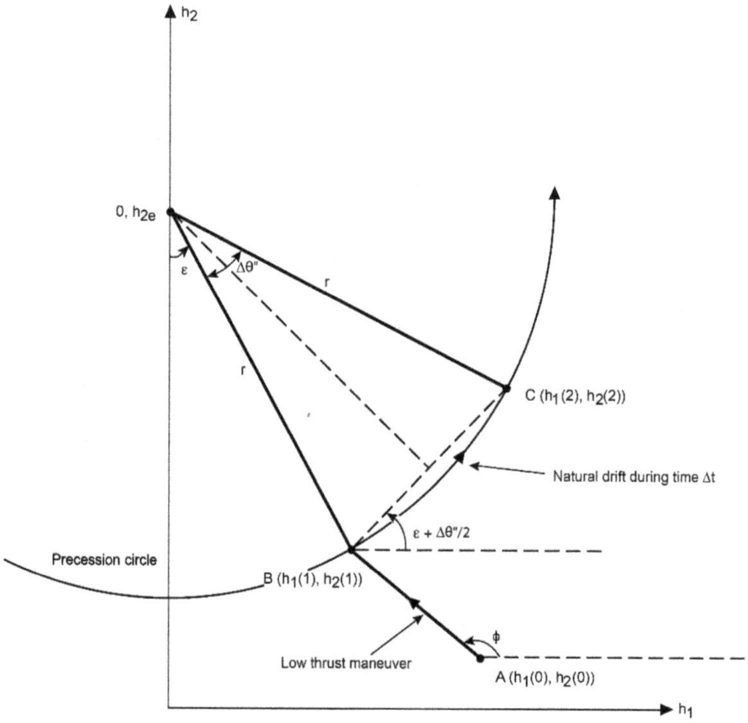

Fig. 5.11 Geometry of low-thrust maneuver and subsequent drift.

solution to compute the coordinates of point B given the initial conditions at point A (Fig. 5.11). Thus

$$h_1(1) = h_1(0) + Vc_\phi \tag{5.67}$$

$$h_2(1) = h_2(0) + Vs_\phi \tag{5.68}$$

The jump from A to B is due to the short low-thrust arc, whereas point C is reached a time Δt later from B by means of the precession circle whose radius is given by

$$r = \{h_1^2(1) + [h_2(1) - h_{2e}]^2\}^{\frac{1}{2}} \tag{5.69}$$

The angle ε is obtained from

$$\varepsilon = \tan^{-1}\left[\frac{h_1(1)}{h_{2e} - h_2(1)}\right] \tag{5.70}$$

and the angular motion $\Delta\theta''$ in time Δt is obtained from $\Delta\theta'' = \omega\Delta t$. The segment BC has length $d = 2r\sin(\Delta\theta''/2)$, and therefore, the coordinates of C are

$$h_1(2) = h_1(1) + d\cos\left(\varepsilon + \frac{\Delta\theta''}{2}\right) \tag{5.71}$$

$$h_2(2) = h_2(1) + d\sin\left(\varepsilon + \frac{\Delta\theta''}{2}\right) \tag{5.72}$$

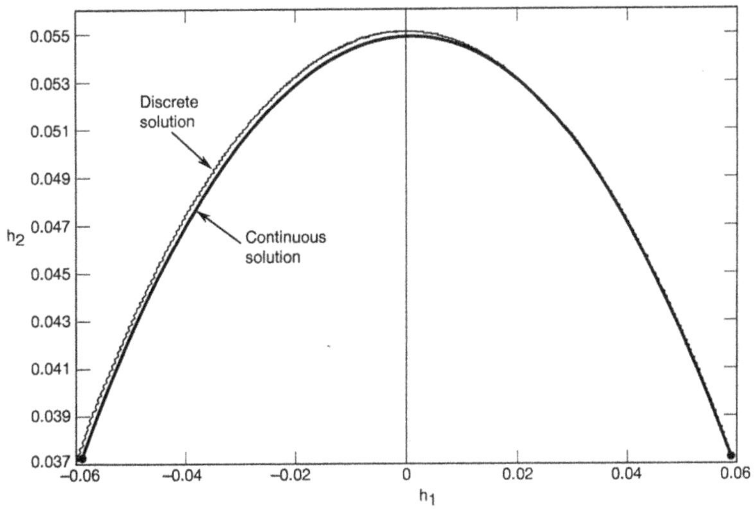

Fig. 5.12 Continuous and discretized optimal return trajectories for a 4-deg inclination deadband.

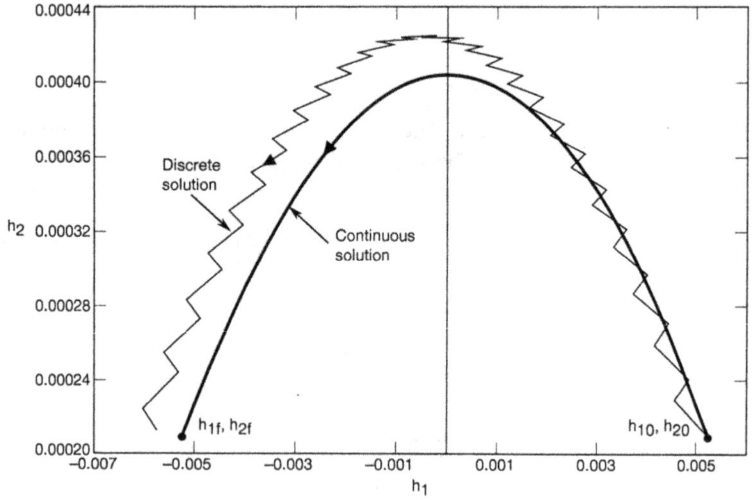

Fig. 5.13 Continuous and discretized optimal return trajectories for a 0.3-deg inclination deadband.

These steps are repeated for the next thrust arc and drift period until the final time τ_f' is reached. Figure 5.12 shows both the continuous optimal trajectory and the discretized equivalent trajectory with a good match for the case of the 4-deg inclination tolerance. The match is still good for the case of a much smaller tolerance, such as in Fig. 5.13, which corresponds to $i_{\max} = 0.3$ deg. In this case, a final trim is needed to match exactly the final conditions h_{1f} and h_{2f}, and this can be accomplished by adjusting the angular position of the final thrust arc along the spacecraft orbit and using a longer thrust arc to reach the target if necessary.

5.7 CONCLUSION

An optimal steering strategy for the north-south stationkeeping of geostationary spacecraft using infrequent low-thrust maneuvers has been devised in this chapter. The optimal drift cycle in inclination and node presented by Kamel and Tibbitts for the zeroth-order analytic solution is employed as an example, as the ideal drift cycle to repeat once the inclination tolerance deadband is violated. The recovery of the ideal initial conditions, or any other desired target conditions, by a single impulse with chemical thrusters requires a series of short-duration low-thrust arcs judiciously positioned along the current spacecraft orbit such that these conditions are reached in minimum time, thus requiring minimum propellant usage for a given common wait period between successive thrust arcs. This optimal transfer design takes into account the natural drift of the orbit plane orientation parameters during the nonmaneuvering coast arcs, by creating intermediate target conditions to be achieved by each thrust arc. This problem has been shown to be an optimal control problem of the Zermelo type for navigation in (i, Ω) space. Further improvements to this analysis are possible by considering the exact numerical description of the free-drift dynamics, as well as variable-length thrust arcs within prescribed bounds, which would also be optimized but still restricted to pure out-of-plane acceleration.

REFERENCES

[1] Frick, R. H., and Garber, T. B., "Perturbations of a synchronous satellite," Report R-399-NASA, Rand Corp., Santa Monica, CA, May 1962.

[2] Allen, R. R., and Cook, G. E., "The long-period motion of the plane of a distant circular orbit," *Proceedings of the Royal Society*, Vol. 280, No. 1380, July 1964, pp. 97–109.

[3] Billik, B. H., "Cross-track sustaining requirements for a 24-hr satellite," *Journal of Spacecraft and Rockets*, Vol. 4, No. 3, March 1967, pp. 297–301.

[4] Balsam, R. E., "A simplified approach for correction of perturbations on a stationary orbit," *Journal of Spacecraft and Rockets*, Vol. 6, No. 7, 1969, pp. 805–811.

[5] Eckstein, M. C., Leibold, A., and Hechler, F., "Optimal autonomous stationkeeping of geostationary satellites," AAS/AIAA Astrodynamics Specialist Conference, AAS Paper 81-206, Lake Tahoe, NV, Aug. 1981.

[6] Slavinskas, D. D., Johnson, G. K., and Benden, W. J., "Efficient inclination control for geostationary satellites," AIAA Paper 85-0216, AIAA 23rd Aerospace Sciences Meeting, Reno, NV, Jan. 1985.

[7] Kamel, A., and Tibbitts, R., "Some useful results on initial node locations for near-equatorial circular satellite orbits," *Celestial Mechanics*, Vol. 8, Aug. 1973, pp. 45–73.

[8] Kechichian, J. A., "Optimal steering for north-south stationkeeping of geostationary spacecraft," AAS Paper 95-118, AAS/AIAA Spaceflight Mechanics Meeting, Albuquerque, NM, Feb. 1995.

[9] Kechichian, J. A., "Optimal steering for north-south stationkeeping of geostationary spacecraft," *Journal of Guidance, Control, and Dynamics*, Vol. 20, No. 3, May–June 1997, pp. 435–444.

[10] Bryson, A. E., Jr., and Ho, Y.-C., *Applied Optimal Control: Optimization, Estimation and Control*, Ginn and Company, Waltham, MA, 1969.

CHAPTER 6

Optimal Thrust Pitch Profiles for Constrained Orbit Control in Near-Circular and Elliptical Orbits

NOMENCLATURE

a = semimajor axis, km
E = eccentric anomaly
e = eccentricity
f_r, f_θ, f_h = components of the perturbation acceleration vector along the radial, perpendicular, and out-of-plane thrust acceleration, km/s^2
h = orbital angular momentum, km^2/s
i = orbit plane inclination
n = mean motion of the satellite orbit, rad/s
p = orbit parameter, $a(1 - e^2)$, km
$\hat{r}$ = unit vector in the radial direction
r = radial distance, km
$\boldsymbol{T}$ = thrust acceleration vector
α = thrust pitch angle
Δa = change in semimajor axis, km
Δe = change in eccentricity
θ^* = true anomaly
μ = gravitational constant of Earth, 398601.3 km^3/s^2
ω = argument of perigee

6.1 INTRODUCTION

In this chapter, optimal thrust pitch profiles that maximize the change in the semimajor axis while constraining the eccentricity change to zero over a revolution, in the presence of Earth shadow, where no thrust is applied, are generated using numerical quadrature methods. The procedure is extended to the dual problem of maximizing the change in eccentricity while constraining the change in the semimajor axis to zero for the case of a near-circular orbit with shadowing. The method is further applied to the more general elliptical case using continuous thrust to circularize a highly elliptical synchronous orbit that remains synchronous during the circularization maneuver. For the simplified problem where the thrust is applied normal to the line of apsides, the use of the average rate of the eccentricity leads to an analytic expression relating the current eccentricity to the accumulated thrust time during the circularization. This expression allows for a quick evaluation of the required velocity change for a given initial elliptical orbit to circularize. Finally, for the near-circular case, the simultaneous circularization and orbit plane rotation using a constant thrust yaw profile and an inertially fixed in-plane thrust orientation, as in the simplified circularization problem, leads to an analytic expression for the required velocity change to effect a given orbit rotation with given initial eccentricity. The constant yaw angle needed for such a transfer is also obtained analytically.

Several of the more fundamental orbit control laws using both impulsive and low-thrust propulsion modes have been summarized by Edelbaum [1], including the important optimal transfer problem between inclined circular orbits that lead to Edelbaum's closed-form expression for the total required velocity change that is widely used in industry for rapid first-order evaluations. Cass [2] extended Edelbaum's analysis to the case of intermittent thrusting under the assumption of circular orbits. Thus, the variable pitch angle is optimized in a direction to keep the orbit circular despite the presence of a shadow arc along the orbit, while simultaneously minimizing the transfer time from the given initial orbit to the final destination orbit and carrying out the required orbit plane rotation in the process.

The theory of maxima is used instead of the calculus of variations to find the optimal pitch profile that maximizes the change in the semimajor axis while constraining the eccentricity change to zero over one revolution in a circular orbit [1–3] for both the continuous- and discontinuous-thrust cases. This maximization technique is applied in the present chapter to the dual problem of maximizing the change in eccentricity while constraining the semimajor axis change to zero for the near-circular case in the presence of shadowing. The method is then extended to the more general elliptical case in the continuous-thrust case only, to circularize a highly elliptical orbit slowly without affecting its orbital energy. This particular example is presently used in the final steps of transferring certain communications satellites to their geosynchronous orbits after their proper placement in an intermediate elliptical synchronous orbit using chemical propulsion. The optimal thrust pitch profile in this continuous-thrust case nearly matches the much simpler Spitzer [4, 5] scheme, where the thrust vector is aligned normal to the orbit line of apsides such that it maintains an inertially fixed orientation in space and is easily implemented. When average rates of change of the eccentricity variable are used, analytic expressions that relate the current eccentricity to the accumulated thrust time and the velocity change required to circularize an initial elliptical orbit are easily obtained for the Spitzer scheme. When the perturbation in the true anomaly is neglected, the change in the eccentricity using Spitzer's suboptimal firing mode is carried out without affecting the orbit semimajor axis, as desired. However, the optimal thrust profile is to be used instead in the discontinuous-thrust case, because, then, the semimajor axis cannot be held constant, as shown in the near-circular examples in this chapter. In the near-circular case, Spitzer's in-plane firing strategy is easily combined with the orbit plane rotation using a constant out-of-plane thrust angle switched between its positive and negative values every one-half revolution, and analytic expressions for the out-of-plane thrust angle and the total velocity change required to simultaneously circularize and rotate the orbit plane are derived for rapid first-order evaluations in the continuous-thrust transfer case. This chapter is based on a previously published symposium paper [6] and journal article [7].

6.2 MAXIMIZATION OF THE CHANGES IN SEMIMAJOR AXIS AND ECCENTRICITY IN NEAR-CIRCULAR ORBIT IN THE PRESENCE OF A SHADOW ARC

Let Fig. 6.1 represent a near-circular orbit whose perigee is at point O and let no thrust be applied on the shadow arc $O'O$. When the true anomaly θ^* is measured from the shadow exit point O and the thrust pitch angle α is measured from the direction $\hat{r}$, the optimal thrust pitch angle profile that maximizes the change in the semimajor axis Δa while constraining the change in eccentricity Δe to zero after

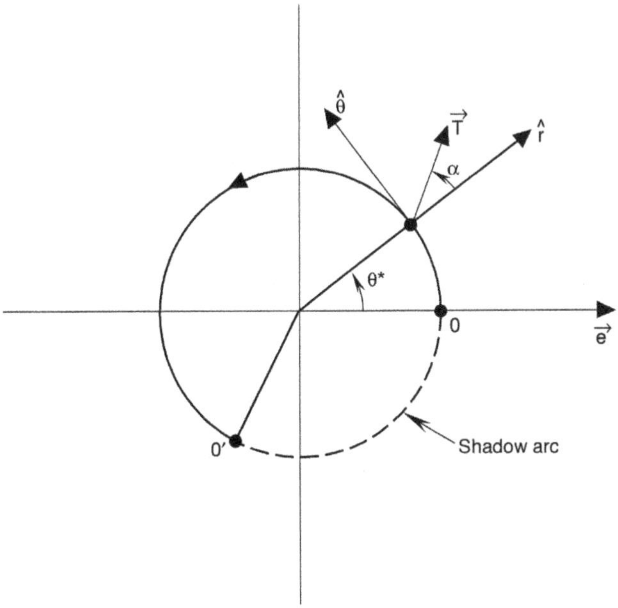

Fig. 6.1 Thrust geometry and shadow arc in a near-circular orbit.

one thrust cycle along OO' can be determined following the approach used by Cass [2].

The variation-of-parameters equations in terms of the radial and normal acceleration components f_r and f_θ, respectively, are given by

$$\dot{a} = (2a^2/h)[es_{\theta^*}f_r + (p/r)f_\theta] \tag{6.1}$$

$$\dot{e} = (1/h)\{ps_{\theta^*}f_r + [(p+r)c_{\theta^*} + re]f_\theta\} \tag{6.2}$$

Here, $f_r = kc_\alpha$ and $f_\theta = ks_\alpha$, where $k = T$, the thrust acceleration magnitude; $s_{\theta^*} = \sin\theta^*$; and $c_{\theta^*} = \cos\theta^*$.

These equations are readily converted to the form [2]

$$\dot{a} = \frac{2es_{\theta^*}}{n(1-e^2)^{1/2}}f_r + \frac{2(1+ec_{\theta^*})}{n(1-e^2)^{1/2}}f_\theta \tag{6.3}$$

$$\dot{e} = \frac{(1-e^2)^{1/2}s_{\theta^*}}{na}f_r + \frac{(1-e^2)^{1/2}}{nae}\left[1 + ec_{\theta^*} - \frac{1-e^2}{1+ec_{\theta^*}}\right]f_\theta \tag{6.4}$$

For near-circular orbits, after one sets $e = 0$, these equations simplify to

$$\dot{a} = 2f_\theta/n \tag{6.5}$$

$$\dot{e} = (s_{\theta^*}/na)f_r + (2c_{\theta^*}/na)f_\theta \tag{6.6}$$

The changes in a and e over one cycle of thrust from $\theta^* = 0$ to $\theta^* = \tau_f$, or from O to O', are given by

$$\Delta a = \frac{2ka^3}{\mu}\int_0^{\tau_f} s_\alpha d\theta^* \tag{6.7}$$

$$\Delta e = \frac{ka^2}{\mu}\int_0^{\tau_f}(c_\alpha s_{\theta^*} + 2s_\alpha c_{\theta^*})d\theta^* \tag{6.8}$$

The maximization of Δa subject to $\Delta e = 0$ is made possible by adjoining the constraint $\Delta e = 0$ to Δa by means of a constant Lagrange multiplier λ such that the performance index is given by

$$I(\alpha) = \int_0^{\tau_f}\left\{\frac{ka^2}{\mu}\left[2as_\alpha + \lambda(c_\alpha s_{\theta^*} + 2s_\alpha c_{\theta^*})\right]\right\}d\theta^* \tag{6.9}$$

Letting F represent the integrand, $\partial F/\partial\alpha = 0$ will yield the optimal α equation as

$$\tan\alpha = \frac{2(a + \lambda c_{\theta^*})}{\lambda s_{\theta^*}} \tag{6.10}$$

such that

$$s_\alpha^2 = \frac{4(a + \lambda c_{\theta^*})^2}{\lambda^2 s_{\theta^*}^2 + 4(a + \lambda c_{\theta^*})^2}$$

$$c_\alpha^2 = \frac{\lambda^2 s_{\theta^*}^2}{\lambda^2 s_{\theta^*}^2 + 4(a + \lambda c_{\theta^*})^2}$$

Substituting the resulting expressions for s_α and c_α into Eq. (6.8) yields

$$\Delta e = \frac{ka^2}{\mu} \int_0^{\tau_f} \frac{\left(4ac_{\theta^*} + \lambda + 3\lambda c_{\theta^*}^2\right)}{\left[4a^2 + 4a\lambda c_{\theta^*} + \lambda\left(\lambda + 3\lambda c_{\theta^*}^2 + 4ac_{\theta^*}\right)\right]^{1/2}} d\theta^* \tag{6.11}$$

where λ is determined numerically by means of a 10-point Gauss–Legendre quadrature of the preceding integral and its value is adjusted slowly until $\Delta e = 0$ is satisfied to within a certain tolerance. The thrust pitch profile is then obtained from Eq. (6.10), and the maximum value of Δa from Eq. (6.7) is obtained by numerical quadrature. Figure 6.2 shows the optimal pitch profiles for $a_0 = 7000$, 10,000, 20,000, 30,000, and 40,000 km using $k = 3.5\ 10^{-7}$ km/s^2 as

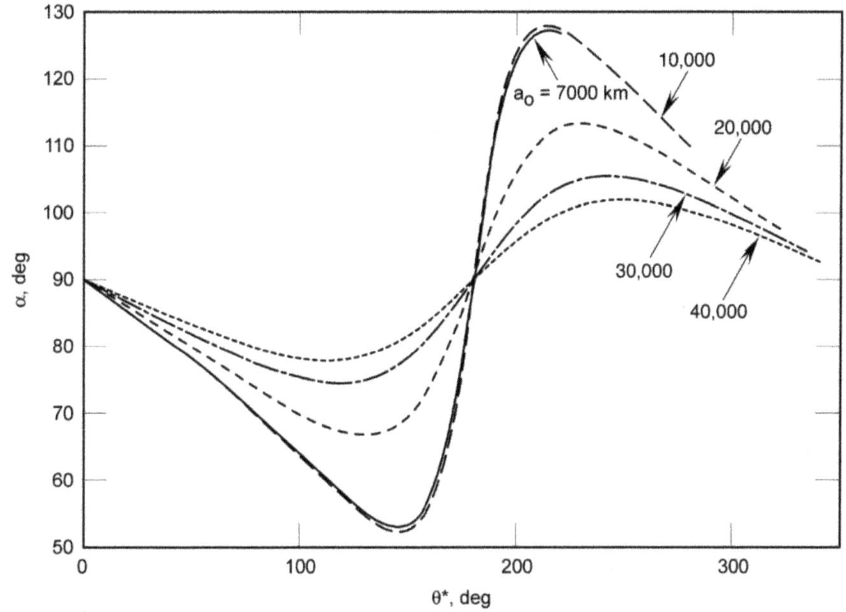

Fig. 6.2 Optimal pitch profiles for constrained maximum Δa with worst-case shadowing in a circular orbit.

functions of θ^* and for the maximum shadow arc length encountered by each circular orbit size. These arcs are such that $\tau_f = 220$, 280, 322, 335, and 341 deg, respectively. These profiles start at $\alpha = 90$ deg, which is also the value for $\theta^* = 180$ deg resulting in the pinch point shown in Fig. 6.2.

The maximization of Δe subject to $\Delta a = 0$ can also be carried out by forming the dual performance index:

$$I(\alpha) = \int_0^{\tau_f} \left\{ \frac{ka^2}{\mu} [(c_\alpha s_{\theta^*} + 2s_\alpha c_{\theta^*}) + \lambda(2as_\alpha)] \right\} d\theta^* \tag{6.12}$$

Letting F now represent the new integrand, the Euler equation $\partial F/\partial\alpha = 0$ leads to the solution

$$\tan\alpha = \frac{2(c_{\theta^*} + a\lambda)}{s_{\theta^*}} \tag{6.13}$$

such that

$$s_\alpha^2 = \frac{4(c_{\theta^*} + a\lambda)^2}{s_{\theta^*}^2 + 4(c_{\theta^*} + a\lambda)^2}$$

$$c_\alpha^2 = \frac{s_{\theta^*}^2}{s_{\theta^*}^2 + 4(c_{\theta^*} + a\lambda)^2}$$

Substituting s_α into Eq. (6.7) for Δa yields

$$\Delta a = 2k\frac{a^3}{\mu}\int_0^{\tau_f} \frac{2(c_{\theta^*} + a\lambda)}{\left[1 + 3c_{\theta^*}^2 + 4a\lambda(a\lambda + 2c_{\theta^*})\right]^{1/2}} d\theta^* = 0 \tag{6.14}$$

As before, λ is determined numerically by adjusting its value until the numerical quadrature of the preceding integral yields a value of zero to within a small tolerance. Δe is then computed by quadrature using the values for s_α, c_α, and λ from the optimal solution in Eq. (6.8). Figure 6.3 shows the optimal pitch profiles for the same five cases, namely, $a_0 = 7{,}000$, 10,000, 20,000, 30,000, and 40,000 km with maximum shadow arcs as functions of θ^*. The varying angle α starts at $\alpha = 90$ deg and crosses the −90 deg value at $\theta^* = 180$ deg, resulting in another pinch point, as shown in Fig. 6.3. The history of the eccentricity is obtained by integrating Eq. (6.6) for $\dot{e}$ with a change of the independent variable from time t to θ^* such that

$$\frac{de}{d\theta^*} = \frac{ka^2}{\mu}(c_\alpha s_{\theta^*} + 2s_\alpha c_{\theta^*}) \tag{6.15}$$

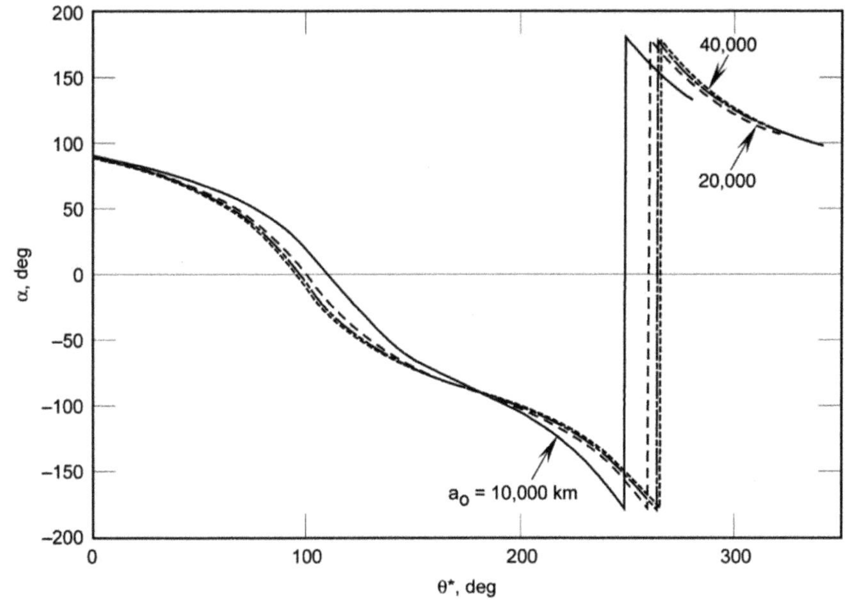

Fig. 6.3 Optimal pitch profiles for constrained maximum Δe with worst-case shadowing in a circular orbit.

In a similar way, $\dot{a}$ in Eq. (6.5) is transformed into

$$\frac{da}{d\theta^*} = 2k\frac{a^3}{\mu}s_\alpha \tag{6.16}$$

using $d\theta^*/dt = n$. The evolution of e for the five cases when a is held as a constant outside the integration sign considered is shown in Fig. 6.4.

Table 6.1 lists the iterated and integrated multiplier and orbit parameters for this analysis. Here, δ is the total maximum shadow angle that occurs when the Sun–Earth line is contained within the spacecraft orbit plane, and $\lambda_{\Delta a}$ and $\lambda_{\Delta e}$ are the iterated values of λ for the maximizations of Δa and Δe, respectively. Δa (optimal) and Δe (optimal) are the corresponding optimal values of Δa and Δe obtained by numerical quadrature of the integrals in Eqs. (6.7) and (6.8), respectively. The final or achieved values of a and e for the two optimization problems are obtained from the integration of Eqs. (6.16) and (6.15), respectively. When Δe is maximized, $(a_f)_{\Delta e}$ very closely matches the values of the initial semimajor axes in all five cases, whereas when Δa is maximized instead, $(e_f)_{\Delta a}$ is effectively very small near the zero value, as desired. Here, $\lambda_{\Delta a}$ is on the order of 10^4 kilometers, and $\lambda_{\Delta e}$ is on the order of 10^{-4} per kilometer.

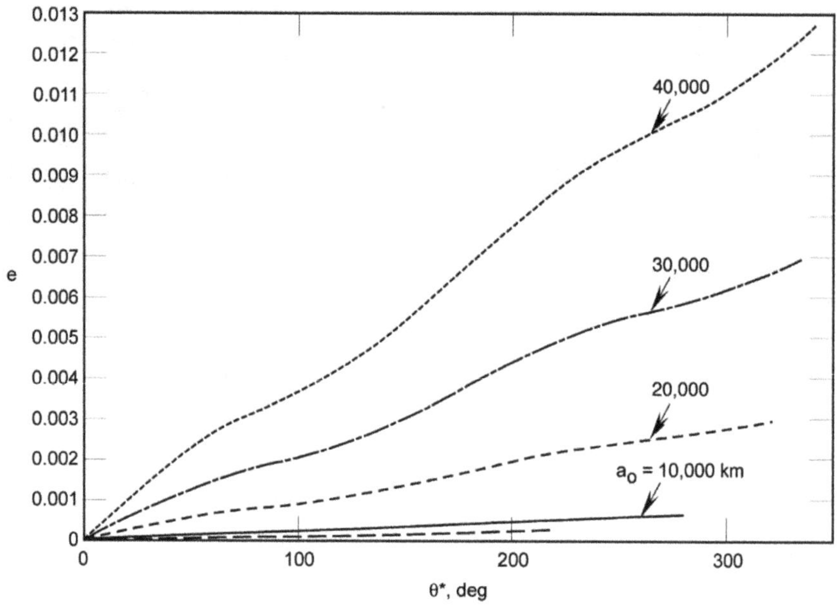

Fig. 6.4 Variation of the eccentricity for the constrained maximum-Δe solutions with worst-case shadowing in a circular orbit.

For continuous thrust over the entire orbit without any shadow arc, the orbit with $a_0 = 7{,}000$ km yields $\lambda_a = 0.155565 \times 10^{-2}$, resulting in a flat α profile at 90 deg with $(e_f)_{\Delta a} = 1.5019 \times 10^{-11}$, which is effectively zero. When Δe is maximized instead, $\lambda_{\Delta e} = 0.185146 \times 10^{-5}$, resulting in $(a_f)_{\Delta e} = 7000.028$ km, effectively constraining Δa [i.e., $(a_f)_{\Delta e} - a$] to zero at the end of one cycle of thrust, with the α profile closely matching the theoretical profile $\tan\alpha = 2/(\tan\theta^*)$ given by Edelbaum [1]. In both optimization cases, λ must be equal to zero, leading to the constant value $\alpha = 90$ deg for the Δa maximization case and to the expression $\tan\alpha = 2/(\tan\theta^*)$ for the Δe maximization case. In the latter case, the total change Δe is given analytically by a complete elliptical integral [1]

$$\Delta e = \frac{8k}{n^2 a}\int_0^{\pi/2}\sqrt{1 - \frac{3}{4}s_{\theta^*}^2}\, d\theta^*$$

$$\Delta e = \frac{8 \times 1.2111k}{n^2 a}$$

which corresponds to a change in velocity ΔV of

$$\Delta V = 0.649 V \Delta e$$

which is directly a function of the orbit semimajor axis and the thrust acceleration.

TABLE 6.1 ITERATED AND INTEGRATED MULTIPLIER AND ORBIT PARAMETERS

a, km	7,000	10,000	20,000	30,000	40,000
τ_f, deg	220	280	322	335	341
δ, deg	140	80	38	25	19
$\lambda_{\Delta a}$	5,823.049823	8,379.265958	13,034.177567	14,569.241955	15,548.465162
$\lambda_{\Delta e}$	0.513697×10^{-4}	0.359287×10^{-4}	0.87259×10^{-5}	0.39454×10^{-5}	0.23495×10^{-5}
Δa (optimal), km	2.107	7.660	75.856	272.102	661.563
Δe (optimal)	0.254313×10^{-3}	0.615390×10^{-3}	0.292109×10^{-2}	0.694068×10^{-2}	0.1264526×10^{-1}
$(a_f)_{\Delta a}$,km	7,002.107	10,007.729	20,075.778	30,272.105	40,661.574
$(a_f)_{\Delta e}$,km	7,000.001	10,000.017	19,999.645	29,999.601	40,000.778
$(e_f)_{\Delta a}$	5.799×10^{-8}	9.113×10^{-6}	6.808×10^{-6}	3.229×10^{-7}	9.045×10^{-7}
$(e_f)_{\Delta e}$	2.542205×10^{-4}	6.149686×10^{-4}	2.923213×10^{-3}	6.943196×10^{-3}	1.264781×10^{-2}

6.3 SPITZER STRATEGY

In the mid-1990s, Spitzer [4, 5] proposed the use of a combination of chemical and electric propulsion schemes to transfer a spacecraft to geosynchronous orbit. The final phase of the transfer consists of applying a low-thrust acceleration in an inertially fixed orientation that is orthogonal to the line of apsides such that a synchronous eccentric orbit is slowly circularized at a constant energy level. In this scheme, the pitch angle α is equal to $\pi/2 - \theta^*$ such that the α profile is not too different from the optimal law tan $\alpha = 2/(\tan\ \theta^*)$. When $s_\alpha = c_{\theta^*}$ and $c_\alpha = s_{\theta^*}$ are used in Eqs. (6.15) and (6.16) and the equations are integrated over θ^* from 0 to τ_f, a linear α profile and histories for a and e as functions of θ^* are obtained. Plots of these results (denoted by the subscript "s") are compared to those corresponding to the constrained Δa and Δe maximization solutions in Figs. 6.5–6.7. The Spitzer strategy yields $(e_f)_s = 2.584017 \times 10^{-4}$, which is slightly higher than the value of $(e_f)_{\Delta e} = 2.542205 \times 10^{-4}$ obtained from the maximum-Δe constrained solutions. However, unlike the constrained case, which effectively keeps the semimajor axis unchanged at $a_f = 7000.001$ km after the thrust cycle, the Spitzer approach yields $(a_f)_s = 6999.612$ km.

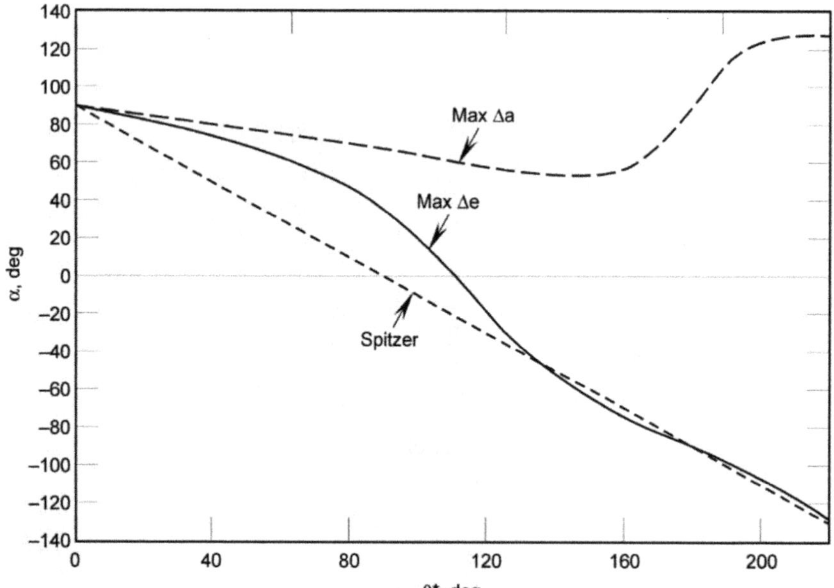

Fig. 6.5 Thrust pitch angle profiles for the maximum-Δe, maximum-Δa, and Spitzer solutions for a circular orbit with $a_0 = 7000$ km and worst-case shadowing.

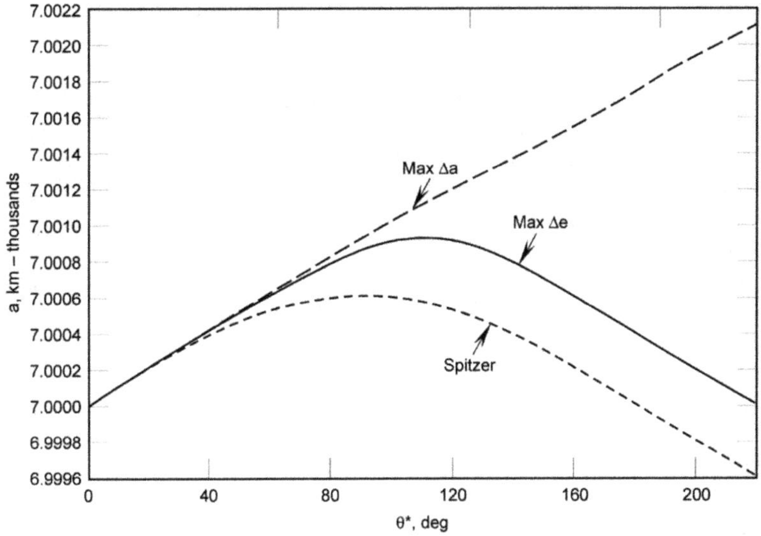

Fig. 6.6 Variation of the semimajor axis for the maximum-Δe, maximum-Δa, and Spitzer solutions for a circular orbit with $a_0 = 7000$ km and worst-case shadowing.

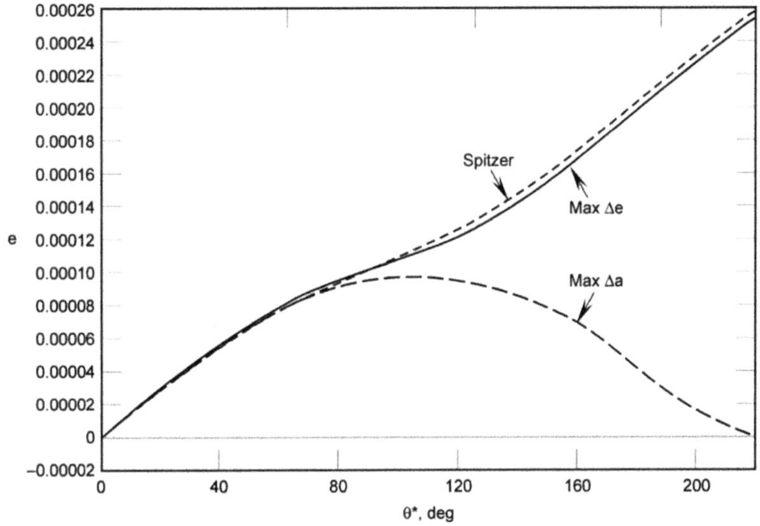

Fig. 6.7 Variation of the eccentricity for the maximum-Δe, maximum-Δa, and Spitzer solutions for a circular orbit with $a_0 = 7000$ km and worst-case shadowing.

The Spitzer and maximum-Δe α profiles cross at $\theta^* = 180$ deg, with $\alpha = -90$ deg exactly in both cases, as shown in Fig. 6.5. If $\alpha = 3\pi/2 - \theta^*$ is used instead in the Spitzer case, which is equivalent to saying that the thrust direction is 180 deg away from the inertially fixed orientation used earlier, namely, $\alpha = \pi/2 - \theta^*$, then the final semimajor axis of $(a_f)_s = 7000.387$ km will be higher than the initial 7000-km value instead of being lower, as earlier.

Figures 6.8–6.10 show the variations in α, a, and e, respectively, for the Spitzer and two optimized schemes for $a_0 = 40{,}000$ km case with $\tau_f = 341$ deg, or the maximum shadow condition. The angle $\alpha = \pi/2 - \theta^*$ is used for the Spitzer case, which yields $(a_f)_s = 39{,}963.408$ km and $(e_f)_{\Delta e} = 1.2647811 \times 10^{-2}$, as opposed to $(a_f)_{\Delta e} = 40{,}000.778$ km and $(e_f)_{\Delta e} = 1.2647811 \times 10^{-2}$ for the constrained maximum-Δe case, which yields a larger change in eccentricity while satisfying the $\Delta a = 0$ constraint. Because of the larger values of eccentricity experienced by thrusting in this larger orbit case, a modified form of Eqs. (6.3) and (6.4) is used because the assumption of a circular orbit that leads to Eqs. (6.15) and (6.16) and has been used in the integrations so far is no longer fully valid. In this case, Eqs. (6.3) and (6.4) are written with θ^* instead of time as the independent variable by neglecting the terms in f_r and f_θ in what follows

$$\frac{d\theta^*}{dt} = \frac{h}{r^2} - \frac{r}{he}[-c_{\theta^*}(1 + ec_{\theta^*})f_r + s_{\theta^*}(2 + ec_{\theta^*})f_\theta]$$

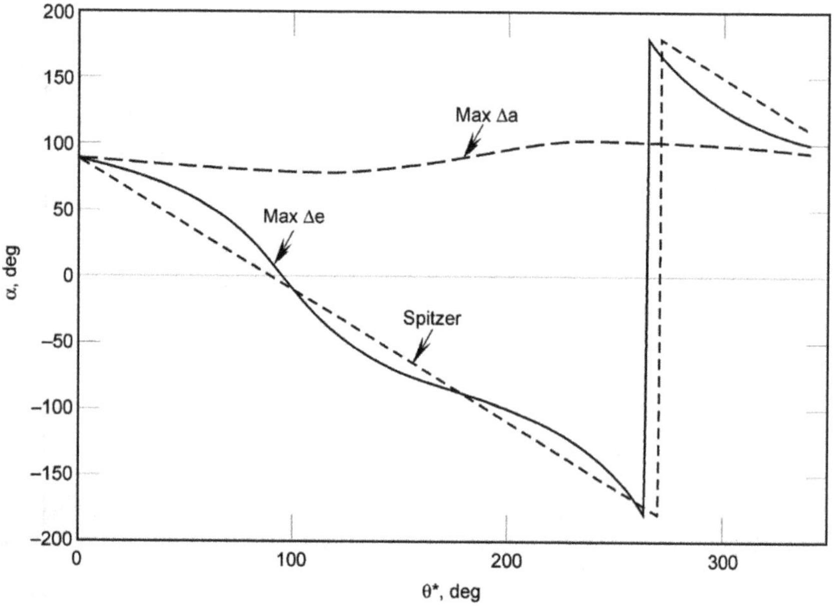

Fig. 6.8 Profiles of the thrust pitch angle for the maximum-Δe, maximum-Δa, and Spitzer solutions for a circular orbit with $a_0 = 40{,}000$ km and worst-case shadowing.

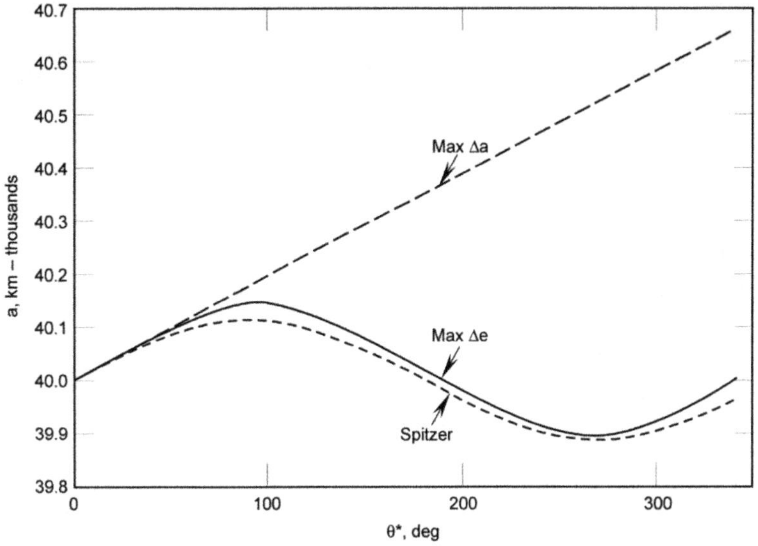

Fig. 6.9 Variation of the semimajor axis for the maximum-Δe, maximum-Δa, and Spitzer solutions for a circular orbit with $a_0 = 40{,}000$ km and worst-case shadowing.

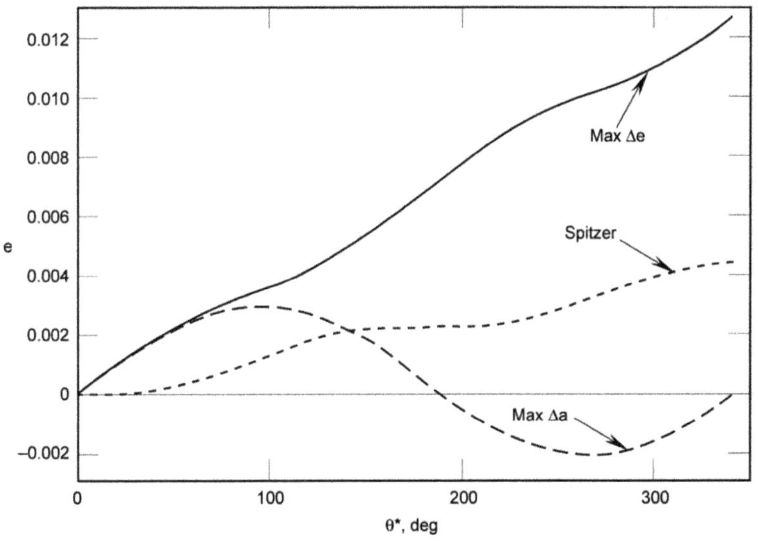

Fig. 6.10 Variation of the eccentricity for the maximum-Δe, maximum-Δa, and Spitzer solutions for a circular orbit with $a_0 = 40{,}000$ km and worst-case shadowing.

such that

$$\frac{d\theta^*}{dt} = \frac{h}{r^2} = \frac{n(1 + ec_{\theta^*})^2}{(1 - e^2)^{3/2}} \tag{6.17}$$

where the relations $h^2 = \mu a(1 - e^2)$ and $r = a(1 - e^2)/(1 + ec_{\theta^*})$ are used. Equations (6.3) and (6.4) are then converted to the form

$$\frac{da}{d\theta^*} = \frac{2es_{\theta^*}(1 - e^2)}{n^2(1 + ec_{\theta^*})^2} kc_\alpha + \frac{2(1 - e^2)}{n^2(1 + ec_{\theta^*})^2} ks_\alpha + \frac{2(1 - e^2)}{n^2(1 + ec_{\theta^*})} ks_\alpha \tag{6.18}$$

$$\frac{de}{d\theta^*} = \frac{(1 - e^2)^2 s_{\theta^*}}{n^2 a(1 + ec_{\theta^*})^2} kc_\alpha + \frac{(1 - e^2)^2}{n^2 ae} \left[\frac{1}{(1 + ec_{\theta^*})} - \frac{(1 - e^2)}{(1 + ec_{\theta^*})^3} \right] ks_\alpha \tag{6.19}$$

These equations are integrated numerically to produce the evolutions of a and e shown in Figs. 6.9 and 6.10, respectively.

As a final note, if $\alpha = 3\pi/2 - \theta^*$ is used instead of $\alpha = \pi/2 - \theta^*$ for the Spitzer case [4, 5], then a final value for the semimajor axis of $(a_f)_s = 40{,}036.591$ km is obtained instead of the original value of 39,963.408 km, indicating that the semimajor axis increases in one case and decreases in the other. The change in eccentricity remains the same, but the location of the final perigee is different.

Incidentally, Eq. (6.16) in the circular case can be integrated analytically for θ^* between 0 and θ^*, yielding

$$a = a_0 \left(1 - 4 \frac{k a_0{}^2}{\mu} s_{\theta^*} \right)^{-1/2} \tag{6.20}$$

where a_0 is the initial semimajor axis. Equation (6.15) can be easily integrated if a is held constant on the right-hand side such that

$$e = e_0 + \frac{ka^2}{\mu} \left(\frac{3}{2} \theta^* + s_{2\theta^*} \right) \tag{6.21}$$

Holding a as constant on the right-hand side of Eq. (6.16) results in the approximation

$$a = a_0 + 2k \frac{a^3}{\mu} s_{\theta^*}$$

Equation (6.20) is exact, but e here is an approximation, and $e_0 = 0$ at the initial time. These examples show that significant errors are associated with the use of the Spitzer strategy [4, 5], when the thrust is off on the shadow arc, as in solar-electric applications.

6.4 CONTINUOUS-THRUST ELLIPTICAL CASE

Equations (6.3) and (6.4) valid in elliptical orbit can be written in terms of the eccentric anomaly E as was done by Burt [8] by using the expression

$$\dot{E} = \left(\frac{\mu}{a}\right)^{1/2}\frac{1}{r} + \frac{1}{es_E}\left(c_E\frac{\mathrm{d}e}{\mathrm{d}t} - \frac{r}{a^2}\frac{\mathrm{d}a}{\mathrm{d}t}\right)$$

Simplifying further by neglecting the $\mathrm{d}e/\mathrm{d}t$ and $\mathrm{d}a/\mathrm{d}t$ contributions gives

$$\dot{E} = \frac{1}{r}\left(\frac{\mu}{a}\right)^{1/2} \tag{6.22}$$

Also, by using the identities

$$\begin{aligned} s_{\theta^*} &= \frac{(1-e^2)^{1/2}s_E}{(1-ec_E)} \\ c_{\theta^*} &= \frac{c_E - e}{(1-ec_E)} \\ r &= p/(1+ec_{\theta^*}) = a(1-ec_E) \\ p &= a(1-e^2) \end{aligned}$$

setting $f_r = ks_{\theta^*}$ and $f_\theta = kc_{\theta^*}$, one can transform Eqs. (6.3) and (6.4) to the Spitzer mode [4, 5]

$$\frac{\mathrm{d}a}{\mathrm{d}E} = \frac{\mathrm{d}a}{\mathrm{d}t}\Big/\frac{\mathrm{d}E}{\mathrm{d}t}$$

$$\frac{\mathrm{d}a}{\mathrm{d}E} = \frac{2a^{7/2}}{\mu(p)^{1/2}}k(1-e^2)c_E \tag{6.23}$$

$$\frac{\mathrm{d}e}{\mathrm{d}E} = \frac{(p)^{1/2}}{\mu}a^{3/2}k\left(1 - 2ec_E + c_E^2\right) \tag{6.24}$$

The change in a after one full cycle of thrust between $E = 0$ and $E = 2\pi$ is

$$\Delta a = \int_{a_0}^{a}\mathrm{d}a = \frac{2a^{7/2}}{\mu(p)^{1/2}}k(1-e^2)\int_0^{2\pi}c_E\,\mathrm{d}E = 0 \tag{6.25}$$

with the semimajor axis a and eccentricity e held constant on the right-hand side.

For the eccentricity change,

$$\Delta e = \int_0^{e}\mathrm{d}e = \frac{(p^{1/2})}{\mu}a^{3/2}k\int_0^{2\pi}(1 - 2ec_E + c_E^2)\,\mathrm{d}E$$

$$\Delta e = 3\pi k a^{3/2} \frac{(p)^{1/2}}{\mu} \tag{6.26}$$

An average rate $(\widetilde{de/dt})$ can be produced from $\Delta e/\Delta T$, where $\Delta T = 2\pi/n$ is the orbit period with $n = \mu^{1/2} a^{-3/2}$ such that

$$\widetilde{\left(\frac{de}{dt}\right)} = \frac{\Delta e}{\Delta T} = \frac{3}{2} k \left(\frac{p}{\mu}\right)^{1/2} \tag{6.27}$$

This expression is integrated from e_0 to e

$$\int_{e_0}^{e} \frac{de}{\sqrt{1 - e^2}} = \frac{3}{2} \frac{k}{\mu^{1/2}} a^{1/2} \int_0^t dt$$

yielding

$$\sin^{-1} e - \sin^{-1} e_0 = \frac{3}{2} \frac{k}{\mu^{1/2}} a^{1/2} t_f \tag{6.28}$$

where t_f is now the total transfer time. The required change in velocity to change the eccentricity from e_0 to e is then given by

$$\Delta V = k t_f = \frac{2}{3} \left(\frac{\mu}{a}\right)^{1/2} (\sin^{-1} e - \sin^{-1} e_0) \tag{6.29}$$

The optimal pitch profile for this continuous-thrust elliptical case is obtained from Eqs. (6.18) and (6.19) for $(da/d\theta^*)$ and $(de/d\theta^*)$ valid in elliptical orbit. Holding a and e constant in these two equations and integrating with respect to θ^* from 0 to 2π yields the changes in Δa and Δe as

$$\Delta a = \frac{2(1 - e^2)k}{n^2} \int_0^{2\pi} \left[\frac{e s_{\theta^*} c_\alpha}{(1 + e c_{\theta^*})^2} + \frac{s_\alpha}{(1 + e c_{\theta^*})} \right] d\theta^* \tag{6.30}$$

$$\Delta e = \frac{(1 - e^2)^2 k}{n^2 a} \int_0^{2\pi} \left[\frac{s_{\theta^*} c_\alpha}{(1 + e c_{\theta^*})^2} + \frac{s_\alpha}{e(1 + e c_{\theta^*})} - \frac{(1 - e^2) s_\alpha}{e(1 + e c_{\theta^*})^3} \right] d\theta^* \tag{6.31}$$

The maximization of Δe subject to the constraint $\Delta a = 0$ leads to the following performance index

$$I(\alpha) = \frac{(1 - e^2)^2 k}{n^2 a} \int_0^{2\pi} \left[\frac{s_{\theta^*} c_\alpha}{(1 + e c_{\theta^*})^2} + \frac{s_\alpha}{e(1 + e c_{\theta^*})} - \frac{(1 - e^2) s_\alpha}{e(1 + e c_{\theta^*})^3} \right] d\theta^*$$
$$+ \lambda \frac{2(1 - e^2)k}{n^2} \int_0^{2\pi} \left[\frac{e s_{\theta^*} c_\alpha}{(1 + e c_{\theta^*})^2} + \frac{s_\alpha}{(1 + e c_{\theta^*})} \right] d\theta^*$$

or

$$I(\alpha) = \frac{(1-e^2)^2 k}{n^2} \int_0^{2\pi} \left\{ \frac{(1-e^2)}{a} \left[\frac{s_{\theta^*} c_\alpha}{(1+ec_{\theta^*})^2} + \frac{s_\alpha}{e(1+ec_{\theta^*})} - \frac{(1-e^2)s_\alpha}{e(1+ec_{\theta^*})^3} \right] \right.$$
$$\left. +2\lambda \left[\frac{es_{\theta^*} c_\alpha}{(1+ec_{\theta^*})^2} + \frac{s_\alpha}{(1+ec_{\theta^*})} \right] \right\} d\theta^* \tag{6.32}$$

Letting F represent the integrand, the optimal angle α is obtained from Euler's equation $\partial F/\partial \alpha = 0$, which gives

$$\tan\alpha = \frac{-(1-e^2)\left[(1+ec_{\theta^*})^2 - (1-e^2)\right] - 2ae\lambda(1+ec_{\theta^*})^2}{e(1+ec_{\theta^*})[-(1-e^2)s_{\theta^*} - 2ae\lambda s_{\theta^*}]} \tag{6.33}$$

In turn, this expression yields

$$s_\alpha{}^2 = \left\{-(1-e^2)\left[(1+ec_{\theta^*})^2 - (1-e^2)\right] - 2ae\lambda((1+ec_{\theta^*})^2\right\}^2 \Big/ K$$
$$c_\alpha{}^2 = e^2(1+ec_{\theta^*})^2\left[-(1-e^2)s_{\theta^*} - 2ae\lambda s_{\theta^*}\right]^2/K$$

where K is given by

$$K = e^2(1+ec_{\theta^*})^2\left[-(1-e^2)s_{\theta^*} - 2ae\lambda s_{\theta^*}\right]^2$$
$$+ \left\{-(1-e^2)\left[(1+ec_{\theta^*})^2 - (1-e^2)\right] - 2ae\lambda(1+ec_{\theta^*})^2\right\}^2$$

By using the expressions for s_α and c_α in Eq. (6.30) and applying some minor manipulations, one can write the Δa integral effectively in terms of θ^* as

$$\Delta a = \frac{2(1-e^2)k}{n^2} \int_0^{2\pi} \frac{\left[-(1-e^2) - 2ae\lambda\right](1+e^2+2ec_{\theta^*}) + (1-e^2)^2}{(1+ec_{\theta^*})K^{1/2}} d\theta^* = 0 \tag{6.34}$$

Because K itself is a function of the constant λ, the latter multiplier's value can be determined numerically using a search method until the preceding quadrature is effectively driven to zero to within a small tolerance. Once λ is thus determined, the Δe value achieved can also be obtained by numerical quadrature using Eq. (6.31). For $a = 40{,}000$ km and $e = 0.7$ with $k = 3.5 \times 10^{-7}$ km/s^2 used as an example, λ is found to be $\lambda = -0.1669322321 \times 10^{-5}$, with the corresponding value $\Delta e = -0.87672449 \times 10^{-2}$ from the integral in Eq. (6.31).

The integration of Eqs. (6.18) and (6.19) for $da/d\theta^*$ and $de/d\theta^*$ using the optimal control law in Eq. (6.33) yields $a_f = 40{,}004.722$ km and $e_f = 0.690371244$, or $\Delta e = 0.7 - e_f = 0.96287557 \times 10^{-2}$, which is slightly larger than the value obtained from the analytic form of Eq. (6.26) valid for the Spitzer [4, 5] firing scheme, which is $\Delta e = 0.945595 \times 10^{-2}$. If $\alpha = 3\pi/2 - \theta^*$ is used to generate the numerically integrated Spitzer trajectory using Eqs. (6.18) and (6.19) by holding a and e as constants on the right-hand sides of these two differential equations, then the final values obtained for a_f and e_f at $\theta^* = 360$ deg are 40,000.000 km and 0.6905440421, respectively, where the latter corresponds to $\Delta e = 0.945595 \times 10^{-2}$, exactly the same as the value from the analytic formula. There is practically no advantage in letting a and e vary on the right-hand sides of Eqs. (6.18) and (6.19) for this very low-thrust acceleration level, especially when the perturbation in θ^* is neglected and the integration is essentially carried out along the initial conic. Upon integration, the fully coupled Eqs. (6.18) and (6.19) yield $a_f = 39{,}999.9992$ km and $e_f = 0.6904825413$. In the absence of any shadow arc, the Spitzer mode [4, 5] yields rather accurate results, effectively holding the orbit energy at its initial value and providing almost the largest change in eccentricity possible with the variable α law of the optimal solution. These facts are verified in Figs. 6.11–6.13, which show the close similarity of the evolutions of a and e for the Spitzer and the optimal solutions.

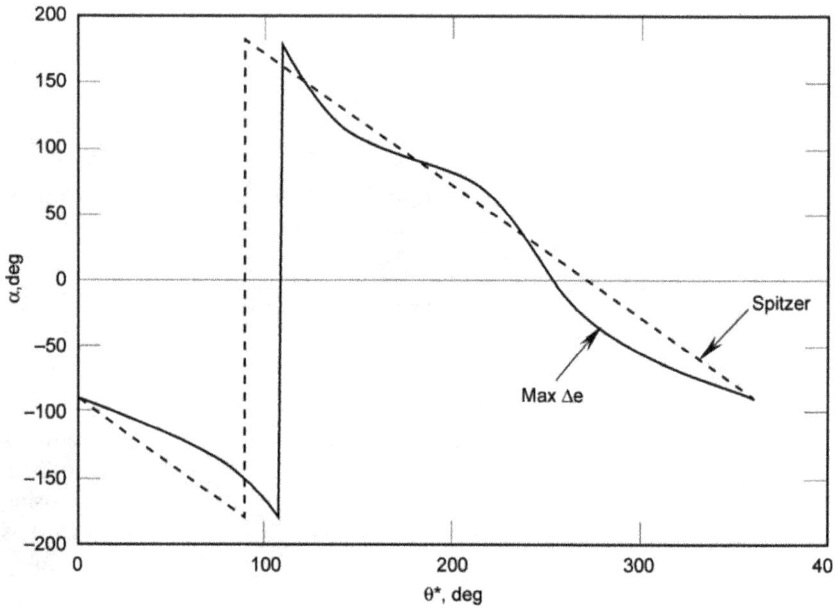

Fig. 6.11 Thrust pitch angle profiles for the maximum-Δe and Spitzer solutions for an elliptical orbit with $a_0 = 40{,}000$ km and $e_0 = 0.7$ without a shadow arc.

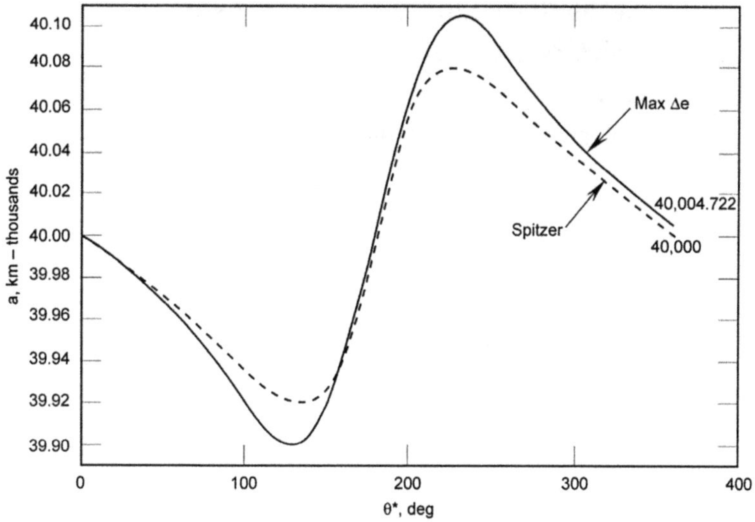

Fig. 6.12 Variation of the semimajor axis for the maximum-Δe and Spitzer solutions for an elliptical orbit with $a_0 = 40{,}000$ km and $e_0 = 0.7$ without a shadow arc.

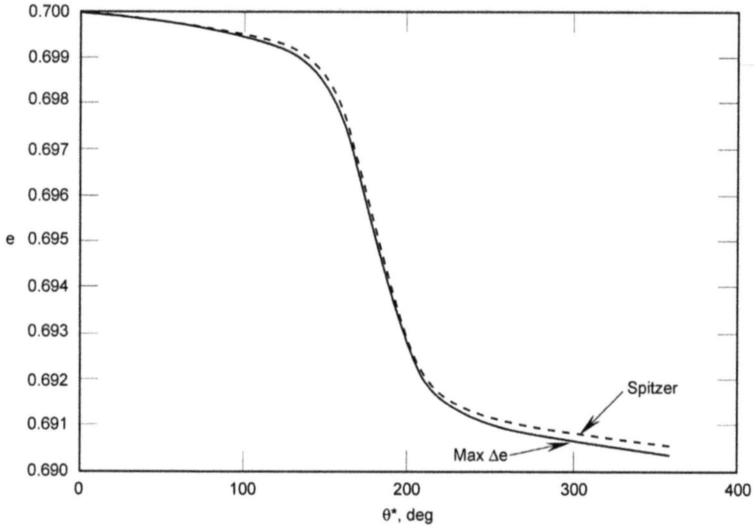

Fig. 6.13 Variation of the eccentricity for the maximum-Δe and Spitzer solutions for an elliptical orbit with $a_0 = 40{,}000$ km and $e_0 = 0.7$ without a shadow arc.

6.5 SIMULTANEOUS CIRCULARIZATION AND ORBIT ROTATION IN THE NEAR-CIRCULAR CASE

The average rate of change of the eccentricity using the in-plane Spitzer [4, 5] firing mode can be rewritten from Eq. (6.27) as

$$\widetilde{\left(\frac{\mathrm{d}e}{\mathrm{d}t}\right)} = \frac{3}{2} k c_\beta \left(\frac{p}{\mu}\right)^{1/2} \tag{6.35}$$

where k has been replaced by kc_β, the in-plane thrust acceleration, and β is the out-of-plane thrust angle or the angle between the thrust vector and its projection onto the orbit plane. An average rate for the inclination is also easily constructed from the exact rate of change $\mathrm{d}i/\mathrm{d}t$ valid in near-circular orbit

$$\frac{\mathrm{d}i}{\mathrm{d}t} = \frac{r}{(\mu p)^{1/2}} \cos(\omega + \theta^*) f_h$$

$$\cong \frac{1}{\left(\frac{\mu}{a}\right)^{1/2}} \cos(\omega + \theta^*) f_h$$

$$\widetilde{\left(\frac{\mathrm{d}i}{\mathrm{d}t}\right)} = \frac{k s_\beta}{(\mu/a)^{1/2}} \frac{1}{2\pi} \times 2 \int_{-\pi/2}^{\pi/2} c_{\theta^*} \, \mathrm{d}\theta^*$$

$$\widetilde{\left(\frac{\mathrm{d}i}{\mathrm{d}t}\right)} = \frac{2}{\pi} \frac{k s_\beta}{(\mu/a)^{1/2}} \tag{6.36}$$

where the out-of-plane acceleration f_h is written as ks_β and where $\omega = 0$ for convenience. This form can also be reached by writing $\mathrm{d}i/\mathrm{d}E$ as $(\mathrm{d}i/\mathrm{d}E)/(\mathrm{d}E/\mathrm{d}t)$ and integrating between the appropriate bounds to generate an expression for Δi, which is then divided by $2\pi/n$ to give the average rate, in accordance with Burt [8].

We can now obtain an expression for the constant out-of-plane angle β by forming

$$\widetilde{\left(\frac{\mathrm{d}i}{\mathrm{d}e}\right)} = \frac{\widetilde{(\mathrm{d}i/\mathrm{d}t)}}{\widetilde{(\mathrm{d}e/\mathrm{d}t)}} = \frac{4}{3\pi} \frac{\tan\beta}{(1-e^2)^{1/2}}$$

such that

$$\int_{i_0}^{i} \mathrm{d}i = \frac{4}{3\pi} \tan\beta \int_{e_0}^{e} \frac{\mathrm{d}e}{(1-e^2)^{1/2}}$$

or

$$(i - i_0) = \frac{4}{3\pi} \tan\beta (\sin^{-1} e - \sin^{-1} e_0)$$

and finally

$$\tan\beta = \frac{\dfrac{3\pi}{4}(i - i_0)}{(\sin^{-1}e - \sin^{-1}e_0)} \tag{6.37}$$

This expression can also be obtained from Eq. (6.28), written as

$$\sin^{-1} e - \sin^{-1} e_0 = \frac{3}{2}\left(\frac{a}{\mu}\right)^{1/2} kc_\beta t_f \tag{6.38}$$

and Eq. (6.36), which, upon integration, gives

$$\Delta i = i - i_0 = \frac{2}{\pi}\left(\frac{a}{\mu}\right)^{1/2} ks_\beta t_f \tag{6.39}$$

Substituting t_f from Eq. (6.39) into Eq. (6.38) yields Eq. (6.37), which is now used to obtain

$$s_\beta = \frac{\dfrac{3}{4}\pi(i - i_0)}{\left[(\sin^{-1}e - \sin^{-1}e_0)^2 + \dfrac{9}{16}\pi^2(i - i_0)^2\right]^{1/2}}$$

This expression for s_β can be inserted into Eq. (6.39) to yield the transfer time t_f, which gives the velocity change as

$$\Delta V = kt_f = \frac{2}{3}\left(\frac{\mu}{a}\right)^{1/2}\left[(\sin^{-1} e - \sin^{-1}e_0)^2 + \frac{9}{16}\pi^2(i - i_0)^2\right]^{1/2} \tag{6.40}$$

6.6 CONCLUSION

The maximization of the change in the semimajor axis subject to the constraint of zero change in the eccentricity has been extended to the dual problem of maximizing the change in the eccentricity subject to zero change in the semimajor axis, in the case of a near-circular orbit with discontinuous thrust over a single revolution. The solution is obtained by direct application of the theory of maxima and minima and through a numerical search for the value of the appropriate constant Lagrange multiplier such that a certain integral is driven to zero. The analysis has also been extended to the general elliptical case using continuous thrust and compared with Spitzer's [4, 5] simple scheme of applying the thrust vector normal to the orbit line of apsides. When the Spitzer firing mode is used, analytic expressions for the current value of the eccentricity and the accumulated velocity change are derived for the circularization problem.

The use of the Spitzer [4, 5] scheme in the presence of a shadow arc leads to a significant change in the semimajor axis that can be effectively constrained only through the use of the optimal solution presented in this chapter.

Finally, simple expressions for the constant out-of-plane thrust angle and total required velocity change to circularize and rotate a near-circular orbit simultaneously are also derived analytically for preliminary analyses.

REFERENCES

[1] Edelbaum, T. N., "Propulsion requirements for controllable satellites," *ARS Journal*, Vol. 31, No. 8, Aug. 1961, pp. 1079–1089.

[2] Cass, J. R., "Discontinuous low thrust orbit transfer," M.S. Thesis, Rept. AFIT/GA/AA/83D-1, School of Engineering, U.S. Air Force Institute of Technology, Wright-Patterson AFB, OH, Dec. 1983.

[3] Kechichian, J. A., "Low-thrust eccentricity-constrained orbit raising," *Journal of Spacecraft and Rockets*, Vol. 35, No. 3, May 1998, pp. 327–335.

[4] Spitzer, A., "Near optimal transfer orbit trajectory using electric propulsion," AAS Paper 95-215, AAS/AIAA Spaceflight Mechanics Meeting, Albuquerque, NM, Feb. 1995.

[5] Spitzer, A., "Novel orbit raising strategy makes low thrust commercially viable," 24th International Electric Propulsion Conference, IEPC Paper 95-212, Moscow, Russia, Sept. 1995.

[6] Kechichian, J. A., "Optimum thrust pitch profiles for certain orbit control problems," 2nd International Symposium on Low Thrust Trajectories, Centre Spatial de Toulouse, France, June 2002.

[7] Kéchichian, J. A., "Optimum thrust pitch profiles for certain orbit control problems," *Journal of Spacecraft and Rockets*, Vol. 40, No. 2, March–April 2003, pp. 253–259.

[8] Burt, E. G. C., "On space manoeuvers with continuous thrust," *Planetary and Space and Science*, Vol. 15, No. 1, Jan. 1967, pp. 103–122.

CHAPTER 7

Constrained Circularization in Elliptical Orbit Using Low Thrust with the Effect of Shadowing

NOMENCLATURE

a_0, a = initial and current orbit semimajor axes, respectively, m
a = orbit semimajor axis, m
b = orbit semiminor axis, m
c_β, s_β = cos β, sin β
E = eccentric anomaly
e = eccentricity
f_r, f_θ = components of the perturbation acceleration vector along the radial and perpendicular directions, respectively
h = orbital angular momentum
k = thrust acceleration, T/m, km/s^2
n = orbit mean motion, $\mu^{1/2}a^{-3/2}$, rad/s
$R_\oplus$ = radius of Earth, 6378.14 km
$\hat{r}$ = unit vector in the radial direction
$\boldsymbol{T}$ = thrust vector
x = direction of the eccentricity vector
α = angle between the radial direction and the thrust vector
β = Sun angle
γ = flight direction angle
θ_1^*, θ_2^* = true anomalies of the shadow entry and exit points
λ = Lagrange multiplier
$\boldsymbol{v}$ = velocity vector

7.1 INTRODUCTION

The optimal thrust pitch angle variation that results in the maximum change in the eccentricity in a general elliptical orbit using continuous, constant, low-thrust acceleration while keeping the orbit energy unchanged after a full thrust cycle is determined by direct use of the theory of maxima and through numerical quadrature and search techniques. The analysis takes into account the presence of a shadow arc arbitrarily positioned along the elliptical orbit, where thrust is cut off. Unlike the well-known nonoptimal scheme that uses a thrust orientation perpendicular to the line of apsides at all times, the present optimal scheme allows for the maximum change in eccentricity for a more efficient orbit circularization. Approximate but highly accurate analytic expressions for the changes in the eccentricity and semimajor axis of a general elliptical orbit, perturbed by a constant low-thrust acceleration applied along the fixed inertial direction normal to the orbit major axis, are also derived for general use and rapid calculations.

The problem of the maximization of the change in the eccentricity of a general elliptical orbit using continuous, constant, low-thrust acceleration while constraining the semimajor axis a to remain constant after one full cycle of thrust is analyzed in this chapter, by also taking into account the presence of Earth shadow arc where the thrust is turned off. Circularization of elliptical orbits with electric thrusters is currently performed by geostationary communications spacecraft, which are initially released into highly elliptical supersynchronous orbits that are subsequently circularized while also maintaining constant orbital energy.

The optimization method is similar to the one used by Edelbaum [1], Cass [2], and McCann [3], which was also applied in an earlier work by the author [4] under the same assumption of an initially circular orbit model. In particular, Cass [2] and McCann [3] considered the problem of transferring a spacecraft between two inclined circular orbits, of different size and inclination, in minimum time, using discontinuous low-thrust acceleration. The two-body thrust-perturbed orbit is thus constrained to remain circular during the transfer, and the optimal control law for the thrust direction derived for the fast-time-scale problem of maximizing the inclination change for a given change in the semimajor axis is used in an averaging procedure to solve the overall slow-time-scale transfer problem. The present analysis addresses the fast-time-scale planar problem in the more general elliptical case by deriving the optimal thrust pitch profile that maximizes the change in eccentricity without changing the semimajor axis and by also extending the analysis to the dual problem of maximizing the orbit semimajor axis while keeping the eccentricity e unchanged after one cycle of intermittent thrust. The change in eccentricity over a single orbit can then be used in conjunction with an inclination change to produce average rates of change in the eccentricity and inclination to solve the overall slow-time-scale minimum-time transfer problem of circularizing and rotating an initial elliptical orbit without changing its orbital energy. The theory of maxima is thus employed in the present planar problem, and the value of a certain constant Lagrange multiplier is determined numerically by means of an iterative scheme such that the corresponding integral constraint evaluated by numerical quadrature is driven to zero. The pitch angle of the optimal thrust vector is then obtained as a direct function of the orbital position of the spacecraft, which is selected here as the true anomaly. Simple orbit circularization schemes were proposed by Burt [5] and Spitzer [6, 7] for the continuous-thrust case. Spitzer [6, 7] employed an inertially fixed firing orientation in which the thrust is applied perpendicular to the line of apsides, providing near-optimal results while also ensuring constancy, to first order, of the semimajor axis in the process.

The presence of a shadow arc disrupts the constancy of the semimajor axis when the inertially fixed firing direction is used. The optimal mode derived here by extending the results obtained in the circular case [8] ensures that the tailored thrust pitch profile not only optimizes the change in the eccentricity but also satisfies the semimajor axis constraint regardless of the size and orbital location of the shadow arc.

Finally, analytic integrations of the differential equations for the variables a and e, which consist of the variational equations of the orbit, are carried out using the inertially fixed firing scheme in the elliptical case, so that the changes in a and e are easily evaluated from their corresponding approximate analytic expressions, valid for about one revolution about the central body, without the need to integrate the differential equations numerically. The analysis of this chapter is based on a previously published conference paper [9] and journal article [10].

7.2 MAXIMIZATION OF THE CHANGE IN ECCENTRICITY SUBJECT TO ZERO CHANGE IN THE SEMIMAJOR AXIS FOR DISCONTINUOUS THRUSTING IN AN ELLIPTICAL ORBIT

Figure 7.1 shows an elliptical orbit with the Earth–Sun line contained in the orbit plane, yielding the worst-case shadowing represented by the arc $E'H'$, where no thrust is applied. The thrust pitch angle α is the angle between the radial direction

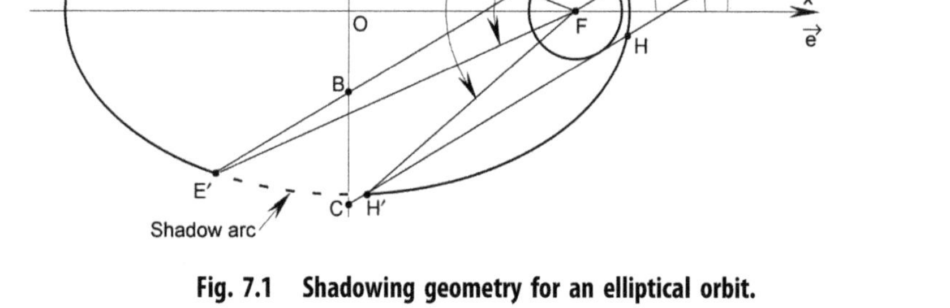

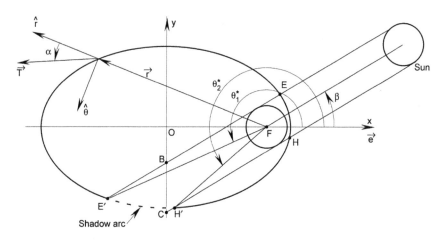

Fig. 7.1 Shadowing geometry for an elliptical orbit.

$\hat{r}$ and the thrust vector $\boldsymbol{T}$, and θ_1^* and θ_2^* are the true anomalies of the shadow entry and exit points E' and H', respectively. The Earth–Sun line is inclined at an angle β with respect to the eccentricity vector orientation, with the sunlight cylinder tangent to the Earth sphere intersecting the orbit at four distinct locations, E, H, E', and H'. Letting $R_\oplus = 6378.14$ km be the radius of Earth, the equations of the lines EE' and HH' are given by

$$y = (x - ae)\tan\beta + \frac{R_\oplus}{c_\beta} \tag{7.1}$$

$$y = (x - ae)\tan\beta - \frac{R_\oplus}{c_\beta} \tag{7.2}$$

where x is along the direction of the eccentricity vector and y is along the semiminor axis; c_β represents $\cos\beta$; and a and b are the orbit semimajor axis and semiminor axis, respectively. The coordinates of the points H and H' are obtained by substituting y from Eq. (7.2) into the equation of the ellipse, namely, $x^2/a^2 + y^2/b^2 = 1$, and solving for the two values of x from the resulting quadratic $Ax^2 + Bx + C = 0$, where

$$A = \frac{1}{a^2} + \frac{\tan^2\beta}{b^2}$$

$$B = -\left(2ae\tan^2\beta + 2\frac{R_\oplus}{c_\beta}\tan\beta\right)\Big/b^2$$

$$C = D - 1$$

$$D = \frac{1}{b^2}\left(\frac{R_\oplus}{c_\beta} + ae\tan\beta\right)^2$$

The coordinates of the intersection points of the line EE' with the ellipse are obtained likewise using Eq. (7.1) and solving for the value of x from the quadratic $Ax^2 + B'x + C'$, with $C' = D' - 1$ and

$$B' = -\left(2ae\tan^2\beta - 2\frac{R_\oplus}{c_\beta}\tan\beta\right)\Big/b^2$$

$$D' = \frac{1}{b^2}\left(\frac{R_\oplus}{c_\beta} - ae\tan\beta\right)^2$$

For $\beta = 90$ deg and 270 deg, the lines HH' and EE' are described by the equations $x = ae + R_\oplus$ and $x = ae - R_\oplus$, respectively, such that the four intersection points are now obtained by substituting these expressions for x into the equation of the ellipse and solving for the values of y as

$$y = \pm\left[b^2 - \frac{b^2}{a^2}(ae + R_{\oplus})^2\right]^{1/2}$$

$$y = \pm\left[b^2 - \frac{b^2}{a^2}(ae - R_{\oplus})^2\right]^{1/2}$$

respectively. Because $x = ac_E$ and $y = bs_E$, the value of the eccentric anomaly E is readily obtained from the coordinates x and y, and finally, the true anomaly of each intersection point is known from the identities

$$s_{\theta}* = \frac{(1 - e^2)^{1/2} s_E}{(1 - ec_E)}$$

$$c_{\theta}* = \frac{(c_E - e)}{(1 - ec_E)}$$

Thus, for a given Sun angle β, the true anomalies of the shadow entry and exit points, θ_1^* and θ_2^*, respectively, are easily obtained.

These angles are reported in Table 7.1 for various angles β covering the whole range $0 \leq \beta \leq$ 360 deg, where θ_0^* represents θ_2^* and θ_f^* represents θ_1^*. Note that 360 deg has been added to θ_f^* where needed to ensure that the thrust is applied from θ_0^* (shadow exit) to θ_f^* (shadow entry), with the angle always increasing for reliable integration.

TABLE 7.1 SHADOW ENTRY AND EXIT TRUE ANOMALIES VS β FOR WORST-CASE SHADOWING WITH a = 40,000 KM AND e = 0.7

β, deg	0 (180)	30 (210)	60 (240)
θ_0^*, deg	185.438741 (30.128698)	218.068229 (55.823269)	254.786809 (80.428847)
θ_f^*, deg	534.561259 (329.871302)	563.566516 (357.902271)	590.083173 (389.834034)
β, deg	**90 (270)**	**120 (300)**	**150 (330)**
θ_0^*, deg	293.593947 (104.863308)	330.165966 (129.916827)	2.097728 (156.433484)
θ_f^*, deg	615.136692 (426.406053)	639.571153 (465.213191)	304.176731 (509.931771)

When $k = T/m$ is the magnitude of the constant thrust acceleration, the $\hat{r}$ and $\hat{\theta}$ components of this acceleration, namely, f_r and f_θ, are given by kc_α and ks_α, respectively. Here, $\hat{\boldsymbol{r}}$ and $\hat{\boldsymbol{\theta}}$ are unit vectors along the radial direction and the direction 90 deg ahead of it, respectively, such that the differential equations valid in an elliptical orbit are given by [11]

$$\dot{a} = \frac{2es_{\theta^*}}{n(1-e^2)^{1/2}} f_r + \frac{2(1+ec_{\theta^*})}{n(1-e^2)^{1/2}} f_\theta \tag{7.3}$$

$$\dot{e} = \frac{(1-e^2)^{1/2}}{na} s_{\theta^*} f_r + \frac{(1-e^2)^{1/2}}{nae}\left(1 + ec_{\theta^*} - \frac{1-e^2}{1+ec_{\theta^*}}\right) f_\theta \tag{7.4}$$

where n represents the orbit mean motion $\mu^{1/2}a^{-3/2}$. Because the magnitude of the thrust acceleration k is very low, we can neglect the contributions of the terms in f_r and f_θ in the differential equation for θ^*, namely

$$\frac{\mathrm{d}\theta^*}{\mathrm{d}t} = \frac{h}{r^2} - \frac{r}{he}\left[-c_{\theta^*}(1+ec_{\theta^*})f_r + s_{\theta^*}(2+ec_{\theta^*})f_\theta\right]$$

such that this general singular form is reduced to the following simple form

$$\frac{\mathrm{d}\theta^*}{\mathrm{d}t} = \frac{h}{r^2} = \frac{n(1+ec_{\theta^*})^2}{(1-e^2)^{3/2}} \tag{7.5}$$

after the replacement of h by $\mu^{1/2}a^{1/2}(1-e^2)^{1/2}$ and of r by $a(1-e^2)/(1+ec_{\theta^*})$. After division by Eq. (7.5), Equations (7.3) and (7.4) can now be written in terms of θ^*, the independent variable (instead of t, the physical time), such that the following approximate forms are obtained:

$$\frac{\mathrm{d}a}{\mathrm{d}\theta^*} = \frac{2es_{\theta^*}(1-e^2)}{n^2(1+ec_{\theta^*})^2} kc_\alpha + \frac{2(1-e^2)}{n^2(1+ec_{\theta^*})} ks_\alpha \tag{7.6}$$

$$\frac{\mathrm{d}e}{\mathrm{d}\theta^*} = \frac{(1-e^2)^2 s_{\theta^*}}{n^2a(1+ec_{\theta^*})^2} kc_\alpha + \frac{(1-e^2)^2}{n^2ae}\left[\frac{1}{1+ec_{\theta^*}} - \frac{(1-e^2)}{(1+ec_{\theta^*})^3}\right] ks_\alpha \tag{7.7}$$

Then, as a result of applying the low-thrust acceleration between θ_0^* and θ_f^*, the changes in semimajor axis and eccentricity over one revolution can be written as follows:

$$\Delta a = \frac{2(1-e^2)k}{n^2}\int_{\theta_0^*}^{\theta_f^*}\left[\frac{es_{\theta^*}c_\alpha}{(1+ec_{\theta^*})^2} + \frac{s_\alpha}{(1+ec_{\theta^*})}\right]\mathrm{d}\theta^* \tag{7.8}$$

$$\Delta e = \frac{(1-e^2)^2k}{n^2a}\int_{\theta_0^*}^{\theta_f^*}\left[\frac{s_{\theta^*}c_\alpha}{(1+ec_{\theta^*})^2} + \frac{s_\alpha}{e(1+ec_{\theta^*})} - \frac{(1-e^2)s_\alpha}{e(1+ec_{\theta^*})^3}\right]\mathrm{d}\theta^* \tag{7.9}$$

To maximize the change in e (i.e., Δe) subject to the constraint $\Delta a = 0$, the following augmented integral is formed with the use of the constant Lagrange multiplier λ that effectively adjoins Δa to the Δe expression in Eq. (7.9), that is, $I(\alpha) = \Delta e + \lambda \Delta a$

$$I(\alpha) = \int_{\theta_0^*}^{\theta_f^*} \frac{(1-e^2)k}{n^2} \left\{ \frac{(1-e^2)}{a} \left[\frac{s_{\theta^*} c_\alpha}{(1+ec_{\theta^*})^2} + \frac{s_\alpha}{e(1+ec_{\theta^*})} - \frac{(1-e^2)s_\alpha}{e(1+ec_{\theta^*})^3} \right] + 2\lambda \left[\frac{es_{\theta^*} c_\alpha}{(1+ec_{\theta^*})^2} + \frac{s_\alpha}{(1+ec_{\theta^*})} \right] \right\} \mathrm{d}\theta^* \tag{7.10}$$

Let F represent the integrand in Eq. (7.10). Then, Euler's equation $\partial F/\partial \alpha = 0$, which is a necessary condition for optimality, provides the optimal thrust pitch law as

$$\tan \alpha = \frac{(1-e^2)\left[(1+ec_{\theta^*})^2 - (1-e^2)\right] + 2ae\lambda(1+ec_{\theta^*})^2}{e(1+ec_{\theta^*})[(1-e^2)+2ae\lambda]s_{\theta^*}} \tag{7.11}$$

In turn, this expression yields

$$s_\alpha^2 = \left\{-(1-e^2)\left[(1+ec_{\theta^*})^2 - (1-e^2)\right] - 2ae\lambda(1+ec_{\theta^*})^2\right\}^2 \Big/ X$$

$$c_\alpha^2 = e^2(1+ec_{\theta^*})^2\left[-(1-e^2)s_{\theta^*} - 2ae\lambda s_{\theta^*}\right]^2 / X$$

where X is given by

$$X = e^2(1+ec_{\theta^*})^2\left[-(1-e^2)s_{\theta^*} - 2ae\lambda s_{\theta^*}\right]^2 + \left\{-(1-e^2)\left[(1+ec_{\theta^*})^2 - (1-e^2)\right] - 2ae\lambda(1+ec_{\theta^*})^2\right\}^2$$

Using these expressions for s_α and c_α in Eq. (7.8) for Δa and performing some manipulations results in the following integral:

$$\Delta a = \frac{2(1-e^2)^2 k}{n^2} \int_{\theta_0^*}^{\theta_f^*} \frac{\left[-(1-e^2) - 2ae\lambda\right](1+e^2+2ec_{\theta^*}) + (1-e^2)^2}{(1+ec_{\theta^*})X^{1/2}} \mathrm{d}\theta^* \tag{7.12}$$

A search is performed on the value of λ, and the integral constraint (7.12) is evaluated for each value of λ by numerical quadrature between the limits θ_0^* and θ_f^*, which is the region where the thrust is on, until the value of Δa is driven to zero to within a small tolerance. A 10-point Gauss–Legendre quadrature [12] of the Δa integral is performed, and the zero crossing of the Δa function is established, such that the van Wijngaarden–Dekker–Brent method [12], which consists of combining root bracketing, bisection, and inverse quadratic interpolation, is used to pinpoint the exact value of λ from the zero-crossing neighborhood. For the example orbit of Table 7.1, λ is on the order of -10^{-6} 1/km, and for the dual problem of maximizing Δa with $\Delta e = 0$, λ is on the order of 10^5 km. Once λ is determined, the angle α that maximizes the change in the eccentricity Δe is obtained from Eq. (7.11) such that Δe itself is now determined by numerical quadrature from Eq. (7.13), that is

$$\Delta e = \frac{(1-e^2)^2 k}{n^2 a} \int_{\theta_0^*}^{\theta_f^*} \left[\frac{s_{\theta^*} c_\alpha}{(1+ec_{\theta^*})^2} + \frac{s_\alpha}{e(1+ec_{\theta^*})} - \frac{(1-e^2)s_\alpha}{e(1+ec_{\theta^*})^3} \right] \mathrm{d}\theta^* \tag{7.13}$$

Because of the squaring operations, the expressions for s_α^2 and c_α^2 provide s_α and c_α with plus-or-minus signs. As the two signs must be the same, because otherwise the expression for tan α will change sign, the λ value that drives Δa to zero is the same in either case. However, the plus sign will yield Δe with pitch angle α, with Δe being either positive or negative, and the minus sign will yield the same absolute value of Δe with the opposite sign, using the thrust pitch angle $\alpha + \pi$, because the thrust direction in the latter case is exactly opposite to the direction that corresponds to the selection of the plus sign. In other words, if Δe is positive with the plus sign selected for the s_α and c_α expressions, then it will be negative with the minus sign selection, and vice versa. If the constraint $\Delta a = 0$ is not enforced, then the maximization of Δe in an elliptical orbit is made possible by simply setting $\lambda = 0$ in Eq. (7.11), resulting in the optimal law

$$\tan \alpha = \frac{(1+ec_{\theta^*})^2 - (1-e^2)}{e(1+ec_{\theta^*})s_{\theta^*}} \tag{7.14}$$

In this case, this angle α also maximizes the instantaneous rate of change of e. In the circular case, the limit of expression (7.14) as e approaches zero will be given by [1]

$$\tan \alpha = \frac{2}{\tan \theta^*} \tag{7.15}$$

which is the well-known optimal thrust steering program to maximize Δe from an initially circular orbit. In practice, when e is near zero, the nonzero average value of e over one revolution is to be used instead, and θ^* is just the angular position measured from the shadow exit point with $\theta_0^* = 0$.

The case that corresponds to $\beta = 180$ deg is now solved with $\theta_0^* = 30.128698$ deg and $\theta_f^* = 329.871302$ deg using $k = 3.5 \times 10^{-7}$ km/s^2 for the constant thrust acceleration and an initial orbit given by $a = 40{,}000$ km and $e = 0.7$, yielding $\lambda = -0.445453 \times 10^{-6}$ 1/km and a corresponding Δe value from Eq. (7.13) of $-0.9042309 \times 10^{-2}$. Equations (7.6) and (7.7) are now integrated numerically from the initial state using $\alpha = f(\theta^*)$ from Eq. (7.11) and holding a, e, and n constant on the right-hand side of each of the two differential equations, over the θ^* range from θ_0^* to θ_f^*. This verification run yields the final state $a_f = 40{,}005.308$ km and $e_f = 0.69060484$, which corresponds to an achieved value for $(\Delta e)_f$ of $-0.9395152 \times 10^{-2}$.

7.3 MAXIMIZATION OF THE CHANGE IN SEMIMAJOR AXIS SUBJECT TO ZERO CHANGE IN THE ECCENTRICITY FOR DISCONTINUOUS THRUSTING IN AN ELLIPTICAL ORBIT

The dual problem of maximizing the orbit energy subject to the constraint $\Delta e = 0$ is carried out by forming the following augmented integral, namely, $I(\alpha) = \Delta a + \lambda\Delta e$:

$$I(\alpha) = \frac{(1-e^2)k}{n^2}\int_{\theta_0^*}^{\theta_f^*}\left\{2\left[\frac{e s_{\theta^*} c_\alpha}{(1+e c_{\theta^*})^2} + \frac{s_\alpha}{(1+e c_{\theta^*})}\right] + \frac{\lambda(1-e^2)}{a}\left[\frac{s_{\theta^*} c_\alpha}{(1+e c_{\theta^*})^2} + \frac{s_\alpha}{e(1+e c_{\theta^*})} - \frac{(1-e^2)s_\alpha}{e(1+e c_{\theta^*})^3}\right]\right\}\mathrm{d}\theta^* \tag{7.16}$$

If F represents the integrand in Eq. (7.16), then, as before, the optimal pitch angle α is obtained as a function of θ^* and the multiplier λ by direct application of the Euler equation $\partial F/\partial\alpha = 0$, leading to

$$\tan\alpha = \frac{\lambda(1-e^2)\left[(1+e c_{\theta^*})^2 - (1-e^2)\right] + 2ae(1+e c_{\theta^*})^2}{e(1+e c_{\theta^*})[\lambda(1-e^2) + 2ae]s_{\theta^*}} \tag{7.17}$$

If the $\Delta e = 0$ constraint is not enforced, then $\lambda = 0$ is selected in Eq. (7.17) for the pitch profile that would maximize Δa for the unconstrained elliptical case, giving

$$\tan\alpha = \frac{(1+e c_{\theta^*})}{e s_{\theta^*}} \tag{7.18}$$

However, the flight direction angle γ, which is the angle between the radius vector $\boldsymbol{r}$ and the velocity vector $\boldsymbol{v}$, is given by [11]

$$s_\gamma = \frac{\mu}{hv}(1+e c_{\theta^*})$$

$$c_\gamma = \frac{\mu}{hv} e s_{\theta^*}$$

where h is the orbital angular momentum, such that

$$\tan\gamma = \frac{(1+e c_{\theta^*})}{e s_{\theta^*}} \tag{7.19}$$

which is identical to Eq. (7.18). Thus, $\alpha = \gamma$, and the thrust vector is then aligned with the velocity vector itself, which is the well-known solution for maximizing the orbit semimajor axis in the unconstrained elliptical case. In the limiting case where e approaches zero, tan α will tend toward infinity, and α will tend toward 90 deg, again pointing the thrust vector along the velocity vector.

Returning to the constrained case, Eq. (7.17) yields

$$s_\alpha^2 = \left\{(1-e^2)\left[\lambda(1+ec_{\theta^*})^2 - \lambda(1-e^2)\right] + 2ae(1+ec_{\theta^*})^2\right\}^2 / X'$$

$$c_\alpha^2 = e^2(1+ec_{\theta^*})^2\left[\lambda(1-e^2)s_{\theta^*} + 2aes_{\theta^*}\right]^2 / X'$$

where X' is the sum of the squares of the numerator and denominator in the expression for tan α. The same discussion concerning the signs of s_α and c_α as was made in Sec. 7.2 also applies here.

With substitution into the Δe expression in Eq. (7.9) and some mathematical manipulation, the following form for that integral is obtained:

$$\begin{aligned}\Delta e = \frac{(1-e^2)^2 k}{n^2 ae}\int_{\theta_0^*}^{\theta_f^*}\left\{\frac{\left[\lambda(1-e^2)+2ae\right](2e^2+2ec_{\theta^*}) - \lambda(1-e^2)^2}{(1+ec_{\theta^*})X'^{1/2}}\right. \\ \left. + \frac{\lambda(1-e^2)^3}{(1+ec_{\theta^*})^3 X'^{1/2}}\right\}\mathrm{d}\theta^*\end{aligned} \tag{7.20}$$

As before, a search is performed on the value of λ, and the integral in Eq. (7.20) is evaluated by numerical quadrature between the bounds θ_0^* and θ_f^*, which is the region where the thrust is on, until $\Delta e = 0$ is achieved to within a small tolerance.

Δa is then obtained from Eq. (7.8), also by quadrature, once λ is determined, and in Eq. (7.17), α is made a function of only the constants a and e and the integration variable θ^*.

The example initial orbit of Sec. 7.2 is used, and the optimal solution that maximizes Δa is obtained as $\lambda = 0.20597067506 \times 10^5$ km with $\Delta a =$ 545.016 km and with integrated values for a_f and e_f obtained using Eqs. (7.6) and (7.7) of 40,529.360 km and 0.699020655, respectively.

The two optimal solutions, namely, the Δe and Δa-maximized values thus generated, are now compared to the simple nonoptimal Spitzer firing scheme [6, 7], for which the thrust vector is directed along a direction normal to the orbit line of apsides, resulting in an inertially fixed firing mode. In this scheme, $c_\alpha = -s_{\theta^*}$ and $s_\alpha = -c_{\theta^*}$ will effectively decrease the orbit eccentricity, whereas $c_\alpha = s_{\theta^*}$ and $s_\alpha = c_{\theta^*}$ will increase it, with the first mode firing opposite the velocity vector at perigee and along the velocity vector at apogee effectively lowering the apogee and raising the perigee to decrease the eccentricity and the second mode giving exactly the reverse result. With use of the first Spitzer mode to decrease the eccentricity, the initial orbit is integrated using Eqs. (7.6) and (7.7) between $\theta_0^* = 30.128698$ and $\theta_f^* = 329.871302$ deg as for the two

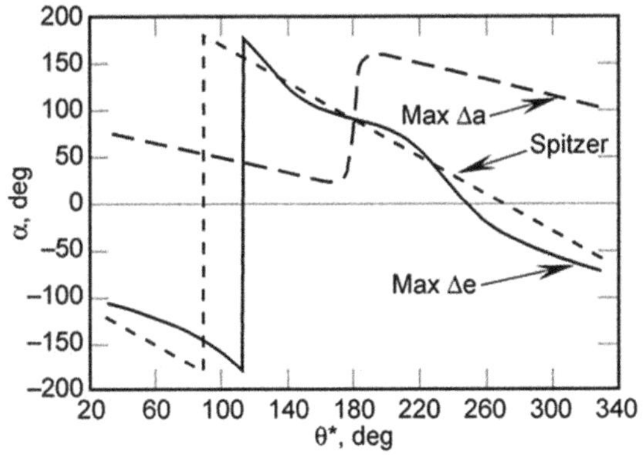

Fig. 7.2 Variation of the thrust pitch angle for the maximum-Δe, maximum-Δa, and Spitzer [6, 7] solutions.

preceding optimal cases, yielding $a_f = 40{,}035.843$ km and $e_f = 0.690812921$, with the latter value being slightly higher than the one generated by the Δa-constrained, maximum-Δe solution at $e_f = 0.69060484$; a_f itself is larger than the desired value of 40,000 km in this case because the Spitzer modes cannot constrain the energy of the orbit.

Figures 7.2–7.4 show the variations of the pitch angle, semimajor axis, and eccentricity, respectively, for the two constrained optimal solutions, as well as the Spitzer [6, 7] solution, showing how the first Spitzer firing mode is nearly identical to the optimal mode that maximizes Δe. Unlike these two modes, which display a sign change in α to accelerate and decelerate the vehicle along its orbit, the maximum-Δa solution maintains a positive α value throughout. Figure 7.3 shows how the Spitzer mode fails to recover the initial semimajor axis after one cycle of thrust even though it closely matches the optimal Δe value, as shown in Fig. 7.4.

7.4 FURTHER NUMERICAL COMPARISONS

A series of runs was performed for various values of the Sun angle β using both the constrained Δe-maximizing strategy and the Spitzer [6, 7] strategy, with the true anomalies of the shadow exit and entry in Table 7.1 for each corresponding β angle. Tables 7.2 and 7.3 list the values of the iterated

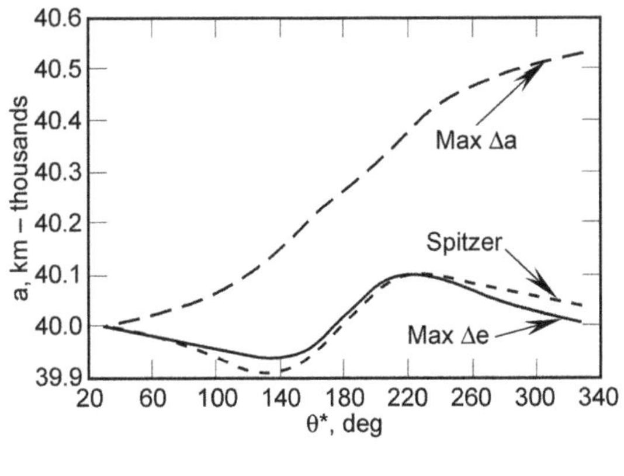

Fig. 7.3 Variation of the semimajor axis for the maximum-Δe, maximum-Δa, and Spitzer [6, 7] solutions.

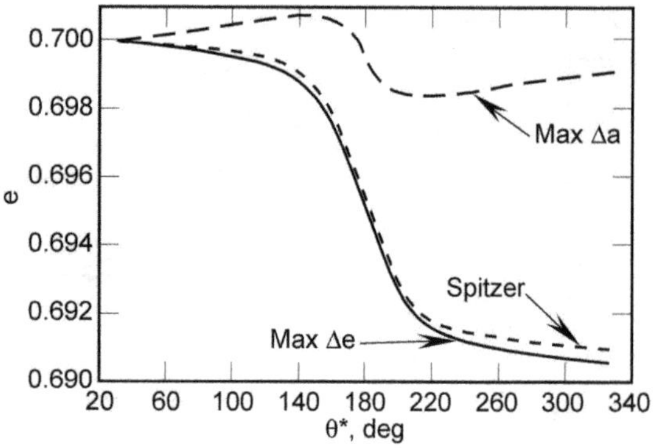

Fig. 7.4 Variation of the eccentricity for the maximum-Δe, maximum-Δa, and Spitzer [6, 7] solutions.

multiplier λ, as well as the $(\Delta e)_{\Delta e}$ values computed from the Δe integral by quadrature and the numerically integrated $(a_f)_{\Delta e}$ and $(e_f)_{\Delta e}$ final parameters of the optimized solutions for each β value. The $(a_f)_s$ and $(e_f)_s$ values of the Spitzer integrated trajectories are also included in Tables 7.2 and 7.3, with $(a_f)_s$ undershooting the desired 40,000-km mark in the β range from 330 to 30 deg and overshooting in the much wider range of $60 < \beta < 300$ deg. When β is either 0 or 180 deg, that is, when the Earth–Sun line is along the orbit line of apsides, the shadow arcs are centered at the orbit apogee and perigee, respectively, resulting in higher variations of $(a_f)_s$ from the 40,000-km value, which, by contrast, is fairly well matched by the constrained solutions to within only a few kilometers, as shown in Tables 7.2 and 7.3.

A higher-order quadrature method than the 10-point Gauss–Legendre method used here might more closely enforce the $\Delta a = 0$ constraint. Nevertheless, Fig. 7.5 shows both the $(a_f)_{\Delta e}$ and $(a_f)_s$ values as functions of β for the entire range $0 < \beta < 360$ deg, with the $(a_f)_{\Delta e}$ value of the optimized solutions closely matching the 40,000-km constraint regardless of the value of β, whereas the Spitzer $(a_f)_s$ bell-shaped curve shows how the constraint is violated except for two narrow ranges centered around β values of 45 and 315 deg.

Finally, in Fig. 7.6, the optimal Δe value and the Δe value of the Spitzer [6, 7] solution are plotted vs β, showing how the optimal solution yields the larger change, as desired, while also complying with the $\Delta a = 0$ constraint for this worst case of shadowing in this particular elliptical orbit example.

7.5 ANALYTIC INTEGRATIONS FOR THE SPITZER SCHEME

Because of the very low level of the low-thrust acceleration provided by the electric thrusters, it is possible to obtain analytic expressions for the changes experienced by the eccentricity and semimajor axis of a given orbit for the case where the constant thrust acceleration is applied in a fixed inertial orientation, such as in Spitzer [6, 7] mode. Because a and e vary by small amounts after one cycle of thrust or one orbit, their values can be held constant on the right-hand side of the variational equations, paving the way for their analytic integration.

TABLE 7.2 FINAL PARAMETERS OBTAINED WITH Δa-CONSTRAINED, Δe-OPTIMIZING, AND SPITZER [6, 7] MODES FOR $\beta = 0-150$ DEG

β, deg	0	30	60
$(\lambda)_{\Delta e}$, 1/km	$-0.25669812 \times 10^{-5}$	$-0.24080373 \times 10^{-5}$	$-0.60456815 \times 10^{-6}$
$(\Delta e)_{\Delta e}$	$-0.80332152 \times 10^{-2}$	$-0.87636901 \times 10^{-2}$	$-0.94523115 \times 10^{-2}$
$(a_f)_{\Delta e}$, km	39998.623	40002.482	39996.023
$(e_f)_{\Delta e}$	0.691958845	0.691218603	0.690659402
$(a_f)_s$, km	39964.157	39985.214	40012.055
$(e_f)_s$	0.692067423	0.691339135	0.690850669

β, deg	90	120	150
$(\lambda)_{\Delta e}$, 1/km	$-0.2563464 \times 10^{-6}$	$-0.2338952 \times 10^{-7}$	$-0.11460911 \times 10^{-6}$
$(\Delta e)_{\Delta e}$	$-0.9314426 \times 10^{-2}$	$-0.9246776 \times 10^{-2}$	$-0.96956425 \times 10^{-2}$
$(a_f)_{\Delta e}$, km	40005.449	39997.453	39997.334
$(e_f)_{\Delta e}$	0.690620353	0.690614399	0.690606069
$(a_f)_s$, km	40026.495	40032.887	40035.272
$(e_f)_s$	0.690785068	0.69079769	0.690809371

TABLE 7.3 FINAL PARAMETERS OBTAINED WITH Δa-CONSTRAINED, Δe-OPTIMIZING, AND SPITZER [6, 7] MODES FOR $\beta = 0-330$ DEG

β, deg	180	210	240
$(\lambda)_{\Delta e}$, 1/km	$-0.4454532 \times 10^{-6}$	$-0.11460911 \times 10^{-6}$	$-0.2338952 \times 10^{-7}$
$(\Delta e)_{\Delta e}$	$-0.90423093 \times 10^{-2}$	$-0.96956425 \times 10^{-2}$	$-0.9246776 \times 10^{-2}$
$(a_f)_{\Delta e}$, km	40,005.308	39,997.334	39,997.453
$(e_f)_{\Delta e}$	0.690604847	0.690606069	0.690614399
$(a_f)_s$, km	40,035.843	40,035.272	40,032.887
$(e_f)_s$	0.69081291	0.690809371	0.690797699
β, deg	**270**	**300**	**330**
$(\lambda)_{\Delta e}$, 1/km	$-0.2563464 \times 10^{-6}$	$-0.60456815 \times 10^{-6}$	$-0.24080373 \times 10^{-5}$
$(\Delta e)_{\Delta e}$	$-0.9314426 \times 10^{-2}$	$-0.94523115 \times 10^{-2}$	$-0.87636901 \times 10^{-2}$
$(a_f)_{\Delta e}$, km	40,005.449	39,996.023	40,002.482
$(e_f)_{\Delta e}$	0.690620353	0.69065940	0.691218603
$(a_f)_s$, km	40,026.495	40,012.055	39,985.214
$(e_f)_s$	0.690785068	0.690850669	0.691339135

Let us first start with the simplified equations of motion valid in a circular orbit, which are obtained from Eqs. (7.6) and (7.7) by neglecting all terms of order e and higher:

$$\frac{da}{d\theta^*} = 2k\frac{a^3}{\mu}s_\alpha \tag{7.21}$$

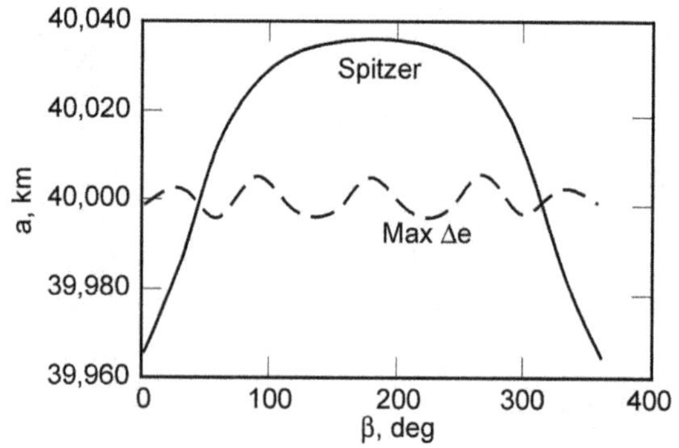

Fig. 7.5 Final semimajor axis values obtained for the maximum-Δe and Spitzer [6, 7] solutions as a function of Sun angle β.

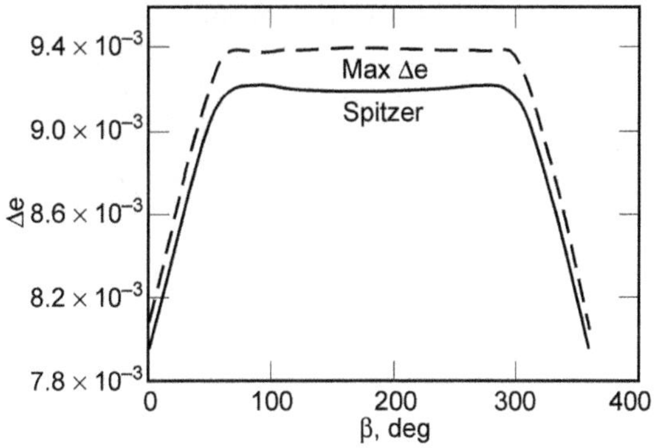

Fig. 7.6 Eccentricity changes for the maximum-Δe and Spitzer [6, 7] solutions as a function of Sun angle β.

$$\frac{\mathrm{d}e}{\mathrm{d}\theta^*} = \frac{ka^2}{\mu}(c_\alpha s_{\theta^*} + 2s_\alpha c_{\theta^*}) \tag{7.22}$$

Using $s_\alpha = c_{\theta^*}$ and $c_\alpha = s_{\theta^*}$ (for increasing e), holding a constant on the right-hand sides of Eqs. (7.21) and (7.22), and integrating between 0 and θ^* yields the simple expressions

$$a = a_0 + 2k\frac{a^3}{\mu}s_{\theta^*} \tag{7.23}$$

$$e = e_0 + \frac{ka^2}{\mu}\left(\frac{3}{2}\theta^* + \frac{s_{2\theta^*}}{4}\right) \tag{7.24}$$

An exact integration is also possible in this case. For instance, rewriting Eq. (7.21) as

$$a^{-3}\mathrm{d}a = 2\frac{k}{\mu}c_{\theta^*}\,\mathrm{d}\theta^*$$

and integrating results in the exact expression

$$a = a_0\left(1 - 4\frac{ka_0^2}{\mu}s_{\theta^*}\right)^{-1/2} \tag{7.25}$$

If this expression is now substituted into Eq. (7.22), then we obtain

$$e - e_0 = \frac{K}{4}\int_0^{\theta^*} \frac{1 + c_{\theta^*}^2}{1 - Ks_{\theta^*}}\,\mathrm{d}\theta^*$$

where $K = 4ka_0^2/\mu$ is a constant and its value is less than 1 for typical low-thrust accelerations used in Earth orbit.

The first part of this integral is easily obtained as

$$\int_0^{\theta^*} \frac{d\theta^*}{(1 - K s_{\theta^*})} = \frac{2}{(1 - K^2)^{1/2}} \tan^{-1}\left[\frac{\tan\frac{\theta^*}{2} - K}{(1 - K^2)^{1/2}}\right]\Bigg|_0^{\theta^*} \tag{7.26}$$

The second part is more complicated, and it is given by MATLAB® in the form

$$\frac{K}{4}\int_0^{\theta^*} \frac{c_{\theta^*}^2}{(1 - K s_{\theta^*})} d\theta^* = -\frac{1}{2K}(1 - K^2)^{1/2} \tan^{-1}\left[\frac{\tan\frac{\theta^*}{2} - K}{(1 - K^2)^{1/2}}\right]\Bigg|_0^{\theta^*} + \frac{\theta^*}{4K}\bigg|_0^{\theta^*} - \frac{1}{2} c_{\theta^*/2}^2\bigg|_0^{\theta^*} \tag{7.27}$$

Therefore, a closed expression for e reads

$$e = e_0 + \frac{K}{2(1 - K^2)^{1/2}} \tan^{-1}\left[\frac{\tan\frac{\theta^*}{2} - K}{(1 - K^2)^{1/2}}\right]\Bigg|_0^{\theta^*} + \frac{\theta^*}{4K}\bigg|_0^{\theta^*} - \frac{1}{2} c_{\theta^*/2}^2\bigg|_0^{\theta^*} - \frac{1}{2K}(1 - K^2)^{1/2} \tan^{-1}\left[\frac{\tan\frac{\theta^*}{2} - K}{(1 - K^2)^{1/2}}\right]\Bigg|_0^{\theta^*} \tag{7.28}$$

The first and last inverse tangent terms could be combined into a single term; however, they are left as is because, unlike the first one, the second inverse tangent term is very sensitive to small K when evaluated at $\theta^* = 0$. For example, for $k = 3.5 \times 10^{-7}$ km/s^2, $a_0 = 7000$ km, $e_0 = 0$, and $\mu = 398601.3$ km^3/s^2, $K = 1.72101796959 \times 10^{-4}$, and $(1 - K^2)^{1/2} = 0.99999998519$, such that sufficient digits must be carried to prevent a less-than-precise calculation of e from Eq. (7.28), because of the already mentioned sensitivity. Also, the proper branch of the inverse tangent function must be selected such that it corresponds to the one on which $\theta^*/2$ lies for any particular choice of θ^*. Otherwise, if the principal branch of the inverse tangent function is always employed in a systematic manner, then the answer will be wrong if the function does not represent the principal value. For example, for $\theta^* = 220$ deg, $\theta^*/2 = 110$ deg, and the branch passing through $\theta^* = 180$ deg must be selected, such that 180 deg must be added to the principal value for accurate evaluations. When these particular values of k, a_0, μ, and $\theta^* = 220$ deg are used, Eqs. (7.23) and (7.25) yield $a = 6999.6127813$ km and $a = 6999.612845$ km, respectively, whereas Eqs.

(7.24) and (7.28) yield $e = 2.58401747 \times 10^{-4}$ and $e = 2.58419725 \times 10^{-4}$, respectively.

The analytic integration of $\mathrm{d}a/\mathrm{d}\theta^*$ and $\mathrm{d}e/\mathrm{d}\theta^*$ in the elliptical case requires that a, e, and n be held constant on the right-hand sides of the corresponding differential equations, namely, Eqs. (7.6) and (7.7), which are rewritten for convenience as

$$\frac{\mathrm{d}a}{\mathrm{d}\theta^*} = \frac{2es_{\theta^*}(1-e^2)}{n^2(1+ec_{\theta^*})^2}kc_\alpha + \frac{2(1-e^2)}{n^2(1+ec_{\theta^*})}ks_\alpha \tag{7.29}$$

$$\frac{\mathrm{d}e}{\mathrm{d}\theta^*} = \frac{(1-e^2)^2 s_{\theta^*}}{n^2a(1+ec_{\theta^*})^2}kc_\alpha + \frac{(1-e^2)^2}{n^2ae}\left[\frac{1}{1+ec_{\theta^*}} - \frac{(1-e^2)}{(1+ec^3_{\theta^*})}\right]ks_\alpha \tag{7.30}$$

For low-thrust acceleration, the assumption of holding a, e, and n constant is an excellent one with negligible effect on the final result when compared to the numerically integrated solution of the fully coupled preceding system over one cycle of thrust or one revolution. Let $c_\alpha = -s_{\theta^*}$ and $s_\alpha = -c_{\theta^*}$ for the Spitzer [6, 7] scheme that reduces the orbital eccentricity after one cycle of thrust; then Eqs. (7.29) and (7.30) lead to

$$\int_{a_0}^{a} \mathrm{d}a = k_1\int_0^{\theta^*}\frac{s^2_{\theta^*}}{(1+ec_{\theta^*})^2}\mathrm{d}\theta^* + k_2\int_0^{\theta^*}\frac{c_{\theta^*}}{(1+ec_{\theta^*})}\mathrm{d}\theta^* \tag{7.31}$$

$$\int_{e_0}^{e} \mathrm{d}e = k_3\int_0^{\theta^*}\frac{s^2_{\theta^*}}{(1+ec_{\theta^*})^2}\mathrm{d}\theta^* + k_4\int_0^{\theta^*}\frac{c_{\theta^*}}{(1+ec_{\theta^*})}\mathrm{d}\theta^*$$
$$+ k_5\int_0^{\theta^*}\frac{c_{\theta^*}}{(1+ec_{\theta^*})^3}\mathrm{d}\theta^* \tag{7.32}$$

where $k_1 = -2e(1-e^2)k/n^2$, $k_2 = k_1/e$, $k_3 = -(1-e^2)^2k/(n^2a)$, $k_4 = k_3/e$, and $k_5 = -k_4(1-e^2) = -k_3(1-e^2)/e$. The values of a and e are the initial values that correspond to $\theta^* = 0$. The integrals involving $s^2_{\theta^*}$ are broken into two integrals after substituting $s^2_{\theta^*} = 1 - c^2_{\theta^*}$ such that the following integrals are needed:

$$\int\frac{\mathrm{d}\theta^*}{(1+ec_{\theta^*})^2},\quad \int\frac{c^2_{\theta^*}}{(1+ec_{\theta^*})^2}\mathrm{d}\theta^*,\quad \int\frac{c_{\theta^*}}{(1+ec_{\theta^*})}\mathrm{d}\theta^*,\quad \text{and} \quad \int\frac{c_{\theta^*}}{(1+ec_{\theta^*})^3}\mathrm{d}\theta^*$$

The first integral is solved through [13]

$$\int \frac{d\theta^*}{(1+ec_{\theta^*})^2} = -\frac{es_{\theta^*}}{(1-e^2)(1+ec_{\theta^*})} + \frac{1}{(1-e^2)}\int \frac{d\theta^*}{(1+ec_{\theta^*})}$$
$$= \frac{-es_{\theta^*}}{(1-e^2)(1+ec_{\theta^*})} + \frac{2}{(1-e^2)^{3/2}}\tan^{-1}\left[\sqrt{\frac{1-e}{(1+e)}}\tan\frac{\theta^*}{2}\right] \tag{7.33}$$

Also

$$\int \frac{c_{\theta^*}}{(1+ec_{\theta^*})}d\theta^* = \frac{\theta^*}{e} - \frac{1}{e}\int \frac{d\theta^*}{(1+ec_{\theta^*})}$$
$$= \frac{\theta^*}{e} - \frac{2}{e(1-e^2)^{1/2}}\tan^{-1}\left[\sqrt{\frac{1-e}{(1+e)}}\tan\frac{\theta^*}{2}\right] \tag{7.34}$$

Note that

$$\tan^{-1}\left[\sqrt{\frac{1-e}{1+e}}\tan\frac{\theta^*}{2}\right] = \frac{E}{2}$$

but that Wintner [14] provided the following more convenient expression relating θ^* and E

$$\tan\left(\frac{\theta^*-E}{2}\right) = \frac{fs_E}{1-fc_E}$$

where

$$f = \frac{(1+e)^{1/2}-(1-e)^{1/2}}{(1+e)^{1/2}+(1-e)^{1/2}}$$

Wintner's [14] expression then leads to

$$\theta^* = E + 2\tan^{-1}\left[\frac{es_E}{1+(1-e^2)^{1/2}-ec_E}\right]$$

and conversely

$$E = \theta^* - 2\tan^{-1}\left[\frac{es_{\theta^*}}{1+(1-e^2)^{1/2}+ec_{\theta^*}}\right]$$

which directly involve the sines and cosines of the angles E and θ^*.

Next, from the general form [13]

$$\int \frac{A + Bc_x}{(a + bc_x)^n} \mathrm{d}x = \frac{1}{(n-1)(a^2 - b^2)} \left[\frac{(aB - bA)s_x}{(a + bc_x)^{n-1}} + \int \frac{(Aa - bB)(n-1) + (n-2)(aB - bA)c_x}{(a + bc_x)^{n-1}} \mathrm{d}x \right]$$

we have

$$\int \frac{c_{\theta^*}}{(1 + ec_{\theta^*})^3} \mathrm{d}\theta^* = \frac{s_{\theta^*}}{2(1 - e^2)(1 + ec_{\theta^*})^2} - \frac{e}{(1 - e^2)} \int \frac{\mathrm{d}\theta^*}{(1 + ec_{\theta^*})^2} + \frac{1}{2(1 - e^2)} \int \frac{c_{\theta^*}}{(1 + ec_{\theta^*})^2} \mathrm{d}\theta^* \tag{7.35}$$

We can use this general form again for the second integral on the right-hand side of Eq. (7.35) to obtain

$$\int \frac{c_{\theta^*}}{(1 + ec_{\theta^*})^2} \mathrm{d}\theta^* = \frac{s_{\theta^*}}{(1 - e^2)(1 + ec_{\theta^*})} - \frac{2e}{(1 - e^2)^{3/2}} \tan^{-1}\left(\sqrt{\frac{1 - e}{1 + e}} \tan \frac{\theta^*}{2} \right) \tag{7.36}$$

In addition, for the remaining integral in Eq. (7.35), the general form [13]

$$\int \frac{\mathrm{d}x}{(a + bc_x)^n} = -\frac{1}{(n-1)(a^2 - b^2)} \left[\frac{bs_x}{(a + bc_x)^{n-1}} - \int \frac{(n-1)a - (n-2)bc_x}{(a + bc_x)^{n-1}} \mathrm{d}x \right]$$

provides

$$\int \frac{\mathrm{d}\theta^*}{(1 + ec_{\theta^*})^2} = -\frac{es_{\theta^*}}{(1 - e^2)(1 + ec_{\theta^*})} + \frac{2}{(1 - e^2)^{3/2}} \tan^{-1}\left(\sqrt{\frac{1 - e}{1 + e}} \tan \frac{\theta^*}{2} \right) \tag{7.37}$$

Finally, after substitution for the pertinent integrals, Eq. (7.35) can be written as follows:

$$\int \frac{c_{\theta^*}}{(1 + ec_{\theta^*})^3} \mathrm{d}\theta^* = \frac{s_{\theta^*}\left[(1 + ec_{\theta^*})(1 + 2e^2) + (1 - e^2)\right]}{2(1 - e^2)^2(1 + ec_{\theta^*})^2} - \frac{3e}{(1 - e^2)^{5/2}} \tan^{-1}\left(\sqrt{\frac{1 - e}{1 + e}} \tan \frac{\theta^*}{2} \right) \tag{7.38}$$

The final integral given next is obtained by means of symbolic integration using MALTAB:

$$\int \frac{c_{\theta^*}^2}{(1+ec_{\theta^*})^2} d\theta^* = \frac{2(2e^2-1)}{e^2(1-e^2)^{3/2}} \tan^{-1}\left(\sqrt{\frac{1-e}{1+e}} \tan\frac{\theta^*}{2}\right) - \frac{s_{\theta^*}}{e(1-e^2)(1+ec_{\theta^*})} + \frac{\theta^*}{e^2} \tag{7.39}$$

Equation (7.31) can now readily be integrated to yield, after regrouping of terms, the following analytic expression for $\Delta a = a - a_0$:

$$\Delta a = \left.\frac{k_1 s_{\theta^*}}{e(1+ec_{\theta^*})}\right|_0^{\theta^*}, \quad \text{where} \quad k_1 = -2e(1-e^2)k/n^2 \tag{7.40}$$

As e approaches zero, because k_1 is linear in e, Eq. (7.40) reduces to

$$\Delta a = -\left.\frac{2k}{n^2} s_{\theta^*}\right|_0^{\theta^*}$$

which is identical to the simple expression in Eq. (7.23) except for the minus sign. The difference arises because, here, we used $s_\alpha = -c_{\theta^*}$ and $c_\alpha = -s_{\theta^*}$ to reduce eccentricity, in contrast to Eq. (7.23), which was obtained using $s_\alpha = c_{\theta^*}$ and $c_\alpha = s_{\theta^*}$, firing in the reverse direction to increase eccentricity.

The integration of Eq. (7.32) can also be carried out, to yield, after regrouping of terms, the following analytic expression for $\Delta e = e - e_0$, as a function of θ^*, valid in general elliptical orbit:

$$\Delta e = \left.\frac{(1+ec_{\theta^*})(1-4e^2)-(1-e^2)}{2e(1-e^2)(1+ec_{\theta^*})^2} k_3 s_{\theta^*}\right|_0^{\theta^*} + \left.\frac{3k_3}{(1-e^2)^{3/2}} \tan^{-1}\left(\sqrt{\frac{1-e}{1+e}}\right) \tan\frac{\theta^*}{2}\right|_0^{\theta^*}, \quad \text{where} \quad k_3 = -\frac{(1-e^2)^2 k}{n^2 a} \tag{7.41}$$

For small e, the leading term in the numerator of the first fraction of this expression is ec_{θ^*} such that, as e approaches zero, Eq. (7.41) collapses to

$$\Delta e = -\left.\frac{k}{n^2 a} \frac{s_{2\theta^*}}{4}\right|_0^{\theta^*} - \left.\frac{3k}{2n^2 a} \theta^*\right|_0^{\theta^*}$$

which is identical to Eq. (7.24) except, once again, for the change in sign due to the reversed firing orientation as compared to the direction assumed in Eq. (7.24).

Thus, the expressions for a and e in Eqs. (7.40) and (7.41), respectively, that are valid in the general elliptical case reduce to the simpler expressions in Eqs. (7.23) and (7.24) as e tends to zero, providing a good check. The integration limits can, of course, be any two true anomaly values and not necessarily 0 and the general θ^* values. The values of a and e appearing on the right-hand sides of all of the equations derived in this chapter correspond to the initial values a_0 and e_0. The subscript zero was left out for ease of writing. If we let $a_0 = 40{,}000$ km and $e_0 = 0.7$ and use the same values of $\mu = 398{,}601.3\ \mathrm{km^3/s^2}$ and $k = 3.5 \times 10^{-7}\ \mathrm{km/s^2}$ as before, the numerical integrations of Eqs. (7.29) and (7.30) with $a = a_0$ and $e = e_0$ held constant on the right-hand sides during integration between the bounds $\theta_0^* = 30.1286983$ deg and $\theta_f^* = 329.871302$ deg, which correspond to the case of $\beta = 180$ deg in Table 7.1, yield $a_f = 40{,}035.8429$ km and $e_f = 0.6908129217$ for the Spitzer [6, 7] Δe strategy.

Equations (7.40) and (7.41) yield $\Delta a = 35.842917$ km and $\Delta e = -9.18707831 \times 10^{-3}$, respectively, or $a_f = 40{,}035.842917$ km and $e_f = 0.690812922$, which are identical to the numerically integrated values, thereby validating the analytic expressions derived here. Finally, these analytic integrations are related to the exact analytic elliptical integrals in Cartesian coordinates [15] for the case of thrust oriented in a constant inertial direction.

7.6 CONCLUSION

The maximization of the change in the eccentricity of a general elliptical orbit in the presence of a shadow arc, where no thrust is applied, subject to the constraint of zero change in the semimajor axis has been carried out by direct application of the theory of maxima. The method has also been applied to the dual problem of maximizing the orbital energy while keeping eccentricity unchanged. A typical high-energy, high-eccentricity orbit was used as an example, and the worst case of shadowing geometry, which occurs when the Sun–Earth line is contained in the spacecraft orbit plane, was selected to compare the optimal changes in the eccentricity obtained for various Sun angles to the suboptimal results obtained with the simpler scheme using a fixed firing orientation. Furthermore, analytic expressions for the changes in the semimajor axis and eccentricity resulting from the latter firing mode were also derived for the general elliptical case.

REFERENCES

[1] Edelbaum, T. N., "Propulsion requirements for controllable satellites," *ARS Journal*, Vol. 31, No. 8, Aug. 1961, pp. 1079–1089.

[2] Cass, J. R., "Discontinuous low thrust orbit transfer," *M.S. thesis, Rept. AFIT/GA/AA/83D-1, School of Engineering, U.S. Air Force Institute of Technology*, Wright-Patterson AFB, OH, Dec. 1983.

[3] McCann, J. M., "Optimal launch time for a discontinuous low thrust orbit transfer," *M.S. Thesis, Rept. AFIT/GA/AA/88D-7, School of Engineering, U.S. Air Force Institute of Technology*, Wright-Patterson AFB, OH, Dec. 1988.

[4] Kechichian, J. A., "Low-thrust eccentricity-constrained orbit raising," *Journal of Spacecraft and Rockets*, Vol. 35, No. 3, May 1998, pp. 327–335.

[5] Burt, E. G. C., "On space manoeuvers with continuous thrust," *Planetary and Space Science*, Vol. 15, No. 1, Jan. 1967, pp. 103–122.

[6] Spitzer, A., "Near optimal transfer orbit trajectory using electric propulsion," *AAS Paper 95-215, AAS/AIAA Spaceflight Mechanics Meeting*, Albuquerque, NM, Feb. 1995.

[7] Spitzer, A., "Novel orbit raising strategy makes low thrust commercially viable," *24th International Electric Propulsion Conference*, IEPC Paper 95-212, Moscow, Russia, Sept. 1995.

[8] Kéchichian, J. A., "Optimum thrust pitch profiles for certain orbit control problems," *Journal of Spacecraft and Rockets*, Vol. 40, No. 2, March–April 2003, pp. 253–259.

[9] Kechichian, J. A., "Constrained circularization in elliptic orbit using low-thrust with shadowing effect," *AIAA/AAS Astrodynamics Specialist Conference, AIAA Paper 2002-4894*, Monterey, CA, Aug. 2002.

[10] Kechichian, J. A., "Constrained circularization in elliptic orbit using low thrust with shadowing effect," *Journal of Guidance, Control, and Dynamics*, Vol. 26, No. 6, Nov.–Dec. 2003, 949–955.

[11] Battin, R. H., *An Introduction to the Mathematics and Methods of Astrodynamics*, AIAA Education Series, AIAA, New York, 1987, pp. 128, 488.

[12] Press, W. H., Flannery, B. P., Teukolsky, S. A., and Vetterling, W. T., *Numerical Recipes (Fortran)*, Cambridge University Press, Cambridge, U.K., 1989.

[13] Gradshteyn, I. S., and Ryzhik, I. M., *Table of Integrals, Series, and Products*, Academic Press, New York, 1980.

[14] Wintner, A., *The Analytical Foundation of Celestial Mechanics*, Princeton University Press, Princeton, NJ, 1947, pp. 201–210.

[15] Grodzovskii, G. L., Ivanov, Y. N., and Tokarev, V. V., "Mechanics of low-thrust spaceflight," translated from Russian by A. Baruch, edited by Y. M. Timant, *NASA TTF-507, TT 68-50301, Israel Program for Scientific Translation, Jerusalem, Israel*, 1969, pp. 431–432.

CHAPTER 8

Efficient Analytic Computation of Fractional Reentering Debris from an Idealized Isotropic Explosion in a General Elliptical Orbit

NOMENCLATURE

a	=	orbit semimajor axis, m
c_θ, s_θ	=	cos θ, sin θ
e	=	eccentricity
h	=	angular momentum of the pre-explosion orbit, $[\mu a/(1-e^2)]^{1/2}$
n_f	=	fraction of fragments whose orbits pass below r_p
p	=	orbit parameter, $a\,(1-e^2)$
p_f	=	fraction of fragments whose orbits remain above r_p
$\boldsymbol{r}$	=	position vector of the spacecraft with respect to the center of Earth
r	=	radial distance
r_e	=	radius of Earth at the equator, 6,378.14 km
r_p	=	user-defined perigee radius required for the fragment postexplosion orbit
V_c	=	orbital velocity of a spacecraft in a general elliptical orbit at time zero
V_s	=	explosion velocity associated with the direction θ_s as measured relative to the V_c orientation
$\dot{X}_0$, $\dot{Y}_0$, $\dot{Z}_0$	=	inertial velocity components of the pre-explosion fragment
γ_0	=	flight path angle of V_c
θ^*	=	true anomaly
θ_1^*, θ_2^*	=	locations on the orbit that intersect the surface of Earth
θ_s	=	angle between the radial direction and the thrust vector
μ	=	gravitational constant of Earth, 398601.3 km^3/s^2

8.1 INTRODUCTION

The efficient computation of the fraction of debris from an isotropic explosion in a general elliptical orbit that would fly in orbits whose perigees are below a certain given altitude is presented in this chapter. Given an explosion velocity, a spacecraft idealized as a sphere breaks up into a myriad of small fragments that either remain in Earth orbit or reenter the atmosphere.

The two angles ϕ and θ_s uniquely define the orientation of the explosion velocity with respect to the pre-explosion velocity in a suitable rotating orbital reference frame. The locus of all points on the surface of the idealized sphere that separate the fragments that reenter the atmosphere from the fragments that remain in orbit was previously obtained as an expression for ϕ in terms of θ_s. Because the counts of debris fragments on either side of the locus curve on the idealized sphere are proportional to the respective areas enclosed by the curve, the area calculations are made possible if θ_s is instead cast in terms of ϕ because, otherwise, the angle ϕ can be multivalued in θ_s, complicating the area evaluations. By contrast, an analytic expression in the form of a quartic in the cosine of θ_s is obtained and solved analytically for those values of ϕ called by the quadrature routine. Two real solution branches for cos θ_s are thus generated, with one branch being valid for a certain range in ϕ and the other branch providing the required solution for the complement of that range.

The percentages of the debris fragments emanating from either side of the separating curve on the idealized sphere are thus evaluated rapidly through a few calls to the quartic solver routine. This analytic method provides percentage counts that are identical in accuracy to those obtained by a purely numerical method, and it can be readily extended to analyze nonisotropic explosions as well.

An algorithm was developed by Sorge [1] to determine the fraction of debris from an isotropic explosion whose perigees are below a given altitude. The exploding object in space is assumed to be in a general elliptical orbit and to have a spherical shape. Upon its explosion, the object produces fragments that fly radially out from its center with a uniform relative velocity. The relative velocity vector of each exploding fragment is uniquely described by two angles referred to a suitable orbital rotating frame that is initially attached to the object in space. As the sphere is effectively divided into two parts corresponding to one set of fragments with perigees below the desired value and the remaining set staying higher than that altitude, an analytic expression for one of the angles in terms of the other was found by Sorge [1] that determined the locus of points on the sphere dividing the two regions. As the fraction of fragments on either side of that curve is proportional to the corresponding area on the sphere, it is then sufficient to evaluate the two spherical areas to compute the two fractions of interest. With the original formulation of the locus curve, the dependent angle was found to be multivalued in the independent angle, complicating the area quadrature calculations. In this chapter, the dependent–independent pair of angles is switched such that the dependent angle is single-valued in the independent angle. This new formulation leads to the resolution of a quartic for the new dependent variable, with the coefficients of the quartic being a function as the new independent variable angle, as well as other pertinent parameters related to the object's orbit, namely, its location on that orbit and the explosion velocity. The following two sections provide the details of both the original analysis and the postexplosion orbit calculations leading to the purely numerical debris percentage evaluations, as well as the new analysis with several examples showing the accuracy of the analytic methods used in comparison to purely numerical methods, which are more computation-intensive.

The analysis reported in this chapter is based on a previously published journal article [2] and conference paper [3].

8.2 GENERAL ANALYSIS

Let V_c represent the orbital velocity of a spacecraft in a general elliptical orbit at time zero, and let V_s represent the explosion velocity associated with the direction θ_s as measured relative to the V_c orientation. It is assumed here that the magnitude of V_s is constant and independent of the angle θ_s such that the explosion is isotropic, resulting in a locally spherical dispersion of the fragmented pieces at time zero, the explosion time. Let $(\hat{\boldsymbol{r}}, \hat{\boldsymbol{\theta}}, \hat{\boldsymbol{h}})$ represent the Euler–Hill rotating frame attached to the spacecraft at the explosion time, but flying otherwise on a circular orbit with radius r, where $\boldsymbol{r}$ is the position vector of the spacecraft with respect to the center of Earth at that same time. Let also $(\hat{\boldsymbol{V}}_r, \hat{\boldsymbol{V}}_\theta, \hat{\boldsymbol{V}}_h)$ be an orbital frame attached to the spacecraft at the explosion time, with $\hat{\boldsymbol{V}}_r$ and $\hat{\boldsymbol{V}}_\theta$ in the pre-explosion orbit plane such that $\hat{\boldsymbol{V}}_\theta$ is along $\boldsymbol{V}_c$ and $\hat{\boldsymbol{V}}_h$ is along $\hat{\boldsymbol{h}}$, as shown in Fig. 8.1. The unit vectors $\hat{\boldsymbol{V}}_r$, $\hat{\boldsymbol{V}}_\theta$, and $\hat{\boldsymbol{V}}_h$ are related to the unit vectors $\hat{\boldsymbol{r}}$, $\hat{\boldsymbol{\theta}}$, and $\hat{\boldsymbol{h}}$ through the rotation γ_0, which represents the flight path angle of V_c. This orthonormal transformation is given by

$$\begin{pmatrix} \hat{\boldsymbol{V}}_r \\ \hat{\boldsymbol{V}}_\theta \\ \hat{\boldsymbol{V}}_h \end{pmatrix} = \begin{pmatrix} c\gamma_0 & -s\gamma_0 & 0 \\ s\gamma_0 & c\gamma_0 & 0 \\ 0 & 0 & 1 \end{pmatrix} \begin{pmatrix} \hat{\boldsymbol{r}} \\ \hat{\boldsymbol{\theta}} \\ \hat{\boldsymbol{h}} \end{pmatrix} \tag{8.1}$$

The flight path angle γ of the postexplosion velocity $\boldsymbol{V} = \boldsymbol{V}_c + \boldsymbol{V}_s$ remains to be determined. The projection of V_s onto the $\hat{\boldsymbol{V}}_\theta$ direction is given by $V_s \cos\theta_s$ [i.e., $(\boldsymbol{V}_s \cdot \hat{\boldsymbol{V}}_\theta)\hat{\boldsymbol{V}}_\theta$], whereas $V_s s_\theta$ [i.e., $\boldsymbol{V}_s - (\boldsymbol{V}_s \cdot \hat{\boldsymbol{V}}_\theta)\hat{\boldsymbol{V}}_\theta$] is the component of V_s along the plane normal to $\hat{\boldsymbol{V}}_\theta$ such that this projected vector makes an angle ϕ with respect to the $\hat{\boldsymbol{V}}_r$ direction.

The orientation of V_s in three-dimensional space is then uniquely defined by the two angles θ_s and ϕ, with θ_s sweeping the range $(0, \pi)$ and ϕ sweeping the full $(0, 2\pi)$ range such that the tip of the vector V_s sweeps the entire surface of a sphere whose radius is equal to V_s. The postexplosion velocity V of a fragment is given by

$$V^2 = V_c^2 + V_s^2 + 2V_c V_s c_{\theta_s} \tag{8.2}$$

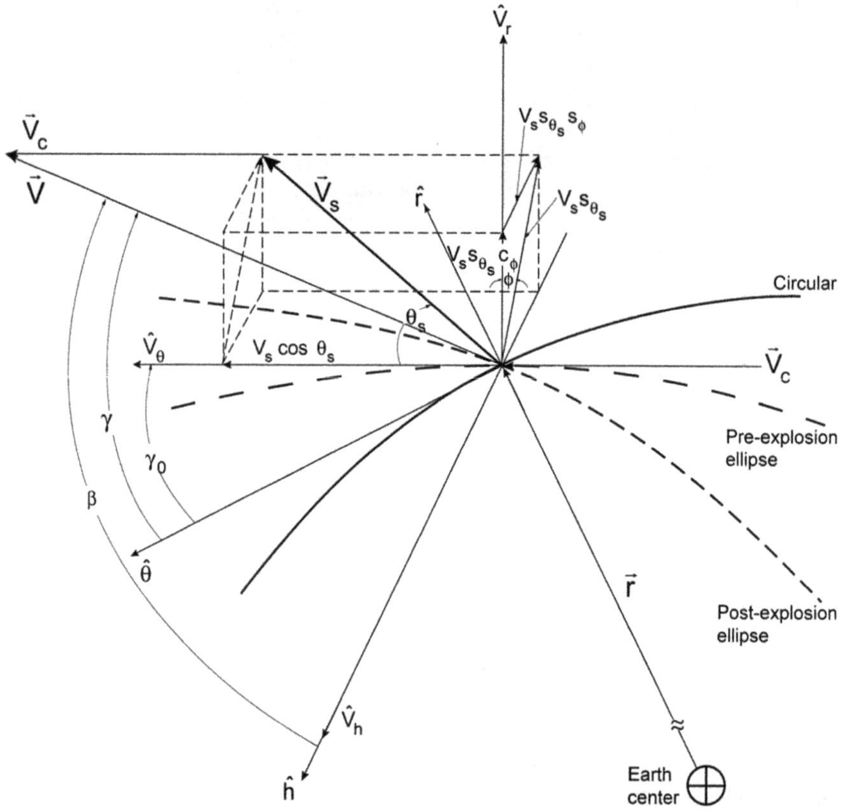

Fig. 8.1 Geometry of the relative coordinate systems.

where orbit parameters a and e are related through the orbit and energy equations and μ is the gravitational constant of Earth

$$r_p = a(1 - e) \tag{8.3}$$

$$\frac{V^2}{2} - \frac{\mu}{r} = -\frac{r}{2a} \tag{8.4}$$

Eliminating the semimajor axis from the previous two equations results in an expression relating the eccentricity to the postexplosion velocity V, the radial distance r, and the perigee radius r_p as well as the gravitational constant of Earth

$$e = 1 + \frac{2r_p\left(\dfrac{V^2}{2} - \dfrac{\mu}{r}\right)}{\mu} \tag{8.5}$$

However, e is also given directly in terms of V, r, and γ by

$$e^2 = \left(\frac{rV^2}{\mu} - 1\right)^2 c_\gamma^2 + s_\gamma^2 \tag{8.6}$$

Substitution of this expression for e into Eq. (8.5) results in

$$c_\gamma^2 = \left(\frac{r_p}{r}\right)^2 \left[\frac{V^2 + 2\mu\left(\frac{1}{r_p} - \frac{1}{r}\right)}{V^2}\right] \tag{8.7}$$

or

$$\gamma = \cos^{-1}\left(\pm\left\{\left(\frac{r_p}{r}\right)^2 \left[\frac{V^2 + 2\mu\left(\frac{1}{r_p} - \frac{1}{r}\right)}{V^2}\right]\right\}^{1/2}\right) \tag{8.8}$$

This expression relates γ to the known quantities V, r, and r_p, where r_p is the perigee radius required for the fragment postexplosion orbit and is therefore a user-defined quantity. γ is also related to θ_s because V is itself a function of θ_s, but it is independent of ϕ. If $\boldsymbol{V}_s$ is expressed in terms of its three components along $\hat{\boldsymbol{V}}_r$, $\hat{\boldsymbol{V}}_\theta$, and $\hat{\boldsymbol{V}}_h$, then it can be written as

$$\boldsymbol{V}_s = V_s s_{\theta_s} c_\phi \hat{\boldsymbol{V}}_r + V_s c_{\theta_s} \hat{\boldsymbol{V}}_\theta + V_s s_{\theta_s} s_\phi \hat{\boldsymbol{V}}_h \tag{8.9}$$

and using Eq. (8.1), as

$$\boldsymbol{V}_s = \begin{pmatrix} V_s s_{\theta_s} c_\phi c_{\gamma_0} + V_s c_{\theta_s} s_{\gamma_0} \\ -V_s s_{\theta_s} c_\phi s_{\gamma_0} + V_s c_{\theta_s} c_{\gamma_0} \\ -V_s s_{\theta_s} s_\phi \end{pmatrix}^T \begin{pmatrix} \hat{\boldsymbol{r}} \\ \hat{\boldsymbol{\theta}} \\ \hat{\boldsymbol{h}} \end{pmatrix} \tag{8.10}$$

Finally, with

$$\boldsymbol{V}_c = V_c s_{\gamma_0} \hat{\boldsymbol{r}} + V_c c_{\gamma_0} \hat{\boldsymbol{\theta}} \tag{8.11}$$

$\boldsymbol{V} = \boldsymbol{V}_c + \boldsymbol{V}_s$ can be written as $\boldsymbol{V} = V_r \hat{\boldsymbol{r}} + V_\theta \hat{\boldsymbol{\theta}} + V_h \hat{\boldsymbol{h}}$ such that

$$\boldsymbol{V} = \begin{pmatrix} V_c s_{\gamma_0} + V_s(s_{\theta_s} c_\phi c_{\gamma_0} + c_{\theta_s} s_{\gamma_0}) \\ V_c c_{\gamma_0} + V_s(c_{\theta_s} c_{\gamma_0} - s_{\theta_s} c_\phi s_{\gamma_0}) \\ -V_s s_{\theta_s} s_\phi \end{pmatrix}^T \begin{pmatrix} \hat{\boldsymbol{r}} \\ \hat{\boldsymbol{\theta}} \\ \hat{\boldsymbol{h}} \end{pmatrix} \tag{8.12}$$

Because the flight path angle γ can be obtained from $\boldsymbol{V} \cdot \hat{\boldsymbol{r}} = V s_\gamma$, it is now possible to solve for s_γ, yielding

$$s_\gamma = \frac{V_c s_{\gamma_0} + V_s(s_{\theta_s} c_\phi c_{\gamma_0} + c_{\theta_s} s_{\gamma_0})}{V} \tag{8.13}$$

In turn, Eq. (8.13) allows c_ϕ to be expressed in terms of s_γ:

$$c_\phi = \frac{Vs_\gamma - V_c s_{\gamma_0} - V_s c_{\theta_s} s_{\gamma_0}}{V_s s_{\theta_s} c_{\gamma_0}} \tag{8.14}$$

If γ and V from Eqs. (8.8) and (8.2), respectively, are used in Eq. (8.14), ϕ will be given in terms of θ_s and the input quantities r_p, V_c, V_s, r, μ, and γ_0. As θ_s is swept from 0 to π, the corresponding angle ϕ that will result in a given perigee radius can be computed, and a continuous curve can be traced on the surface of the sphere in the form of $\phi = f(\theta_s)$ that effectively acts as the boundary between the region on the sphere whose fragments end up in orbits with perigee radii less than r_p and the region whose fragments end up in orbits that have perigee radii greater than r_p.

Edgecombe et al. [4] provided a derivation of the locus of vectors in velocity space that will result in orbits with a given perigee radius r_p. The locus is a hyperboloid of revolution with its axis of rotation parallel to the radial velocity vector. By extending these results, Sorge [1] made a comparison to Eq. (8.14). The locus curve of Eq. (8.14) is the intersection of the hyperboloid of revolution from Edgecombe et al. [4] and the $\boldsymbol{V}_s$ sphere. To verify these results using the conclusions of Edgecombe et al. [4], the hyperboloid in the work by Sorge [1] was shifted into a reference frame centered at the $\boldsymbol{V}_s$ sphere. The equation of the hyperboloid was then converted into spherical coordinates using the angles θ_s and ϕ. The radial coordinate in the spherical system was set to V_s, resulting in an equation for the intersection between the hyperboloid and the $\boldsymbol{V}_s$ sphere. By rearranging constants, Sorge [1] showed that this equation is identical to Eq. (8.14).

It can be verified that $V_r^2 + V_\theta^2 + V_h^2$ in Eq. (8.12) can be expressed as $V^2 + V_c^2 + V_s^2 + 2V_cV_s \cos\theta_s$. Moreover, even though it is correct to write $V_r = Vs_\gamma$, from which we obtained the expression for s_γ in Eq. (8.13), it would be wrong to write $V_\theta = Vc_\gamma$ to also obtain an expression for c_γ, because $\boldsymbol{V}$ is no longer in the $\hat{\boldsymbol{r}}-\hat{\boldsymbol{\theta}}$ plane but is instead in the rotated $\hat{\boldsymbol{r}}_n-\hat{\boldsymbol{\theta}}_n$ plane, which corresponds to the postexplosion orbit plane. Therefore, $\boldsymbol{V}\cdot\hat{\boldsymbol{\theta}} = Vc_\gamma$ does not hold. Instead, we have $\boldsymbol{V}\cdot\hat{\boldsymbol{\theta}}_n = Vc_\gamma$, but $\hat{\boldsymbol{\theta}}_n$ must first be evaluated before c_γ can be extracted. It is also true that $\boldsymbol{V}\cdot\hat{\boldsymbol{r}} = V_r = Vs_\gamma$ holds only because $\boldsymbol{V}\cdot\hat{\boldsymbol{r}} = \boldsymbol{V}\cdot\hat{\boldsymbol{r}}_n$, which, in turn, is true because $\hat{\boldsymbol{r}} = \hat{\boldsymbol{r}}_n$, and that is why s_γ can be extracted from Eq. (8.12).

Now, given a pre-explosion orbit with the parameters a_0, e_0, i_0, Ω_0, ω_0, and θ_0^*, where θ^* represents the true anomaly $r_e = 6{,}378.14$ km represents the radius of Earth at the equator, one can obtain the two angles θ_1^* and θ_2^* for the two locations on that orbit that intersect the surface of Earth for the case where the orbit is of the intersecting type, from the equation

$$r_e = a_0(1 - e_0^2)/(1 + e_0 c_{\theta^*})$$

or

$$c_{\theta^*} = \frac{1}{e_0}\left[\frac{a_0(1-e_0^2)}{r_e} - 1\right] \tag{8.15}$$

which gives θ_1^* between 0 and π, from which $\theta_2^* = 2\pi - \theta_1^*$ is obtained by symmetry. The range $\theta_1^* \leq \theta^* \leq \theta_2^*$ corresponds to the portion of the orbit in vacuum. One can sweep this θ^* range at equal intervals of $\Delta\theta^*$ to determine the locus curve at each point along the orbit by evaluating the three quantities

$$h_0 = (\mu p_0)^{1/2} = [\mu a_0(1-e_0^2)]^{1/2}$$

$$V_c = \left[\frac{\mu}{p_0}(1 + 2e_0 c_{\theta^*} + e_0^2)\right]^{1/2}$$

$$r = a_0(1-e_0^2)/(1+e_0 c_{\theta^*})$$

for each θ^*. The flight path angle γ_0 is obtained each time from

$$s_{\gamma_0} = \frac{\mu}{h_0 V_c}(e_0 s_{\theta^*}) \quad \text{and} \quad c_{\gamma_0} = \frac{\mu}{h_0 V_c}(1 + e_0 c_{\theta^*})$$

It is also true that

$$V_r = \frac{\mu}{h_0} e_0 s_{\theta^*} \quad \text{and} \quad V_\theta = \frac{\mu}{h_0}(1 + e_0 c_{\theta^*})$$

with V_r and V_θ along the $\hat{\boldsymbol{r}}$ and $\hat{\boldsymbol{\theta}}$ directions of the circular rotating reference frame, which are, of course, different from the $\hat{\boldsymbol{V}}_r$ and $\hat{\boldsymbol{V}}_\theta$ unit vectors, as seen in Fig. 8.1.

The locus can be generated for each location on the orbit (i.e., each θ^* value) by sweeping the angle θ_s between 0 and 180 deg. For each θ_s, V is first computed from Eq. (8.2) such that the values of

$$a^+ = \frac{\mu}{\left(\dfrac{2\mu}{r} - V^2\right)} \quad \text{and} \quad e^+ = 1 + \frac{2r_p\left(\dfrac{V^2}{2} - \dfrac{\mu}{r}\right)}{\mu}$$

for the postexplosion orbit are evaluated.

The flight path angle γ is evaluated next from Eq. (8.8) using first the plus sign. γ is returned in the range from 0 to 90 deg, after it has been ensured that c_γ lies in the interval (0, 1), because, otherwise, θ_s would be rejected and the next θ_s value in the sweep would be selected. From the expression $h = rVc_\gamma = [\mu a(1-e^2)]^{1/2}$

$$e^2 = 1 - \frac{r^2 V^2 c_\gamma^2}{\mu a}$$

is evaluated with $a = a^+$, to get e^+ and $r_p^+ = a^+(1 - e^+)$, and the condition $r_p^+ < r_p$ is checked to once again reject that particular value of θ_s. Next, the postexplosion true anomaly θ_n^* is computed, after

$$c_{\theta_n^*} = \frac{\dfrac{a^+\left(1 - e^{+2}\right)}{r} - 1}{e^+}$$

and confirmed to lie in the $(-1, 1)$ range because, otherwise, the value of θ_s would once again be rejected.

The first solution $\theta_{n_1}^*$ is in the interval $(0, \pi)$ as $\gamma_1 = \gamma$, and these two variables are related through the well-known relations [5]

$$s_\gamma = \frac{e s_{\theta^*}}{\left(1 + 2e c_{\theta^*} + e^2\right)^{1/2}} \tag{8.16}$$

$$c_\gamma = \frac{1 + e c_{\theta^*}}{\left(1 + 2e c_{\theta^*} + e^2\right)^{1/2}} \tag{8.17}$$

γ_1 is now used in Eq. (8.14) for c_ϕ, which is first checked to ensure that c_ϕ lies in the range $-1 \le c_\phi \le 1$ because, otherwise, θ_s would once again be rejected. ϕ is now returned in the interval $(0, \pi)$ as the first solution ϕ_{11} corresponding to $0 \le \gamma_1 \le \pi/2$, with the second solution $\phi_{12} = -\phi_{11}$ also valid for the same γ_1 range of $(0, \pi/2)$. The next set of solutions for ϕ are obtained in the same manner by starting now with $\gamma_2 = -\gamma_1$, where $-\pi/2 < \gamma_2 < 0$ corresponds to the second solution $\theta_{n_2}^* = -\theta_{n_1}^*$ with $\pi \le \theta_{n_2}^* \le 2\pi$. This value of γ_2 results in the two ϕ solutions from Eq. (8.14), namely, $\phi_{21} = \phi$ and $\phi_{22} = -\phi_{21}$ with $0 \le \phi_{21} \le \pi$ and $\pi \le \phi_{22} \le 2\pi$ as before, because c_ϕ returns $\pm\phi$ as its two solutions each time, as is the case for c_γ, which returns γ_1 and γ_2, or $\pm\gamma$. It is then clear that up to four solutions for ϕ, namely, ϕ_{11}, ϕ_{12}, ϕ_{21}, and ϕ_{22}, are possible for a given value of θ_s, and it is necessary to reject any value of θ_s that would result in no solution, because depending on the flown pre-explosion orbit, the location of θ^* on that orbit, and the values of V_s and r_p, ϕ will have either no solution; two solutions ϕ_{11} and ϕ_{12} or ϕ_{21} and ϕ_{22}; or even up to four solutions ϕ_{11}, ϕ_{12}, ϕ_{21}, and ϕ_{22}.

Now, to compute the postexplosion fragment orbits individually, it is necessary that, for each feasible solution, the inertial position and velocity components be computed, after which, the corresponding orbit elements can be evaluated through a Kepler routine. Thus, given a pre-explosion orbit $(a, e, i, \Omega, \omega, \theta^*)$, the position vector $\boldsymbol{r}$ components are obtained from

$$X_0 = r(c_\Omega c_\theta - s_\Omega s_\theta c_i) = rR_{11} \tag{8.18}$$

$$Y_0 = r(s_\Omega c_\theta + c_\Omega s_\theta c_i) = rR_{21} \tag{8.19}$$

$$Z_0 = r s_\theta s_i = rR_{31} \tag{8.20}$$

where $r = a(1-e^2)/(1+ec_{\theta^*})$, $\theta = \omega + \theta^*$, and the rotation matrix R involving the three Eulerian angles Ω, i, and θ is given by

$$R = \begin{pmatrix} c_\Omega c_\theta - s_\Omega s_\theta c_i & -c_\Omega s_\theta - s_\Omega c_i c_\theta & s_\Omega s_i \\ s_\Omega c_\theta + c_\Omega c_i s_\theta & -s_\Omega s_\theta + c_\Omega c_i c_\theta & -c_\Omega s_i \\ s_i s_\theta & s_i c_\theta & c_i \end{pmatrix} = \begin{pmatrix} R_{11} & R_{12} & R_{13} \\ R_{21} & R_{22} & R_{23} \\ R_{31} & R_{32} & R_{33} \end{pmatrix} \tag{8.21}$$

Thus, the inertial velocity components of the postexplosion fragment at position $\boldsymbol{r}$ with relative velocity $\boldsymbol{V}_s$ defined by the two angles θ_s and ϕ are given by

$$\dot{X} = \dot{X}_0 + R_{11}(\boldsymbol{V}_s \cdot \hat{\boldsymbol{r}}) + R_{12}(\boldsymbol{V}_s \cdot \hat{\boldsymbol{\theta}}) + R_{13}(\boldsymbol{V}_s \cdot \hat{\boldsymbol{h}}) \tag{8.22}$$

$$\dot{Y} = \dot{Y}_0 + R_{21}(\boldsymbol{V}_s \cdot \hat{\boldsymbol{r}}) + R_{22}(\boldsymbol{V}_s \cdot \hat{\boldsymbol{\theta}}) + R_{23}(\boldsymbol{V}_s \cdot \hat{\boldsymbol{h}}) \tag{8.23}$$

$$\dot{Z} = \dot{Z}_0 + R_{31}(\boldsymbol{V}_s \cdot \hat{\boldsymbol{r}}) + R_{32}(\boldsymbol{V}_s \cdot \hat{\boldsymbol{\theta}}) + R_{33}(\boldsymbol{V}_s \cdot \hat{\boldsymbol{h}}) \tag{8.24}$$

where $\dot{X}_0, \dot{Y}_0$, and $\dot{Z}_0$ are the inertial velocity components of the pre-explosion fragment given by

$$\dot{X}_0 = -\frac{\mu}{h}[c_\Omega(s_\theta + es_\omega) + s_\Omega(c_\theta + ec_\omega)c_i] \tag{8.25}$$

$$\dot{Y}_0 = -\frac{\mu}{h}[s_\Omega(s_\theta + es_\omega) - c_\Omega(c_\theta + ec_\omega)c_i] \tag{8.26}$$

$$\dot{Z}_0 = \frac{\mu}{h}[(c_\theta + ec_\omega)s_i] \tag{8.27}$$

and where

$$\boldsymbol{V}_s \cdot \hat{\boldsymbol{r}} = V_s s_{\theta_s} c_\phi c_{\gamma_0} + V_s c_{\theta_s} s_{\gamma_0} \tag{8.28}$$

$$\boldsymbol{V}_s \cdot \hat{\boldsymbol{\theta}} = -V_s s_{\theta_s} c_\phi s_{\gamma_0} + V_s c_{\theta_s} c_{\gamma_0} \tag{8.29}$$

$$\boldsymbol{V}_s \cdot \hat{\boldsymbol{h}} = -V_s s_{\theta_s} s_\phi \tag{8.30}$$

as can be seen from Eq. (8.12). Here, $h = [\mu a(1-e^2)]^{1/2}$ represents the angular momentum of the pre-explosion orbit, and ϕ is, of course, one of the possible four solutions associated with that particular value of θ_s, as described earlier. The angle γ can be verified to be the correct one by reproducing it using the equations

$$r = (X_0^2 + Y_0^2 + Z_0^2)^{1/2} \tag{8.31}$$

$$V = (\dot{X}^2 + \dot{Y}^2 + \dot{Z}^2)^{1/2} \tag{8.32}$$

$$h = [(Y_0\dot{Z} - Z_0\dot{Y})^2 + (X_0\dot{Z} - Z_0\dot{X})^2 + (X_0\dot{Y} - Y_0\dot{X})^2]^{1/2} \tag{8.33}$$

$$s_\gamma = (X_0\dot{X} + Y_0\dot{Y} + Z_0\dot{Z})/(rV) \tag{8.34}$$

$$c_{\gamma} = \frac{h}{rV} \tag{8.35}$$

Similarly, the components of the eccentricity vector along the inertial axes can be obtained from

$$e_X = \frac{1}{\mu}\left[\left(V^2 - \frac{\mu}{r}\right)X_0 - (\boldsymbol{r} \cdot \boldsymbol{V})\dot{X}\right] \tag{8.36}$$

$$e_Y = \frac{1}{\mu}\left[\left(V^2 - \frac{\mu}{r}\right)Y_0 - (\boldsymbol{r} \cdot \boldsymbol{V})\dot{Y}\right] \tag{8.37}$$

$$e_Z = \frac{1}{\mu}\left[\left(V^2 - \frac{\mu}{r}\right)Z_0 - (\boldsymbol{r} \cdot \boldsymbol{V})\dot{Z}\right] \tag{8.38}$$

and hence, $e = (e_X^2 + e_Y^2 + e_Z^2)^{1/2}$. These equations were derived directly from the vectorial expression of

$$\boldsymbol{e} = \frac{1}{\mu}\left[(\boldsymbol{V} \times \boldsymbol{h}) - \frac{\mu}{r}\boldsymbol{r}\right]$$

by observing that $\boldsymbol{h} = \boldsymbol{r} \times \boldsymbol{V}$ and $\boldsymbol{V} \times \boldsymbol{r} \times \boldsymbol{V} = (\boldsymbol{V} \cdot \boldsymbol{V})\boldsymbol{r} - (\boldsymbol{r} \cdot \boldsymbol{V})\boldsymbol{V} = V^2\boldsymbol{r} - (\boldsymbol{r} \cdot \boldsymbol{V})\boldsymbol{V}$, through direct application of the triple vector product rule. However, the Kepler routine provides all six elements of the postexplosion orbit, a^+, e^+, i^+ Ω^+, ω^+, and θ^{*+}, from which all of the pertinent desired quantities can be generated, say, for verification purposes. Thus,

$$r_p^+ = a^+(1 - e^+), \quad r_a^+ = a^+(1 + e^+), \quad p^+ = a^+(1 - e^{+^2})$$

$$V^2 = \frac{\mu}{p^+}(1 + 2e^+ c_{\theta^{*+}} + e^{+^2}), \quad h^{+^2} = \mu a^+(1 - e^{+^2})$$

$$s_{\gamma^+} = \left(\frac{\mu}{h^+ V}\right)e^+ s_{\theta^{*+}}, \quad c_{\gamma^+} = \left(\frac{\mu}{h^+ V}\right)(1 + e^+ c_{\theta^{*+}}), \quad r = a^+\left(\frac{1 - e^{+^2}}{1 + e^+ c_{\theta^{*+}}}\right)$$

Moreover, the three components of the velocity vector $\boldsymbol{V}$ along the new $(\hat{\boldsymbol{r}}_n, \hat{\boldsymbol{\theta}}_n, \hat{\boldsymbol{h}}_n)$ directions of the circular reference frame corresponding to the post-explosion orbit can be obtained as

$$V_{r_n} = \left(\frac{\mu}{h^+}\right)e^+ s_{\theta*+} \tag{8.39}$$

$$V_{\theta_n} = \left(\frac{\mu}{h^+}\right)(1 + e^+ c_{\theta*+}) \tag{8.40}$$

$$V_{h_n} = 0 \tag{8.41}$$

verifying, once again, that $V^2 = V_{r_n}^2 + V_{\theta_n}^2$ is satisfied. Furthermore, $\Delta r_p = r_p^+ - r_p$ should also be verified to be equal to zero or, in practice, a small number such as 1 m or less to certify the accuracy of the ϕ solutions. Figure 8.2 was generated using an initial orbit given by (a_0, e_0, i_0 Ω_0,

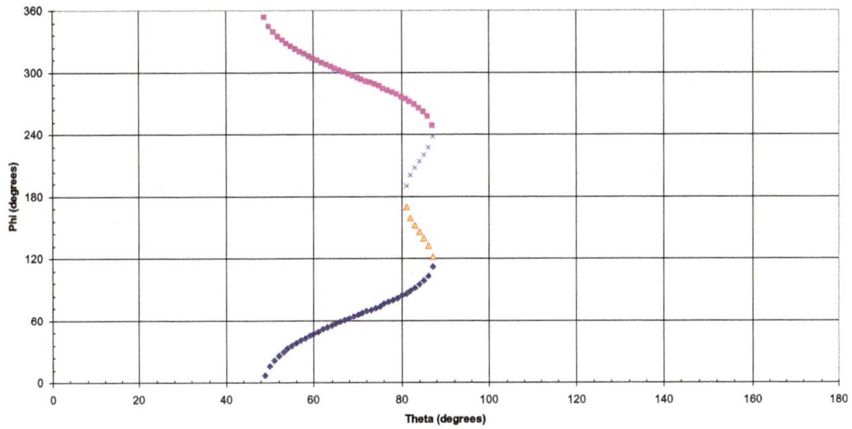

Fig. 8.2 Locus of ϕ angle solutions vs θ_s for θ^* = 122.142 deg.

ω_0, θ_0^*) = (6600 km, 0.1, 0 deg, 0 deg, 0 deg, 0 deg), with $V_s = 1$ km/s and θ_s swept at 1 deg intervals from 0 to 180 deg. The portion of the orbit in vacuum extended from $\theta_1^* = 98.998$deg to $\theta_2^* = 261.001$deg, and a value of $\theta^* = 122.142$ deg was chosen for the location along this orbit where the isotropic velocity V_s was applied. It is clear that the ϕ solutions were not necessarily obtained in a contiguous manner, because of the necessity of computing the multiple ϕ solutions emanating from the various cosine expressions for γ and ϕ. Furthermore, the evaluations of the areas to the left and right of the locus curve were complicated by the multivalued aspect of the $\phi = f(\theta)$ locus curve around the $\theta_s = 90$ deg value, with the left side corresponding to fragment orbits for which the perigees remain above r_p and the right side corresponding to fragment orbits with perigees below r_p.

Figure 8.2 could also be generated by sweeping the angle ϕ numerically, say, at 1 deg intervals between 0 and 360 deg, for each value of θ_s in the interval (0, 180 deg). In this case, after computing the first γ solution from Eq. (8.8) (i.e., $0 \leq \gamma \leq \pi/2$), corresponding to the $0 \leq \theta_n^* \leq \pi$ solution of $c_{\theta_n^*}$, one would sweep the angle ϕ at 1-deg intervals starting from 0, compute the post-explosion orbit as before, and obtain the resulting flight path angle γ_{itr} from the corresponding components of the inertial position and velocity vectors as in Eqs. (8.34) and (8.35). As soon as an interval in ϕ was found for which $\Delta\gamma = \gamma - \gamma_{\text{itr}}$ changed sign, a search would be carried out for the value of ϕ that resulted in $\Delta\gamma = 0$. This first solution would then be used as the starting point for ϕ, which would be further swept by 1 deg at a time, to pinpoint the second ϕ solution that satisfied $\Delta\gamma = 0$. These numerical searches would be repeated for the second γ solution, corresponding to $\pi \leq \theta_n^* \leq 2\pi$, and the other two solutions, if they existed, would also be found using the same search procedure. These calculations would be repeated for each θ_s value until $\theta_s = \pi$ was reached, and the locus curve $\phi = f(\theta_s)$ would thus be generated. This numerical solution could be used to verify the results obtained analytically for the angles ϕ.

The fractions of fragments resulting in orbits with $r_p^+ > r_p$ and with $r_p^+ < r_p$ can be found using this numerical search method by computing the postexplosion orbit of each fragment represented by a point in (θ_s, ϕ) space, with each adjacent point distant by 1 deg in θ_s and ϕ. This is equivalent to partitioning (θ_s, ϕ) space (i.e., $0 \leq \theta_s \leq \pi, 0 \leq \phi \leq 2\pi$) into a rectangular grid, where each point is distant by 1 deg in either θ_s or ϕ from an adjacent point. The area on the sphere covered by an infinitesimal change in both θ_s and ϕ is given by

$$dS = V_s^2 s_{\theta_s} \, d\theta_s \, d\phi \tag{8.42}$$

because the infinitesimal area on the sphere of radius V_s at θ_s is equal to $V_s s_{\theta_s} \, ds \, d\phi$, where $ds = V_s \, d\theta_s$ is the length element generated by $d\theta_s$ for constant ϕ (see Fig. 8.3). Thus, each point corresponding to a given θ_s value must be weighted by a factor of s_{θ_s} regardless of its ϕ value to account for the effective area on the spherical surface. Thus, each (θ_s, ϕ) pair is counted as a count P of

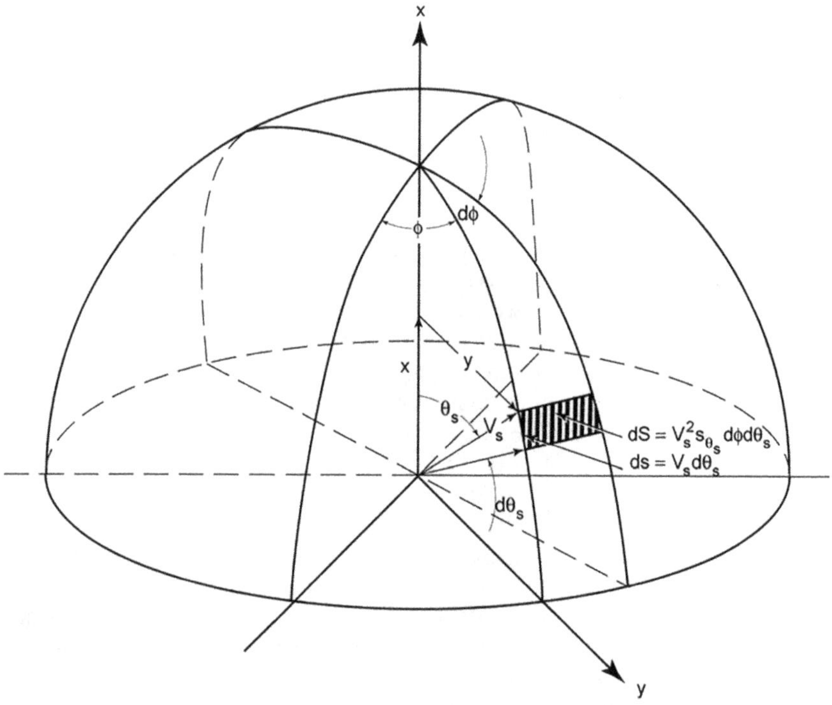

Fig. 8.3 Infinitesimal area on a sphere of radius V_s defined by infinitesimal changes in both angles θ_s and ϕ.

one multiplied by the weight s_{θ_s} if the resulting orbit has $r_p^+ >_{r_p}$, whereas if its orbit has $r_p^+ <_{r_p}$, the pair is counted as a count N of one also multiplied by its s_{θ_s} weight. When all such weighted counts are summed as the (θ_s, ϕ) rectangular grid is traversed 1 deg at a time in θ_s and ϕ, the fraction of the fragments that remain above r_p is given by the ratio $p_f = P/(P+N)$, whereas the remaining fraction with $r_p^+ <_{r_p}$ is given by the ratio $n_f = N/(P+N)$. For our example orbit with a true anomaly of $\theta^* = 122.142$ deg, one obtains $p_f = 31.93\%$ and $n_f = 68.07\%$ after multiplication by 100.

8.3 ANALYTIC INVERSE SOLUTION

Returning to Eq. (8.13) from which the solution for ϕ in terms of θ_s was obtained, the inverse solution for θ_s in terms of ϕ cannot be obtained in a straightforward manner as the direct solution, because V is itself a function of θ_s, namely

$$V^2 = V_c{}^2 + V_s{}^2 + 2V_cV_sc_{\theta_s} \tag{8.43}$$

Equation (8.13) can be rewritten as

$$Vs_\gamma - V_cs_{\gamma_0} = V_s(s_{\theta_s}c_\phi c_{\gamma_0} + c_{\theta_s}s_{\gamma_0})$$

and then squared to yield

$$\begin{aligned}\frac{(Vs_\gamma - V_cs_{\gamma_0})^2}{V_s^2} &= (s_{\theta_s}^2c_\phi^2c_{\gamma_0}^2 + c_{\theta_s}^2s_{\gamma_0}^2 + 2s_{\theta_s}c_{\theta_s}c_\phi s_{\gamma_0}c_{\gamma_0})\\ &= c_\phi^2c_{\gamma_0}^2 + c_{\theta_s}^2(s_{\gamma_0}^2 - c_\phi^2c_{\gamma_0}^2) + s_{2\theta_s}c_\phi s_{\gamma_0}c_{\gamma_0}\end{aligned}$$

such that

$$\begin{aligned}V^2s_\gamma^2 = {}& V_c^2s_{\gamma_0}^2 + V_s^2[c_\phi^2c_{\gamma_0}^2 + c_{\theta_s}^2(s_{\gamma_0}^2 - c_\phi^2c_{\gamma_0}^2) + s_{2\theta_s}c_\phi s_{\gamma_0}c_{\gamma_0}]\\ &+ 2V_cV_ss_{\gamma_0}(s_{\theta_s}c_\phi c_{\gamma_0} + c_{\theta_s}s_{\gamma_0})\end{aligned}$$

and finally, we obtain the form

$$\begin{aligned}\frac{s_\gamma^2}{V_s^2}(V_c^2 + V_s^2 + 2V_cV_sc_{\theta_s}) = {}& s_{\theta_s}^2c_\phi^2c_{\gamma_0}^2 + c_{\theta_s}^2s_{\gamma_0}^2 + \frac{V_c^2}{V_s^2}s_{\gamma_0}^2\\ &+ 2s_{\theta_s}c_{\theta_s}c_\phi s_{\gamma_0}c_{\gamma_0} + 2s_{\theta_s}c_\phi c_{\gamma_0}s_{\gamma_0}\frac{V_c}{V_s} + 2c_{\theta_s}s_{\gamma_0}^2\frac{V_c}{V_s}\end{aligned} \tag{8.44}$$

We can obtain an expression in powers of tan θ if we first use Eq. (8.7) to eliminate $s_\gamma^2 = 1 - c_\gamma^2$ in the Eq. (8.44) after writing c_γ^2 as

$$c_\gamma^2 = \left(\frac{r_p}{r}\right)^2 + \frac{(r_p/r)^2 2\mu(1/r_p - 1/r)}{V_c^2 + V_s^2 + 2V_cV_sc_{\theta_s}} \tag{8.45}$$

and replacing s_{θ_s}, c_{θ_s}, and $c^2_{\theta_s} = 1 - s^2_{\theta_s}$ with the equivalent expressions in terms of $\tan\,\theta_s$, namely

$$s_{\theta_s} = \frac{\tan\,\theta_s}{\pm(1+\tan^2\theta_s)^{1/2}}$$

$$c_{\theta_s} = \frac{1}{\pm(1+\tan^2\theta_s)^{1/2}}$$

$$c^2_{\theta_s} = \left[1 + \frac{1-\tan^2\theta_s}{1+\tan^2\theta_s}\right]\Big/2$$

However, it is preferable to obtain polynomial expressions in terms of s_{θ_s} and c_{θ_s} to solve for the θ_s value without ambiguity.

Returning to Eq. (8.44) and using Eq. (8.45) as before to eliminate s^2_γ yields

$$\frac{s^2_\gamma}{V^2_s}\left(V^2_c + V^2_s + 2V_c V_s c_{\theta_s}\right) = a c_{\theta_s} + b + c \tag{8.46}$$

where

$$a = \frac{2V_c}{V_s}\left[1 - \left(\frac{r_p}{r}\right)^2\right] \tag{8.47}$$

$$b = \frac{\left(V^2_c + V^2_s\right)}{V_s^2}\left[1 - \left(\frac{r_p}{r}\right)^2\right] \tag{8.48}$$

$$c = -2\mu\left(\frac{r_p}{r}\right)^2\left(\frac{1}{r_p} - \frac{1}{r}\right)/V_s^2 \tag{8.49}$$

Equation (8.46) is the result of writing

$$\begin{aligned}\frac{s^2_\gamma}{V^2_s}\left(V^2_c + V^2_s + 2V_c V_s c_{\theta_s}\right) = {} & \frac{\left(V^2_c + V^2_s + 2V_c V_s c_{\theta_s}\right)}{V^2_s} \\ & - \left(\frac{r_p}{r}\right)^2\frac{\left(V^2_c + V^2_s + 2V_c V_s c_{\theta_s}\right)}{V^2_s} \\ & - 2\mu\left(\frac{r_p}{r}\right)^2\left(\frac{1}{r_p} - \frac{1}{r}\right)/V^2_s\end{aligned}$$

Equation (8.44) can now be written as

$$\pm\left(1 - s^2_{\theta_s}\right)^{1/2} a + b + c = d s^2_{\theta_s} + e \pm \left(1 - s^2_{\theta_s}\right)^{1/2} f s_{\theta_s} + g s_{\theta_s} \pm \left(1 - s^2_{\theta_s}\right)^{1/2} h \tag{8.50}$$

or, in a more compact form

$$\pm\left(1-s_{\theta_s}^2\right)^{1/2}(j-fs_{\theta_s}) = ds_{\theta_s}^2 + gs_{\theta_s} + i \tag{8.51}$$

where

$$d = c_\phi^2 c_{\gamma_0}^2 - s_{\gamma_0}^2 \tag{8.52}$$

$$e = s_{\gamma_0}^2 + \frac{V_c{}^2}{V_s{}^2} s_{\gamma_0}^2 \tag{8.53}$$

$$f = 2c_\phi s_{\gamma_0} c_{\gamma_0} \tag{8.54}$$

$$g = 2c_\phi s_{\gamma_0} c_{\gamma_0} \frac{V_c}{V_s} \tag{8.55}$$

$$h = 2s_{\gamma_0}^2 \frac{V_c}{V_s} \tag{8.56}$$

$$j = a - h \tag{8.57}$$

$$i = e - b - c \tag{8.58}$$

After squaring both sides of Eq. (8.51) to eliminate the square root and regrouping terms, the following quartic in s_{θ_s} is obtained

$$s_{\theta_s}^4 + a_q(\phi)s_{\theta_s}^3 + b_q(\phi)s_{\theta_s}^2 + c_q(\phi)s_{\theta_s} + d_q(\phi) = 0 \tag{8.59}$$

where

$$a_q(\phi) = \frac{2(dg - jf)}{(f^2 + d^2)} \tag{8.60}$$

$$b_q(\phi) = \frac{j^2 + g^2 - f^2 + 2di}{(f^2 + d^2)} \tag{8.61}$$

$$c_q(\phi) = \frac{2(jf + gi)}{(f^2 + d^2)} \tag{8.62}$$

$$d_q(\phi) = \frac{i^2 - j^2}{(f^2 + d^2)} \tag{8.63}$$

Now, returning to Eq. (8.44) once more and writing it in terms of c_{θ_s} instead gives

$$ac_{\theta_s} + b + c = -dc_{\theta_s}^2 + e' \pm \left(1 - c_{\theta_s}^2\right)^{1/2} fc_{\theta_s} \pm \left(1 - c_{\theta_s}^2\right)^{1/2} g + hc_{\theta_s} \tag{8.64}$$

with

$$d = c_\phi^2 c_{\gamma_0}^2 - s_{\gamma_0}^2 \tag{8.65}$$

$$e' = c_\phi^2 c_{\gamma_0}^2 + \frac{V_c{}^2}{V_s{}^2} s_{\gamma_0}^2 \tag{8.66}$$

The expression in Eq. (8.64) can be written as

$$dc_{\theta_s}^2 + jc_{\theta_s} + i' = \pm\left(1 - c_{\theta_s}^2\right)^{1/2}\left(fc_{\theta_s} + g\right) \tag{8.67}$$

where

$$j = a - h \tag{8.68}$$

$$i' = b + c - e' \tag{8.69}$$

Equation (8.67) is squared to eliminate the square root, yielding a quartic in c_{θ_s} this time, namely

$$c_{\theta_s}^4 + a'_q(\phi)c_{\theta_s}^3 + b'_q(\phi)c_{\theta_s}^2 + c'_q(\phi)c_{\theta_s} + d'_q(\phi) = 0 \tag{8.70}$$

with

$$a'_q(\phi) = \frac{2(dj + fg)}{(f^2 + d^2)} \tag{8.71}$$

$$b'_q(\phi) = \frac{j^2 + g^2 - f^2 + 2di'}{(f^2 + d^2)} \tag{8.72}$$

$$c'_q(\phi) = \frac{2(ji' - fg)}{(f^2 + d^2)} \tag{8.73}$$

$$d'_q(\phi) = \frac{i'^2 - g^2}{(f^2 + d^2)} \tag{8.74}$$

Because the angle θ_s is confined to the interval $(0, \pi)$, it is sufficient to solve the quartic in c_{θ_s} only, because θ_s will be returned in that interval without ambiguity. Equation (8.70) can be written as

$$x^4 + a'_q x^3 + b'_q x^2 + c'_q x + d'_q = 0 \tag{8.75}$$

with $x = c_{\theta_s}$, and its resolvent cubic is given by [6]

$$y^3 + p_q y^2 + q_q y + r_q = 0 \tag{8.76}$$

where

$$p_q = -b'_q$$

$$q_q = a'_q c'_q - 4d'_q$$

$$r_q = -a'^2_q d'_q + 4b'_q d'_q - c'^2_q$$

The substitution $y = z - (p_q/3)$ will reduce the preceding cubic to the simpler form [6]

$$z^3 + a_c z + b_c = 0 \tag{8.77}$$

where

$$a_c = \frac{1}{3}\left(3q_q - p_q^2\right) \tag{8.78}$$

$$b_c = \frac{1}{27}\left(2p_q^3 - 9p_q q_q + 27r_q\right) \tag{8.79}$$

The three solutions are given by

$$z_1 = A + B \tag{8.80}$$

$$z_2 = -\frac{(A+B)}{2} + \frac{(A-B)}{2}\sqrt{-3} \tag{8.81}$$

$$z_3 = -\frac{(A+B)}{2} - \frac{(A-B)}{2}\sqrt{-3} \tag{8.82}$$

where [6]

$$A = \sqrt[3]{\frac{-b_c}{2} + \sqrt{\frac{b_c^2}{4} + \frac{a_c^3}{27}}} \tag{8.83}$$

$$B = \sqrt[3]{\frac{-b_c}{2} - \sqrt{\frac{b_c^2}{4} + \frac{a_c^3}{27}}} \tag{8.84}$$

If the value of the discriminant is such that

$$\left|\frac{b_c^2}{4} + \frac{a_c^3}{27}\right| < 10^{-7}$$

then it is assumed to be equal to zero, in which case A and B are both real. Moreover, they are both negative if $-b_c/2 < 0$, and positive otherwise. If the discriminant is negative, then the real and imaginary parts of $A = (x + iy)^{1/3}$ lead to one of the three solutions

$$z_1 = 2r^{1/3}\cos\left(\frac{\eta}{3}\right) \tag{8.85}$$

with $r = (x^2 + y^2)^{1/2}$ and $\eta = \tan^{-1}(a/b)$ through the effective use of the ATAN2 routine . This solution results from the expressions

$$A = r^{1/3}\left(\cos\frac{\eta}{3} + i\sin\frac{\eta}{3}\right)$$

$$B = r^{1/3}\left(\cos\frac{\eta}{3} - i\sin\frac{\eta}{3}\right)$$

$$z_1 = A + B$$

If the previously mentioned discriminant is positive, then A and B are real, and depending on the positive or negative sign of the quantity

$$\frac{-b_c}{2} + \left(\frac{b_c^2}{4} + \frac{a_c^3}{27}\right)^{1/2}$$

they are easily obtained as either positive or negative real numbers, respectively, from which, once again, a real solution $z_1 = A + B$ is obtained. Therefore, a solution of the resolvent cubic in Eq. (8.76) is also available as $y_1 = z_1 - p_q/3$.

Once y_1 has been obtained, the discriminant $a_q'^2/4 - b_q' + y_1$ of $R = (a_q'^2/4 - b_q' + y_1)^{1/2}$ is tested for its sign, and if it is negative, R is then the imaginary number $R_{\text{im}} = i\left(\left|a_q'^2/4 - b_q' + y_1\right|\right)^{1/2}$, which leads to the complex number

$$D = \left(\frac{3a_q'^2}{4} + |R_{\text{im}}|^2 - 2b_q' - i\frac{4a_q' b_q' - 8c_q' - a_q'^3}{4|R_{\text{im}}|}\right)^{1/2} = (x + iy)^{1/2} \qquad (86)$$

with real and imaginary parts given by

$$x = \frac{3a_q'^2}{4} + |R_{\text{im}}|^2 - 2b_q'$$

$$y = -\frac{4a_q' b_q' - 8c_q' - a_q'^3}{4|R_{\text{im}}|}$$

The two solutions of D, namely, D_1 and D_2, are then complex and given by

$$D_1 = r^{1/2}(c_{\alpha/2} + is_{\alpha/2}) \qquad (8.87)$$

$$D_2 = r^{1/2}(c_{(\alpha+2\pi)/2} + is_{(\alpha+2\pi)/2}) = -r^{1/2}(c_{\alpha/2} + is_{\alpha/2}) \qquad (8.88)$$

with $r = (x^2 + y^2)^{1/2}$and $\alpha = \tan^{-1}(x/y)$ such that

$$D_1 = D_{1_{\text{re}}} + iD_{1_{\text{im}}} \qquad (8.89)$$

$$D_2 = D_{2_{\text{re}}} + iD_{2_{\text{im}}} = -D_{1_{\text{re}}} - iD_{1_{\text{im}}} = -D_1 \qquad (8.90)$$

If the discriminant of R is positive, then R is readily obtained as a real number, and the discriminant of D is evaluated from

$$D_{\text{dis}} = \frac{3a_q'^2}{4} - R^2 - 2b_q' + \frac{4a_q' b_q' - 8c_q' - a_q'^3}{4R} \qquad (8.91)$$

If this expression is positive, then $D = (D_{\text{dis}})^{1/2}$ is real, whereas if the expression is negative, then D is imaginary: $D = D_{\text{im}} = i(|D_{\text{dis}}|)^{1/2}$. If R is real, then there are either two real or two complex roots of Eq. (8.75) according to whether D is real or imaginary; they are given by

$$x_{1,2} = \frac{-a'_q}{4} + \frac{R}{2} \pm \frac{D}{2} \tag{8.92}$$

Conversely, if R is imaginary, then there are two complex roots of Eq. (8.75), given by

$$x_{1,2} = \frac{-a'_q}{4} \pm \frac{D_{1_{\text{re}}}}{2} + i\left(\frac{|R_{\text{im}}|}{2} \pm \frac{D_{1_{\text{im}}}}{2}\right) \tag{8.93}$$

Identical calculations are carried out to extract the remaining two roots of Eq. (8.75) by first evaluating

$$E = \left(\frac{3a'^2_q}{4} + |R_{\text{im}}|^2 - 2b'_q + i\frac{4a'_q b'_q - 8c'_q - a'^3_q}{4|R_{\text{im}}|}\right)^{1/2} = (x + iy)^{1/2} \tag{8.94}$$

for the case where R is imaginary. The two solutions of E are given by

$$E_1 = r^{1/2}(c_{\beta/2} + is_{\beta/2}) \tag{8.95}$$

$$E_2 = r^{1/2}(c_{(\beta+2\pi)/2} + is_{(\beta+2\pi)/2}) = -r^{1/2}(c_{\beta/2} + is_{\beta/2}) \tag{8.96}$$

with $r = (x^2 + y^2)^{1/2}$ and $\beta = \tan^{-1}(x/y)$ such that

$$E_1 = E_{1_{\text{re}}} + iE_{1_{\text{im}}} \tag{8.97}$$

$$E_2 = E_{2_{\text{re}}} + iE_{2_{\text{im}}} = -E_{1_{\text{re}}} - iE_{1_{\text{im}}} = -E_1 \tag{8.98}$$

If R is real, the discriminant of E is evaluated from

$$E_{\text{dis}} = \frac{3a'^2_q}{4} - R^2 - 2b'_q - \frac{4a'_q b'_q - 8c'_q - a'^3_q}{4R} \tag{8.99}$$

and if R is also positive, then $E = (E_{\text{dis}})^{1/2}$ is real. If E_{dis} is negative, then $E = E_{\text{im}} = i(E_{\text{dis}})^{1/2}$ is imaginary. The two remaining solutions of the quartic in Eq. (8.75) are then obtained as two real roots if E is real or as two complex roots if E is imaginary, namely

$$x_{3,4} = \frac{-a'_q}{4} - \frac{R}{2} \pm \frac{E}{2} \tag{8.100}$$

if E is real and

$$x_{3,4} = \frac{-a'_q}{4} \pm \frac{E_{1_{\text{re}}}}{2} + i\left(-\frac{|R_{\text{im}}|}{2} \pm \frac{E_{1_{\text{im}}}}{2}\right) \tag{8.101}$$

if R is imaginary, in which case E is complex with solutions E_1 and $E_2 = -E_1$, or $\pm E_1$, as was the case with D_1 and $D_2 = -D_1$, or $\pm D_1$, for Eq. (8.93). Each square root in the expressions

$$R = \left(\frac{a'^2_q}{4} - b'_q + y_1\right)^{1/2} \tag{8.102}$$

$$D = \left(\frac{3a'^2_q}{4} - R^2 - 2b'_q + \frac{4a'_q b'_q - 8c'_q - a'^3_q}{4R}\right)^{1/2} \tag{8.103}$$

$$E = \left(\frac{3a'^2_q}{4} - R^2 - 2b'_q - \frac{4a'_q b'_q - 8c'_q - a'^3_q}{4R}\right)^{1/2} \tag{8.104}$$

leads to $\pm R$ or $\pm R_{\text{im}}$ for real and imaginary R, respectively, giving rise to the two forms D and E, with D corresponding to $+R$ or $+R_{\text{im}}$ and E corresponding to $-R$ or $-R_{\text{im}}$, respectively. In these expressions for D and E, $+R$ or $+R_{\text{im}}$ is to be used only because the sign change in E effectively accounts for the $-R$ and $-R_{\text{im}}$ cases.

For our example orbit, Fig. 8.4 shows the two real solutions for c_{θ_s}, for the case $\theta^* = 122.142$ deg. Four solutions are obtained for each value of ϕ, and they are of type I (x_1 and x_2 real, x_3 and x_4 complex) or of type II (x_1 and x_2 complex,

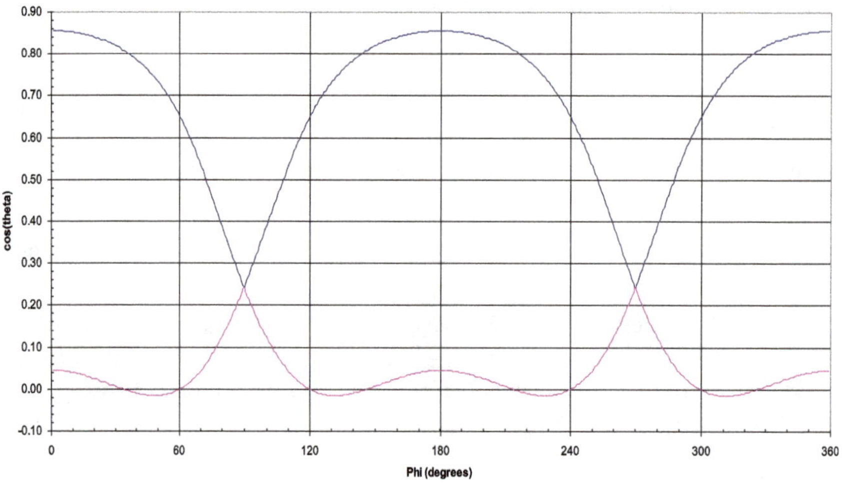

Fig. 8.4 c_{θ_s} **solutions for** $\theta^* = 122.142$ **deg.**

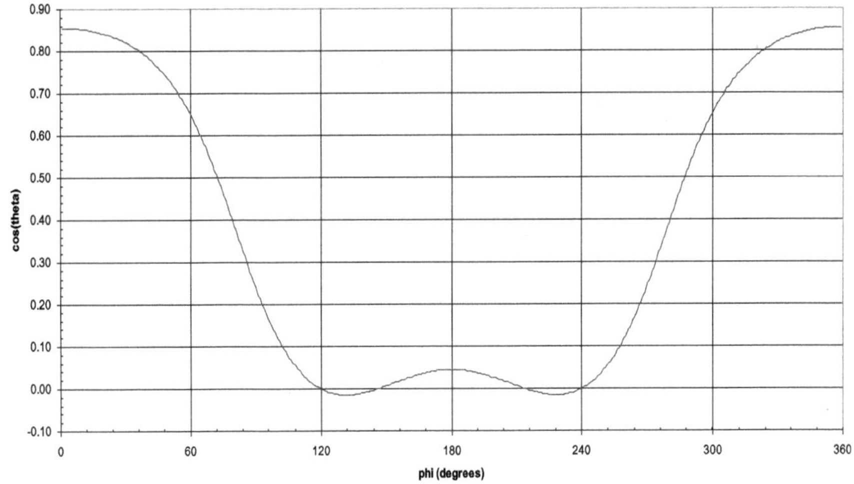

Fig. 8.5 c_{θ_s} **solution for** $\theta^* = 122.142$ **deg.**

x_3 and x_4 real). If the solutions are of type I, $x = x_1$ is chosen for $-\pi/2 \le \phi \le \pi/2$, and $x = x_2$ is chosen for $\pi/2 < \phi < 3\pi/2$, provided that $\theta^* < \pi$. For $\theta^* > \pi$, $x = x_2$ is chosen for $-\pi/2 \le \phi \le \pi/2$, and $x = x_1$ is chosen for $\pi/2 < \phi < 3\pi/2$.

If the solutions are of type II, $x = x_3$ is chosen for $-\pi/2 \le \phi \le \pi/2$, and $x = x_4$ is selected for $\pi/2 < \phi < 3\pi/2$, provided that $\theta^* < \pi$; otherwise, for $\theta^* > \pi$, $x = x_4$ is selected for the interval $-\pi/2 \le \phi \le \pi/2$, and $x = x_3$ is selected for $\pi/2 < \phi < 3\pi/2$.

The top curve in Fig. 8.4 corresponds to x=x_1 or x=x_3, and the bottom curve corresponds to $x = x_2$ or $x = x_4$. The solutions x_3 and x_4 exist in narrow ranges of about ± 5 deg centered at $\phi = \pi/2$ and $\phi = 3\pi/2$. Figure 8.5 selects the correct value of x from either x_1 or x_2 or from either x_3 or x_4 according to the value of ϕ, as described earlier. In this example, there are two real roots at any given value of ϕ, with the remaining two roots being complex and therefore inapplicable, because a real solution must always exist because of the physics of the problem. For examples of orbit and V_s parameters that result in either all fragments entering the atmosphere or all fragments remaining in orbit, a locus curve cannot exist, and therefore, the solutions of the quartic will all be complex, leading to their rejection.

The type I solutions are such that x_1 and x_2 are given by Eq. (8.92) with both R and D real, whereas x_3 and x_4 are given by Eq. (8.100) with R real and E imaginary. For type II solutions , x_1 and x_2 are still given by Eq. (8.92) with R real and D imaginary, whereas x_3 and x_4 are real and are given by Eq. (8.100) with R and E real. However, it is true that, if R is imaginary, then all four solutions are complex, and therefore, a locus curve cannot exist.

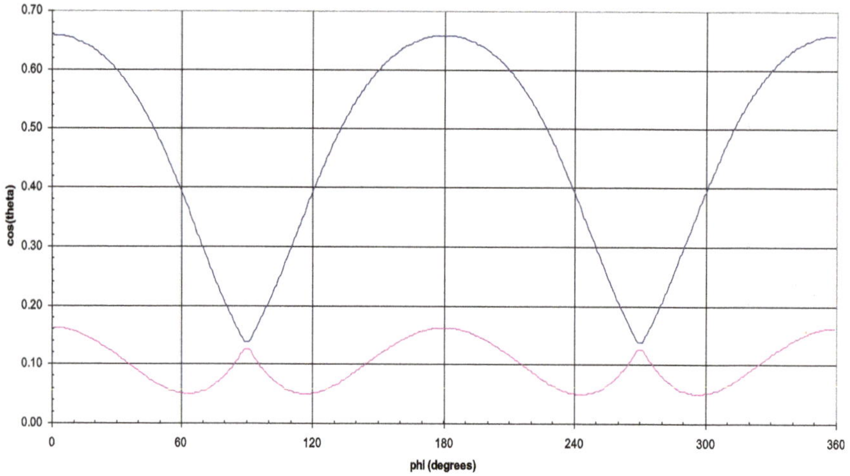

Fig. 8.6 c_{θ_s} **solutions for** $\theta^* = 214.714$ **deg.**

Figure 8.6 is for the example orbit with $\theta^* = 214.714$ deg, this time using the $\theta^* > \pi$ solution tree. Once again, the top curve represents x_1 or x_3, with x_3 confined to small intervals in ϕ around $\phi = \pi/2$ and $3\pi/2$, whereas the bottom curve is for x_2 or x_4, with x_4 also confined to the same small ϕ intervals around $\phi = \pi/2$ and $3\pi/2$. Figure 8.7 shows the x solution that corresponds to the desired locus curve. Figure 8.8 shows how the desired solutions for both $\theta^* = 122.142$ deg and $\theta^* = 214.714$ deg switch between the four real solutions, with two real

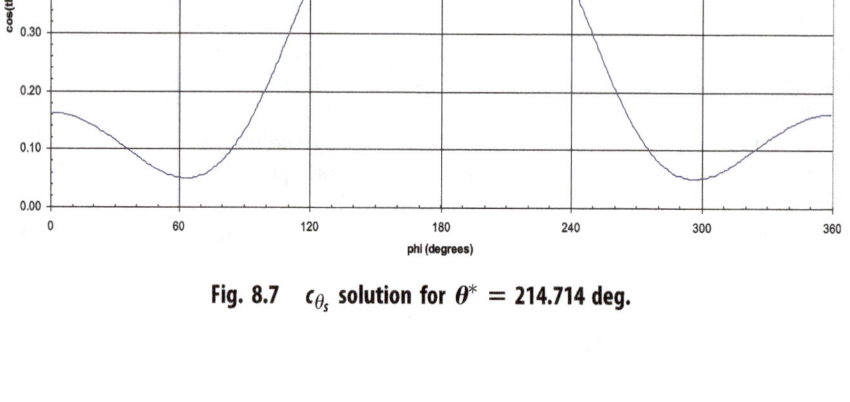

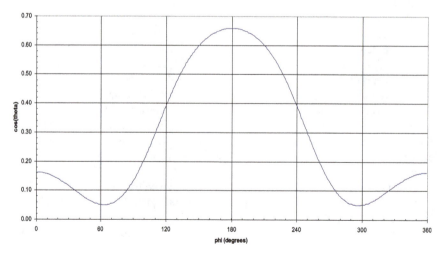

Fig. 8.7 c_{θ_s} **solution for** $\theta^* = 214.714$ **deg.**

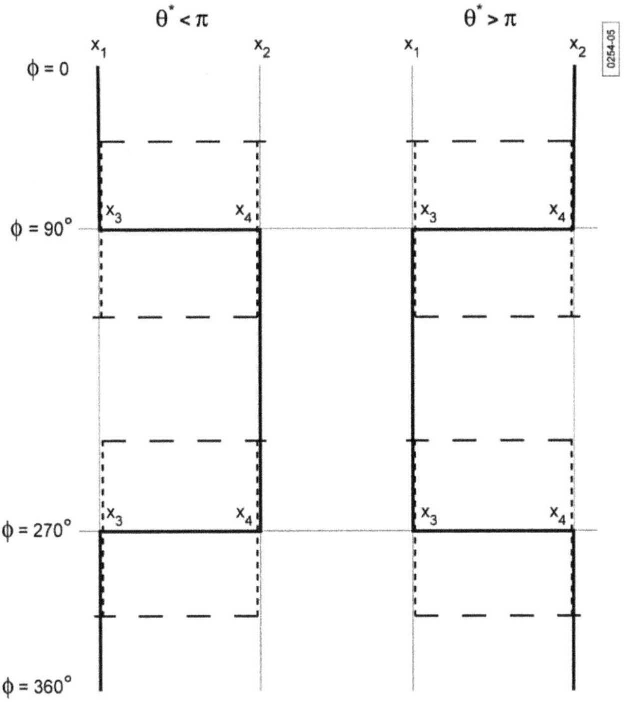

Fig. 8.8 Solution trees for θ^* < π and θ^* > π.

solutions present at a time for a given value of ϕ. The fragment ratios can now be easily computed by quadrature.

Figure 8.9 shows the locus curve $c_{\theta_s} = f(\phi)$ for our first example with $\theta^* = 122.142$ deg. It corresponds to Fig. 8.10, which depicts $\theta_s = f(\phi)$ as the locus curve, with, once again, the area below this curve corresponding to postexplosion orbits that remain in vacuum and the area above the curve corresponding to orbits that have perigees lower that r_p. In Fig. 8.9, c_{θ_s} is plotted between the ± 1 limits corresponding to the $0 \leq \theta_s \leq \pi$ interval, and the area calculations are carried out in this c_{θ_s} space rather than in the θ_s space because the area ratios are identical to the area ratios on the V_s sphere itself, only in (c_{θ_s}, ϕ) space.

This can be shown heuristically by assuming that θ_s is constant for all values of ϕ in the interval $0 \leq \phi \leq 2\pi$. On the sphere of radius V_s, a constant value of θ will effectively cut the sphere into two shells with relative areas of $S_{s_1} = 2\pi V_s^2(1 - c_{\theta_s})$ and $S_{s_2} = 4\pi V_s^2 - 2\pi V_s^2(1 - c_{\theta_s}) = 2\pi V_s^2(1 + c_{\theta_s})$. The ratio is then

$$r_s = \frac{2\pi V_s^2(1 - c_{\theta_s})}{2\pi V_s^2(1 + c_{\theta_s})} = \frac{(1 - c_{\theta_s})}{(1 + c_{\theta_s})}$$

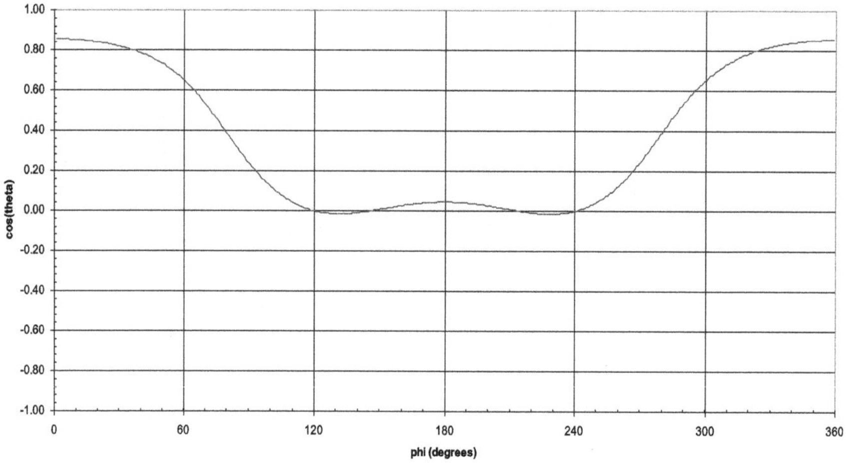

Fig. 8.9 $c_{\theta_s} = f(\phi)$ **locus curve for** $\theta^* = 122.142$ **deg.**

In the $c_{\theta_s} - \phi$ plane, the first area S_{s_1} on the sphere corresponds to the region confined between c_{θ_s} and 1, whose area is given by $S_{b_1} = 2\pi(1 - c_{\theta_s})$.

The remaining area in the $c_{\theta_s} - \phi$ planar box is given by $S_{b_2} = 4\pi - 2\pi(1 - c_{\theta_s})$, such that the ratio of the two areas is

$$r_b = \frac{(1 - c_{\theta_s})}{(1 + c_{\theta_s})}$$

which is identical to r_s. Therefore, the area S_1 below the c_{θ_s} curve in Fig. 8.9 corresponds to the area above the θ_s locus curve of Fig. 8.10 such that all points in these areas result in orbits that reenter the atmosphere.

Once S_1 has been computed by numerical quadrature using a 10-point Gauss–Legendre method with function calls to the quartic solver, the fractions p_f and n_f are obtained as

$$p_f = \frac{S_1}{4\pi} \times 100$$
$$n_f = \frac{4\pi - S_1}{4\pi} \times 100$$

Figure 8.9 yields $p_f = 31.97\%$ and $n_f = 68.03\%$, which are almost identical to the values obtained numerically as 31.93% and 68.07% reported in Sec. 8.2.

As a final example, let $(a_0, e_0, i_0, \Omega_0, \omega_0, \theta_0^*) =$ (6600 km, 10^{-5}, 0, 0, 0, 75.855 deg), be our pre-explosion orbit, which, in this case, is effectively circular. Using $V_s = 1$ km/s as before, numerical instability is avoided when eccentricity is near zero, and invalid roots of the quartic in c_{θ_s} are thus prevented from occurring by setting the term $b_c^2/4 + a_c^3/27$ equal to zero whenever it becomes negative.

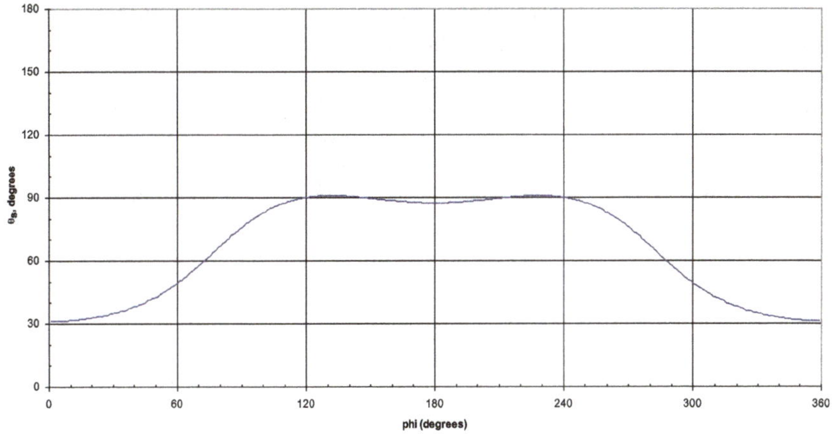

Fig. 8.10 θ_s **vs** ϕ **for** $\theta^* = 122.142$ **deg.**

Because $-b_c/2$ is a positive quantity and because it is much larger than $(b_c^2/4 + a_c^3/27)^{1/2}$, it is necessary to set this square root equal to zero because, otherwise, A and B in Eqs. (8.83) and (8.84) would be complex conjugates, leading to a z_1 root that, in turn, generates four erroneous real roots of the c_{θ_s} quartic that are not feasible, meaning that $|c_{\theta_s}| > 1$ for each one. In contrast, setting $b_c^2/4 + a_c^3/27$ equal to zero whenever it is negative will avoid the numerical instability and

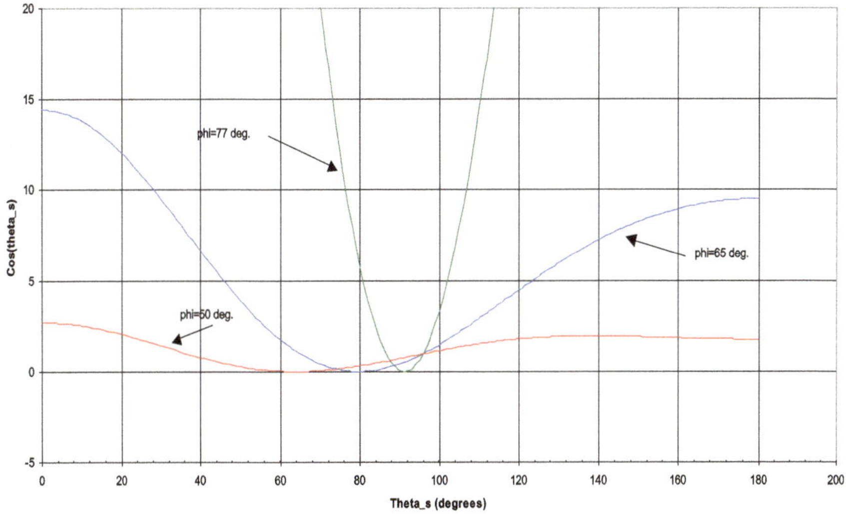

Fig. 8.11 c_{θ_s} **polynomials for the example with** $a = 6600$ **km;** $e = 1.0 \times 10^{-5}$**;** $\theta^* = 75.885$ **deg; and** $\phi = 50$, **65, and 77 deg.**

ensure that A and B remain real with a z_1 solution that is also real, leading to a correct c_{θ_s} solution of the quartic. The two real roots of the quartic become closer as $e \to 0$ and become equal as the quartic reaches its zero value at its tangency point with the θ_s axis. Figure 8.11 shows the quartics in this example for three values of ϕ, namely, $\phi = 50$, 65, and 77 deg. This quartic crosses the θ_s axis in two distinct points when e becomes larger, as the $b_c^2/4 + a_c^3/27$ term remains clearly positive for all values of ϕ, leading to stable numerics. This is shown in Fig. 8.12, which is for the same example except with $e = 0.1$, $\theta^* = 122.142$ deg, and $\phi = 59$ deg. As a final note, the term $(f^2 + d^2)$ appearing as the denominator in Eqs. (8.71 – 8.74) that define $a'_q(\phi)$, $b'_q(\phi)$, $c'_q(\phi)$, and $d'_q(\phi)$ will go to zero for $\phi = 90$ and 270 deg and $\gamma_0 = 0$ deg, introducing an inherent singularity at those two ϕ values for the case where $e \to 0$.

To prevent the solutions from the quartic to jump between its solution branches, R, D, and E are set equal to zero if the discriminants are less than 10^{-7}. When R is set to zero in this way, then D and E can be evaluated instead from [6]

$$D = \sqrt{\frac{3a'^2_q}{4} - 2b'_q + 2\sqrt{y_1^2 - 4d'_q}} \tag{8.105}$$

$$E = \sqrt{\frac{3a'^2_q}{4} - 2b'_q - 2\sqrt{y_1^2 - 4d'_q}} \tag{8.106}$$

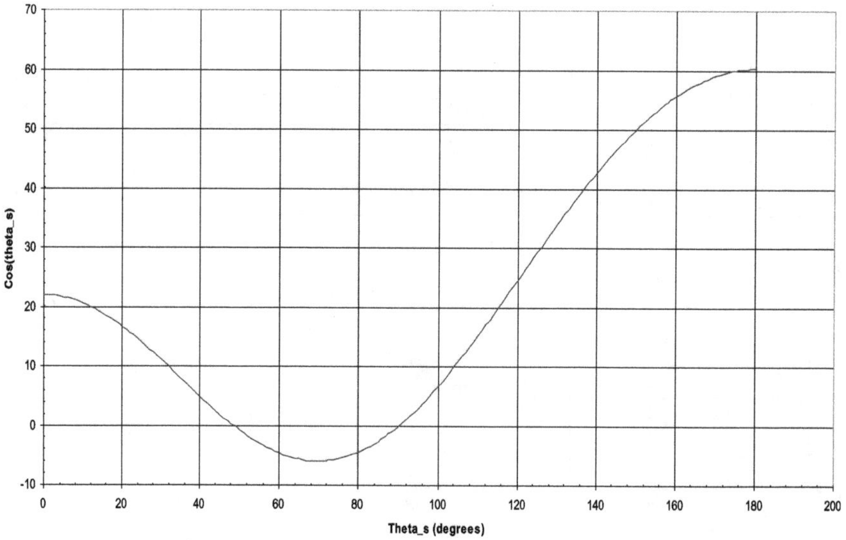

Fig. 8.12 c_{θ_s} **polynomial for the example with $a = 6600$ km, $e = 0.1$, $\theta^* = 122.142$ deg, and $\phi = 59$ deg.**

which yields either real or complex numbers as before. The inner square root is also set equal to zero for those values of ϕ for which $y_1^2 - 4d'_q$ is again less than 10^{-7}. Otherwise, as before, the square roots are on the order of 10^{-14}, and if they are not set equal to zero, they will introduce unwanted numerical instability in addition to jumping between solution branches.

8.4 CONCLUSION

The efficient computation of the fraction of debris from an isotropic explosion in general elliptical orbit that would fly in orbits whose perigees are below a certain given altitude, which usually corresponds to the interface with the atmosphere, has been analyzed in this chapter. Given an explosion velocity, a spacecraft idealized as a sphere is shown to break up into small fragments that either remain in Earth orbit or intersect the interface with the atmosphere, resulting in reentry. The locus of points on the surface of the sphere that separates the two groups of debris was obtained analytically in the form of a quartic in the cosine of one of the two angles that uniquely define the orientation of the explosion velocity vector with respect to the pre-explosion velocity vector, and the corresponding areas were evaluated by Gaussian quadrature for accurate percentage counts. The accuracy of the percentage counts from this analytic method was shown to be identical to that obtained from a purely numerical method based on fine grid characterization of the exploding sphere surface.

REFERENCES

[1] Sorge, M. E., "Determination of the fraction of fragments from an orbital explosion which pass below a given altitude," ATM Technical Memorandum No. *90(5909-02)-3*, Aerospace Corporation, El Segundo, CA, 1990.

[2] Kéchichian, J. A., and Sorge, M. E., "The efficient analytic computation of fractional reentering debris from an idealized isotropic explosion in general elliptic orbit," *Acta Astronautica*, Vol. 61, Nos. 7–8, Oct. 2007, pp. 590–603.

[3] Kéchichian, J. A., "The efficient analytic computation of fractional reentering debris from an idealized isotropic explosion in general elliptic orbit," *AAS/AIAA Space Flight Mechanics Conference*, AAS Paper 06-236, Tampa, FL, Jan. 2006.

[4] Edgecombe, D. S., Fisher, N. H., and Reynolds, R. C., "A model for the evolution of the on-orbit manmade debris environment," 33rd International Astronautical Federation Congress, IAA Paper 82-155, Paris, France, 1982.

[5] Escobal, P. R., *Methods of Orbit Determination*, Krieger Publishing Company, Inc., Malabar, FL, 1976, p. 119.

[6] *CRC Standard Mathematical Tables*, 21st ed., The Chemical Rubber Co., Cleveland, OH, 1973, pp. 103–106.

CHAPTER 9

Optimal Low-Thrust Transfer in a General Circular Orbit Using Analytic Averaging of the System Dynamics

NOMENCLATURE

a	=	orbit semimajor axis, m
c_α, s_α	=	$\cos\alpha$, $\sin\alpha$
E	=	eccentric anomaly
e	=	eccentricity
f	=	thrust acceleration vector
f_t, f_n, f_h	=	components of the thrust acceleration vector along the tangent, normal, and out-of-plane directions, respectively
i	=	relative inclination
J_2	=	second zonal harmonic of Earth's potential
α	=	mean angular position, $\omega + M$
θ^*	=	true anomaly
θ_c	=	instantaneous angular position of the relative line of nodes
μ	=	gravitational constant of Earth, 398601.3 km^3/s^2
n	=	mean motion, rad/s

9.1 INTRODUCTION

Edelbaum's classic problem of minimum-time low-thrust transfer between inclined circular orbits is analyzed in this chapter within the context of the additional perturbation due to Earth's oblateness. The original analytic theory using as variables only the orbital velocity V and relative inclination i, sufficient to describe the transfers between conic orbits, is extended by also considering the right ascension of the ascending node Ω needed to account for the precession of the instantaneous orbit during the transfer due to the perturbation of the second zonal harmonic of Earth's potential J_2. Analytic averaging of the dynamic and adjoint differential equations using a piecewise-constant thrust angle is carried out for the thrust-perturbation-only case within the framework of the three-state description to emulate the purely closed-form Edelbaum solution. For the more general case of a precessed orbit plane, an identical analytic averaging is carried out to rotate the orbit plane around the instantaneous line of nodes to generate the set of averaged differential equations that do not require numerical quadratures during their integration. The suboptimal results are compared to the purely numerical solutions using precision integration on a four-state system description by the addition of the mean angular position α as the fourth state variable, with a continuously varying thrust vector orientation for the unaveraged system dynamics, and additionally using numerical quadrature for the averaged system dynamics. The analysis of the optimal low-thrust orbit transfer between circular orbits of general size and orientation has benefitted from several earlier contributions [1–5]. In particular, Edelbaum [1] provided his classical analytic expression for the velocity change ΔV needed to transfer between two circles of uneven size and relative inclination in the minimum time while applying continuous, constant thrust acceleration. In the absence of the effect of Earth's oblateness due mainly to the harmonic J_2, it is convenient to consider the relative inclination of one orbit with respect to the other and, thus, to ignore the right ascension of the ascending node Ω in the system of dynamic equations, which then reduces to only two state variables, namely, the orbital velocity V and the relative inclination i. The differential equations in V and i can be averaged by considering a simple thrust vectoring strategy that essentially holds the thrust angle piecewise-constant during each revolution, switching signs at the antinodes. The angular position can thereby be removed from these dynamic equations, enabling a closed-form solution for the state variables V and i as functions of time.

Transfer examples were also analyzed by Bryson and Ho [6], and Edelbaum et al. [7] first introduced the use of equinoctial elements in general transfer between any elliptical or circular orbits. This author has also provided several contributions to solve the general problem using various sets of nonsingular elements. This chapter concentrates on the important circular application and assumes that the instantaneous orbit remains circular during the transfer such that the problem can be further investigated using a reduced set of dynamic variables instead of the full six-state system needed in the more general elliptical transfer. Typical long-duration precision integration transfers spanning several months with the full six-state dynamics and continuous thrust vectoring from LEO to GEO remain essentially circular. It is only when high accelerations are used, resulting in fast subday transfers, that the eccentricities of the intermediate orbits can be substantial, not unlike the case of an impulsive geostationary transfer orbit (GTO). The analytic Edelbaum velocity-change estimates for LEO-to-GEO transfers are only less than 3% higher than their exact counterparts, and the transfers are much easier to implement in an operational sense.

In a first analysis, the general circular problem is revisited in the absence of any oblateness-induced perturbation by considering the three state variables V, i, and Ω and by also using the assumption of a piecewise-constant thrust yaw to arrive at a system of averaged equations that can be solved numerically. The results must then mimic Edelbaum's findings. An unaveraged dynamic system, this time using all four state variables including the mean angular position α defined in the next section and continuously orienting the thrust vector along the instantaneous optimal direction, is also solved numerically for comparison purposes and in preparation for the inclusion of the important J_2 perturbation. The J_2 perturbation is then taken into account, both in the "exact" or unaveraged sense with optimal thrust vectoring and within the context of an analytic averaging that uses the piecewise-constant yaw strategy mentioned earlier. However, unlike the thrust-only problem, the addition of J_2 causes the precession of the orbit plane and initiates changes in the orientation of the line of nodes, which effectively remained fixed in the thrust-only problem. Therefore, the instantaneous line of nodes must be continuously updated to continue the process of analytic averaging in a reliable manner. Numerical comparisons are provided with the exact solution that does not average out the angular position, which, in turn, makes the search for the exact solutions more difficult, as discussed in the text. The effect of the J_2 perturbation using the full six-state dynamics was treated by the author in 2000 [8], as well as in later conference presentations [9, 10] and corresponding journal articles [11, 12].

The analysis presented in this chapter is based on a previously published symposium paper [13, 14].

A book [15] which treats the optimal low-thrust orbit transfer problem was published in 2018.

9.2 ANALYTIC AVERAGING OF THE THREE-STATE DYNAMICS FOR THE THRUST-ONLY CASE (i^* THEORY)

In this section, Edelbaum's analytic description of the circle-to-inclined-circle transfer using only the orbital velocity V and relative inclination i is reformulated within the context of the three-state dynamics description, with the right ascension of the ascending node Ω as the third state variable. In the absence of the J_2 perturbation, it is convenient to assume that the initial orbit plane is effectively the equatorial plane such that the relative inclination of the final orbit with respect to the initial orbit represents effectively the equatorial inclination angle itself. Thus, it is sufficient to consider only the state variables V and i.

Starting from the full set of the Gaussian form of the Lagrange planetary equations for near-circular orbits, namely

$$\mathrm{d}a/\mathrm{d}t = 2af_t/V \tag{9.1}$$

$$\mathrm{d}e_x/\mathrm{d}t = 2f_t c_\alpha/V - f_n s_\alpha/V \tag{9.2}$$

$$\mathrm{d}e_y/\mathrm{d}t = 2f_t s_\alpha/V + f_n c_\alpha/V \tag{9.3}$$

$$\mathrm{d}i/\mathrm{d}t = f_h c_\alpha/V \tag{9.4}$$

$$\mathrm{d}\Omega/\mathrm{d}t = f_h s_\alpha/(V s_i) \tag{9.5}$$

$$\mathrm{d}\alpha/\mathrm{d}t = n + 2f_n/V - f_h s_\alpha/(V \tan i) \tag{9.6}$$

where s_α and c_α represent sin α and cos α, respectively; $\alpha = \omega + M$ is the mean angular position given in terms of the classical elements; a is the orbit semimajor axis; and $e_x = ec_\omega$ and $e_y = es_\omega$ are used to replace e and ω, as the latter is poorly defined in near-circular orbit. The mean motion is given by $n = (\mu/a^3)^{1/2}$, where μ is the gravitational constant of Earth, and the thrust acceleration $\boldsymbol{f}$ resolved along the tangent, normal, and out-of-plane directions is given by f_t, f_n, and f_h, with the normal direction oriented toward the center of attraction. Assume that only tangential and out-of-plane acceleration occur (i.e., $f_n = 0$) and that the orbit remains circular during the transfer. Also, let β represent the thrust yaw angle such that $f_t = fc_\beta$ and $f_h = fs_\beta$, and let $\alpha = \omega + M = \omega + \theta^* = \theta$, because $e = 0$, where $\theta = nt$ and θ^* is the true anomaly. Then, the original differential equations (9.1–9.6) are further reduced to the following form:

$$\mathrm{d}a/\mathrm{d}t = 2af_t/V \tag{9.7}$$

$$\mathrm{d}i/\mathrm{d}t = c_\theta f_h/V \tag{9.8}$$

$$\mathrm{d}\Omega/\mathrm{d}t = s_\theta f_h/(Vs_i) \tag{9.9}$$

$$\mathrm{d}\theta/\mathrm{d}t = n - s_\theta f_h/(V \tan i) \tag{9.10}$$

The assumption $f_n = 0$ is validated by the observation that circle-to-circle transfers remain essentially circular, as discussed earlier.

Letting β be piecewise-constant, switching sign at the orbital antinodes, the $s_\theta f_h$ terms will give zero contribution such that Eq. (9.10) is replaced by

$$\mathrm{d}\theta/\mathrm{d}t = n \tag{9.11}$$

Averaging out the angle θ and using the relation $V = na = (\mu/a)^{1/2}$, one can reduce the averaged system equations to the following form:

$$\dot{\tilde{i}} = \frac{1}{2\pi}\int_0^{2\pi}\left(\frac{\mathrm{d}i}{\mathrm{d}t}\right)\mathrm{d}\theta = \frac{2fs_\beta}{2\pi V}\int_{-\pi/2}^{\pi/2} c_\theta\,\mathrm{d}\theta = \frac{2fs_\beta}{\pi V} \tag{9.12}$$

$$\dot{\tilde{V}} = -fc_\beta \tag{9.13}$$

$$\dot{\tilde{\Omega}} = \frac{1}{2\pi}\int_0^{2\pi}\left(\frac{\mathrm{d}\Omega}{\mathrm{d}t}\right)\mathrm{d}\theta = \frac{2fs_\beta}{2\pi V s_i}\int_{-\pi/2}^{\pi/2} s_\theta\,\mathrm{d}\theta = 0 \tag{9.14}$$

The transfer can now be analyzed using only the variables $\tilde{V}$ and $\tilde{i}$, leading to the analytic solutions reported previously by Edelbaum [1, 2] and the author [4, 5]. When the three-state system is considered, the orbit inclination is relative to the equator, as shown in Fig. 9.1, where the orientations of the initial and final orbits are described by the (i_1, Ω_1) and (i_2, Ω_2) pairs, where θ_i and θ_f are the angular positions of the relative node at A and i^*_{tot} is the total relative inclination. Holding β piecewise-constant with a sign change at the antinodes of the relative

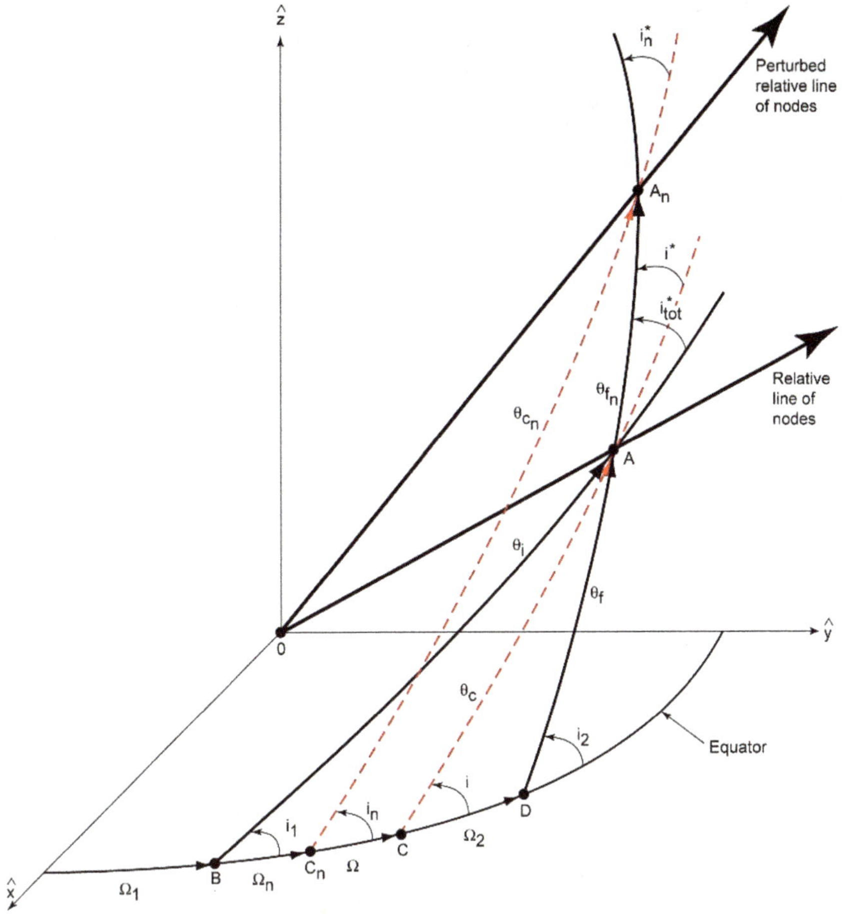

Fig. 9.1 Orbit geometry.

node at A, the averaged equations are now given by

$$\dot{\tilde{V}} = -fc_{\beta} \tag{9.15}$$

$$\dot{\tilde{i}} = \frac{1}{2\pi}\int_0^{2\pi}\left(\frac{\mathrm{d}i}{\mathrm{d}t}\right)\mathrm{d}\theta = \frac{2fs_{\beta}}{2\pi V}\int_{\theta_c-\pi/2}^{\theta_c+\pi/2} c_{\theta}\,\mathrm{d}\theta = \frac{2fs_{\beta}}{\pi V}c_{\theta_c} \tag{9.16}$$

$$\dot{\tilde{\Omega}} = \frac{1}{2\pi}\int_0^{2\pi}\left(\frac{\mathrm{d}\Omega}{\mathrm{d}t}\right)\mathrm{d}\theta = \frac{2fs_{\beta}}{2\pi V s_i}\int_{\theta_c-\pi/2}^{\theta_c+\pi/2} s_{\theta}\,\mathrm{d}\theta = \frac{2fs_{\beta}}{\pi V s_i}s_{\theta_c} \tag{9.17}$$

where θ_c is the instantaneous angular position of A, sweeping the interval from θ_i to θ_f during the transfer, whereas the relative inclination i^* sweeps the interval from i^*_{tot} to 0. As in the original two-state theory, the angle β can now be considered to be a continuous quantity even though, in practice, its value must change from one orbit to the next whereas it is held constant within each revolution. The relative line of nodes remains fixed in space even though the orbit size is continuously changing during the transfer. From spherical trigonometry, we have

$$c_{i^*_{\text{tot}}} = c_{\Omega_1 - \Omega_2} s_{i_1} s_{i_2} + c_{i_1} c_{i_2} \tag{9.18}$$

The angles θ_i and θ_f are also computed from

$$s_{\theta_i} = s_{i_2} s_{\Omega_2 - \Omega_1} / s_{i^*_{\text{tot}}} \tag{9.19}$$

$$c_{\theta_i} = \left(c_{i^*_{\text{tot}}} c_{i_1} - c_{i_2} \right) / \left(s_{i^*_{\text{tot}}} s_{i_1} \right) \tag{9.20}$$

$$s_{\theta_f} = s_{i_1} s_{\Omega_2 - \Omega_1} / s_{i^*_{\text{tot}}} \tag{9.21}$$

$$c_{\theta_f} = \left(c_{i_1} - c_{i_2} c_{i^*_{\text{tot}}} \right) / \left(s_{i^*_{\text{tot}}} s_{i_2} \right) \tag{9.22}$$

The instantaneous angle i^* is then given by

$$s_{i^*} = s_{\Omega_2 - \Omega} s_i / s_{\theta_f} \tag{9.23}$$

$$c_{i^*} = c_{i_1} c_{i_2} + s_{i_1} s_{i_2} c_{\Omega_2 - \Omega} \tag{9.24}$$

This angle can be extracted directly and without ambiguity from Eq. (9.24) because $0 \leq i^* \leq \pi$.

Finally, the angle θ_c is given by

$$s_{\theta_c} = s_{i_2} s_{\theta_f} / s_i \tag{9.25}$$

$$c_{\theta_c} = c_{\Omega_2 - \Omega}\, c_{\theta_f} - s_{\Omega_2 - \Omega} s_{\theta_f} c_{i_2} \tag{9.26}$$

This angle is continuously updated during the numerical integration of the averaged system in Eqs. (9.15–9.17). All variables are thus considered to be averaged values. The minimum-time solution requires the generation of the adjoint or multiplier differential equations for $\tilde{\lambda}_V$, $\tilde{\lambda}_i$, and $\tilde{\lambda}_\Omega$ obtained from the averaged Hamiltonian $\tilde{H}$, given by

$$\tilde{H} = 1 + \tilde{\lambda}_V \dot{\tilde{V}} + \tilde{\lambda}_i \dot{\tilde{i}} + \tilde{\lambda}_\Omega \dot{\tilde{\Omega}} = 1 - \tilde{\lambda}_V f c_\beta + \tilde{\lambda}_i \left(\frac{2 f s_\beta}{\pi V} c_{\theta_c} \right) + \tilde{\lambda}_\Omega \left(\frac{2 f s_\beta}{\pi V s_i} s_{\theta_c} \right) \tag{9.27}$$

such that

$$\dot{\tilde{\lambda}}_V = -\partial\tilde{H}/\partial V = \frac{2fs_\beta}{\pi V^2}\left(\tilde{\lambda}_i c_{\theta_c} + \frac{\tilde{\lambda}_\Omega}{s_i} s_{\theta_c}\right) \tag{9.28}$$

$$\dot{\tilde{\lambda}}_i = -\partial\tilde{H}/\partial i = \tilde{\lambda}_\Omega \frac{2fs_\beta}{\pi V}\frac{c_i}{s_i^2} s_{\theta_c} \tag{9.29}$$

$$\dot{\tilde{\lambda}}_\Omega = -\partial\tilde{H}/\partial\Omega = 0 \tag{9.30}$$

The optimal thrust yaw angle β is obtained from the optimality condition

$$\partial\tilde{H}/\partial\beta = \tilde{\lambda}_V f s_\beta + \tilde{\lambda}_i \frac{2fc_\beta}{\pi V} c_{\theta_c} + \tilde{\lambda}_\Omega \frac{2fc_\beta}{\pi V s_i} s_{\theta_c} = 0 \tag{9.31}$$

An anonymous reviewer of the original journal paper [14] suggested that the strengthened Legendre–Clebsch condition $\partial^2\tilde{H}/\partial\beta^2 > 0$, which is a second-order condition for a local minimum, can be satisfied by the local minimum ensuring thrust direction given by $c_\beta = a(a^2+b^2)^{-1/2}$ and $s_\beta = -b(a^2+b^2)^{-1/2}$ because $\partial\tilde{H}/\partial\beta$ has the form $as_\beta + bc_\beta$.

Equation (9.31) leads to

$$\tan\beta = -\left(\tilde{\lambda}_i c_{\theta_c} + \frac{\tilde{\lambda}_\Omega}{s_i} s_{\theta_c}\right)\frac{2}{\pi V \tilde{\lambda}_V} = \frac{s_\beta}{c_\beta} \tag{9.32}$$

Given initial values V_0, i_0, and Ω_0; final values V_f, i_f, and Ω_f; and a thrust acceleration magnitude f, the two-point boundary-value problem is solved by guessing the initial values of the multipliers and the total flight time t_f and integrating the system of Eqs. (9.15–9.17) and adjoint Eqs. (9.28–9.30) using the optimal thrust angle β given in Eq. (9.32), from time zero t_0 until the final time t_f and iterating on the guessed values until V_f, i_f, and Ω_f, as well as $H_f = 0$, are satisfied.

Note that the Hamiltonian $\tilde{H}$ is not constant here, even though it is not an explicit function of time, because of the formulation that involves the angle θ_c. Even though the relative line of nodes remains fixed in space, the angle θ_c varies from θ_i to θ_f during the transfer such that

$$\frac{d\tilde{H}}{dt} = \frac{\partial\tilde{H}}{\partial t} + \frac{\partial\tilde{H}}{\partial\tilde{x}}\dot{\tilde{x}} + \frac{\partial\tilde{H}}{\partial\tilde{\lambda}}\dot{\tilde{\lambda}} + \frac{\partial\tilde{H}}{\partial\theta_c}\dot{\theta}_c = \frac{\partial\tilde{H}}{\partial\theta_c}\dot{\theta}_c \neq 0$$

Because $\partial\tilde{H}/\partial t = 0$ and $\partial\tilde{H}/\partial\tilde{x} = -\dot{\tilde{\lambda}}$, $\partial\tilde{H}/\partial\tilde{\lambda} = \dot{\tilde{x}}$. However, if we replace c_{θ_c} and s_{θ_c} by their expressions as functions of $\tilde{\Omega}$ and $\tilde{i}$, then the form of $\dot{\tilde{\lambda}}_V = -\partial\tilde{H}/\partial\tilde{V}$ remains unchanged, whereas $\dot{\tilde{\lambda}}_\Omega = -\partial\tilde{H}/\partial\tilde{\Omega}$ and $\dot{\tilde{\lambda}}_i = -\partial\tilde{H}/\partial\tilde{i}$ will have different forms, and the Hamiltonian $\tilde{H}$ will be effectively constant, with tan β also unchanged. The Hamiltonian $\tilde{H}$ decreases from 0.14249 at time zero to essentially zero at t_f for the example considered in the next section.

9.3 PRECISION INTEGRATION OPTIMIZATION USING THE FOUR-STATE SYSTEM

Returning to Eqs. (9.7–9.10), using the fact that $n = V^3/\mu$ for a circular orbit, and once again employing only the $f_t = fc_\beta$ and $f_h = fs_\beta$ accelerations, we have the four-state dynamic equations given by

$$\dot{V} = -fc_\beta \tag{9.33}$$

$$\dot{i} = fs_\beta c_\alpha / V \tag{9.34}$$

$$\dot{\Omega} = fs_\beta s_\alpha / (V s_i) \tag{9.35}$$

$$\dot{\alpha} = V^3/\mu - fs_\beta s_\alpha / (V \tan i) \tag{9.36}$$

For minimum-time solutions, the Hamiltonian of the system is given by

$$\begin{aligned} H &= 1 + \lambda_V \dot{V} + \lambda_i \dot{i} + \lambda_\Omega \dot{\Omega} + \lambda_\alpha \dot{\alpha} \\ &= 1 + \lambda_V(-fc_\beta) + \lambda_i \left(\frac{fs_\beta c_\alpha}{V}\right) + \lambda_\Omega \left(\frac{fs_\beta s_\alpha}{V s_i}\right) + \lambda_\alpha \left(\frac{V^3}{\mu} - \frac{fs_\beta s_\alpha}{V \tan i}\right) \end{aligned} \tag{9.37}$$

This equation leads to the following multiplier equations:

$$\dot{\lambda}_V = -\frac{\partial H}{\partial V} = \lambda_i \frac{fs_\beta c_\alpha}{V^2} + \lambda_\Omega \frac{fs_\beta s_\alpha}{V^2 s_i} - \lambda_\alpha \left(\frac{3V^2}{\mu}\right) - \lambda_\alpha \frac{fs_\beta s_\alpha}{V^2 \tan i} \tag{9.38}$$

$$\dot{\lambda}_i = -\frac{\partial H}{\partial i} = \lambda_\Omega \frac{fs_\beta s_\alpha c_i}{V s_i^2} - \lambda_\alpha \frac{fs_\beta s_\alpha}{V s_i^2} \tag{9.39}$$

$$\dot{\lambda}_\Omega = -\frac{\partial H}{\partial \Omega} = 0 \tag{9.40}$$

$$\dot{\lambda}_\alpha = -\frac{\partial H}{\partial \alpha} = \lambda_i \frac{fs_\beta s_\alpha}{V} - \lambda_\Omega \frac{fs_\beta c_\alpha}{V s_i} + \lambda_\alpha \frac{fs_\beta c_\alpha}{V \tan i} \tag{9.41}$$

The thrust yaw angle β is obtained from the optimality condition

$$\frac{\partial H}{\partial \beta} = \lambda_V fs_\beta + \lambda_i \frac{fc_\beta c_\alpha}{V} + \lambda_\Omega \frac{fc_\beta s_\alpha}{V s_i} - \lambda_\alpha \frac{fc_\beta s_\alpha}{V \tan i} = 0 \tag{9.42}$$

leading to

$$\tan\beta = \frac{-\lambda_i \dfrac{c_\alpha}{V} - \lambda_\Omega \dfrac{s_\alpha}{Vs_i} + \lambda_\alpha \dfrac{s_\alpha}{V\tan i}}{\lambda_V} = \frac{s_\beta}{c_\beta} \tag{9.43}$$

To find the overall minimum-time solution for transfer between given initial and final states, namely, V_0, i_0, Ω_0 and V_f, i_f, Ω_f, respectively, it is necessary to optimize the corresponding initial and final locations on the initial and final orbits, instead of selecting them arbitrarily. The optimal solution, therefore, requires that $(\lambda_\alpha)_0 = (\lambda_\alpha)_f = 0$. Thus, given V_0, i_0, Ω_0, $(\lambda_\alpha)_0 = 0$, and V_f, i_f, Ω_f, $(\lambda_\alpha)_f = 0$, the initial values $(\lambda_V)_0$, $(\lambda_i)_0$, $(\lambda_\Omega)_0$, $(\alpha)_0$, and t_f are guessed, and state and adjoint Eqs. (9.33–9.36) and (9.38– 9.41) are integrated forward from $t_0 = 0$ to t_f using the optimal firing angle β in Eq. (9.43). The guessed values are adjusted by an optimizer program until V_f, i_f, Ω_f, $(\lambda_\alpha)_f = 0$, and $H_f = 0$ are closely matched to within a small tolerance.

An example is now used to generate the relevant optimal transfers, namely, $a_0 = 6563.14$ km, $V_0 = 7.7931587$ km/s, $i_0 = 10$ deg, $\Omega_0 = 20$ deg, and $a_f = 6878$ km, corresponding to $V_f = 7.6126921$ km/s, $i_f = 5$ deg, and $\Omega_f = 10$ deg, with $f = 3.5 \times 10^{-6}$ km/s^2. Because the Hamiltonian system is homogeneous in the multipliers, we can scale the initial values of the multipliers to obtain $H_f = -1$ instead of $H_f = 0$, provided that $H = \boldsymbol{\lambda}^T\dot{\mathbf{x}}$ is used instead of $H = 1 + \boldsymbol{\lambda}^T\dot{\mathbf{x}}$. In this precision-integrated four-state system, $H_f = -1$ is thereby enforced as a boundary condition. Also, H is a constant of the motion, as it is not an explicit function of time. Table 9.1 lists the solutions obtained by the four-state theory, as well as those from the three-state i^* theory of Sec. 9.2 and the purely analytic Edelbaum theory [1, 4]. The two-state (V, i) analytic theory requires the computation of the single relative inclination change i^*, which is directly obtained from the (V_0, i_0) and (V_f, i_f) pairs through Eq. (9.18). Thus, $(\lambda_i)_0$ effectively represents $(\lambda_{i^*})_0$ in the analytic theory.

In the Edelbaum run, i^* goes from 0 to 5.148939835 deg, instead of the reverse, such that $(\lambda_i)_0$ has the opposite sign with respect to the $(\lambda_i)_0$ values of the other two theories.

Table 9.2 lists the initial and final parameters for the three cases.

The $(\alpha)_0$ and $(\alpha)_f$ angular positions are optimized such that they represent the departure and arrival locations on the initial and final orbits, respectively. Also, $(\lambda_\alpha)_f = 0$ is matched closely as a boundary condition, whereas the condition $H_f = 0$ instead of $H_f = -1$ is enforced for the three-state and four-state cases. The histories of the angle β for the three cases are shown in Fig. 9.2, with the analytic and averaged curves being essentially identical because β must switch sign at each antinode to mimic the oscillation of the exact solution. The analytic and averaged β curves have opposite signs because i^* goes from 0 to 5.148939835 deg, as mentioned earlier. However, the two curves are essentially identical. Figure 9.3 shows the evolution of the velocity with a small oscillatory behavior in the exact case, whereas Figs. 9.4 and 9.5 show the evolutions of the inclination and the node.

TABLE 9.1 SOLUTIONS FROM THE THREE-STATE-AVERAGED, FOUR-STATE PRECISION-INTEGRATED, AND ANALYTIC EDELBAUM THEORIES

	$(\lambda_V)_0$ (s/km/s)	$(\lambda_i)_0$ (s/rad)	$(\lambda_\Omega)_0$ (s/rad)	t_f (s)	ΔV (km/s)
Three-state-averaged	0.5915208891×10^5	0.2547555258×10^7	0.4112381940×10^6	3.146527652×10^5	1.1012846
Four-state-unaveraged	0.162483798×10^5	0.240312782×10^7	0.071394913×10^6	3.12638781×10^5	1.0942357
Analytic Edelbaum	0.6646589168×10^5	$-0.3401606421 \times 10^7$	NA	3.146467816×10^5	1.1012637

TABLE 9.2 INITIAL AND FINAL PARAMETERS OBTAINED FROM THE THREE-STATE-AVERAGED, FOUR-STATE PRECISION-INTEGRATED, AND ANALYTIC EDELBAUM THEORIES

	V_0 (km/s) V_f (km/s)	i_0 (deg) i_f (deg)	Ω_0 (deg) Ω_f (deg)	α_0 (deg) α_f (deg)	$(\lambda_\alpha)_0$ (s/rad) $(\lambda_\alpha)_f$ (s/rad)	H_f
Three-state- averaged	7.7931587 7.6126921	10.0 5.000000074	20.0 10.00000030			6.2246×10^{-9}
Four-state- unaveraged	7.7931587 7.6127009	10.0 4.999983319	20.0 9.999944152	345.4613991 46.85238677	0.0 3.49414×10^{-6}	-0.9999785801
Analytic Edelbaum	7.7931587 7.6126921	$i_0^* = 5.148939835$ $i_f^* = 1.590277 \times 10^{-15}$				-1.78258×10^{-17}

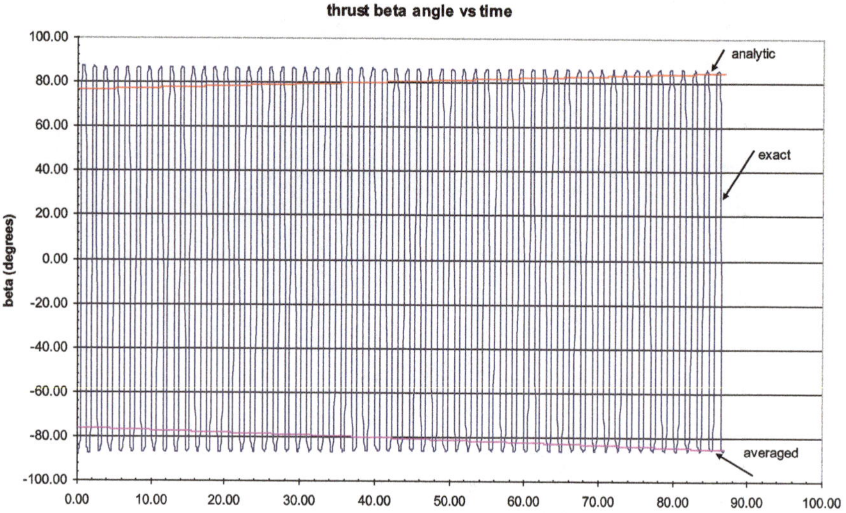

Fig. 9.2 Histories of thrust angle β for the averaged, exact, and Edelbaum theories.

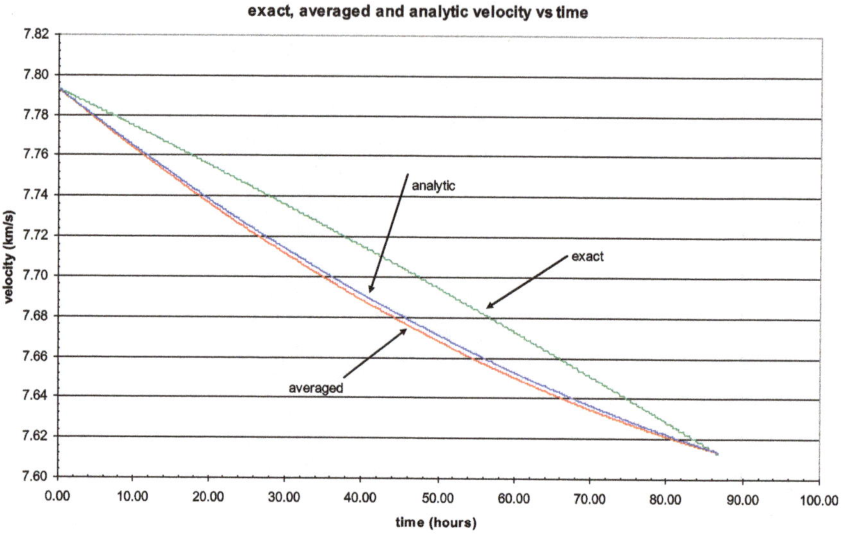

Fig. 9.3 Evolution of orbital velocity for the averaged, exact, and Edelbaum theories.

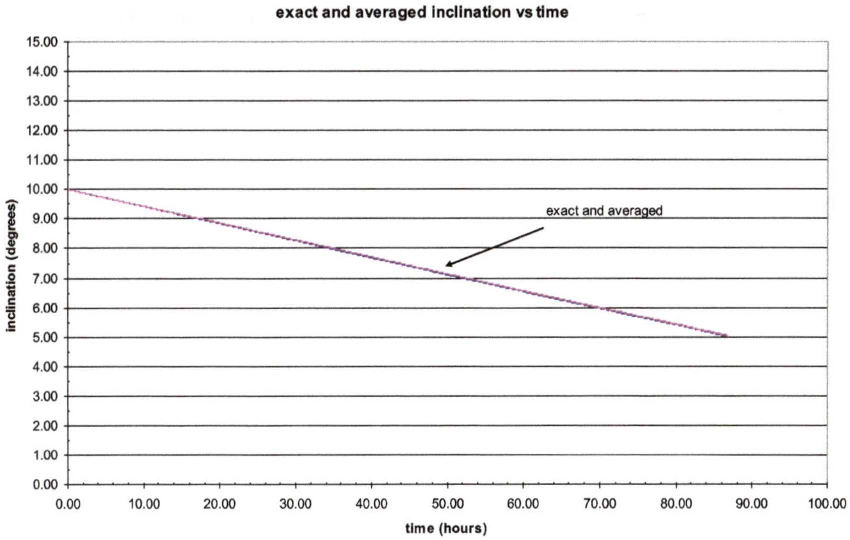

Fig. 9.4 Evolution of orbital inclination for the averaged and exact theories.

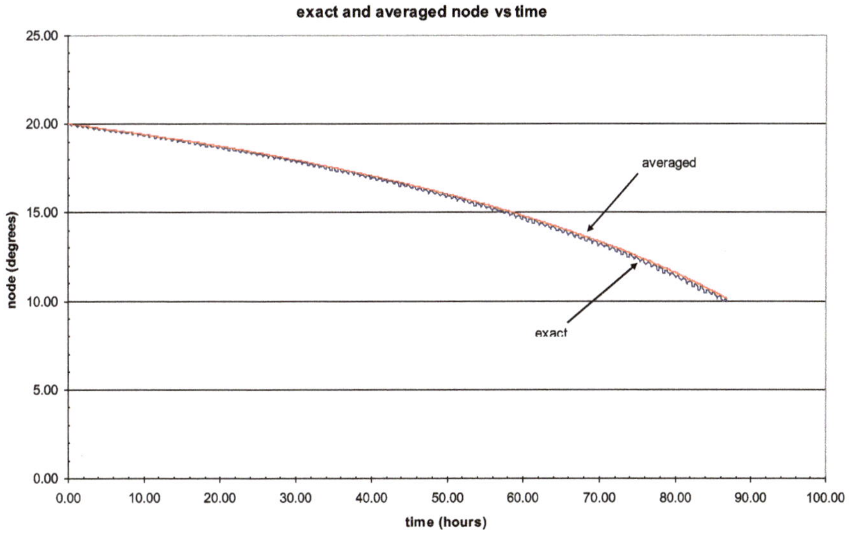

Fig. 9.5 Evolution of the node for the averaged and exact theories.

9.4 PRECISION INTEGRATION OPTIMIZATION USING THE FOUR-STATE SYSTEM UNDER THE INFLUENCE OF J_2

It is well known that, for the case of a circular orbit, for which $r^4 = \mu^4/V^8$ because $V^2 = \mu/r = \mu/a$, the components of the J_2 acceleration along the axes of the rotating Euler–Hill frame $(\hat{\boldsymbol{r}}, \hat{\boldsymbol{\theta}}, \hat{\boldsymbol{h}})$ can be written as

$$(f_r)_{J_2} = -\frac{3\mu J_2 R^2}{2r^4}(1 - 3s_i^2 s_\theta^2) = -\frac{3J_2R^2}{2\mu^3}V^8(1 - 3s_i^2 s_\theta^2) \tag{9.44}$$

$$(f_\theta)_{J_2} = -\frac{3\mu J_2 R^2}{r^4}s_i^2 s_\theta c_\theta = -\frac{3J_2R^2}{\mu^3}V^8 s_i^2 s_\theta c_\theta \tag{9.45}$$

$$(f_h)_{J_2} = -\frac{3\mu J_2 R^2}{r^4}s_i c_i s_\theta = -\frac{3J_2R^2}{\mu^3}V^8 s_i c_i s_\theta \tag{9.46}$$

In Eqs. (9.44–9.46), R represents the radius of the equator, and $(f_r)_{J_2} = -(f_n)_{J_2}$ and $(f_t)_{J_2} = (f_\theta)_{J_2}$ for the circular case. Returning to Eqs. (9.1) and (9.4–9.6), with $\alpha = \theta$, we have

$$\dot{V} = -\left[(f_t)_T + (f_t)_{J_2}\right] = -fc_\beta - (f_\theta)_{J_2} = -fc_\beta + \frac{3J_2R^2}{\mu^3}V^8 s_i^2 s_\alpha c_\alpha \tag{9.47}$$

$$\dot{i} = \left[(f_h)_T + (f_h)_{J_2}\right]\frac{c_\alpha}{V} = \frac{fs_\beta c_\alpha}{V} - \frac{3J_2R^2}{\mu^3}V^7 s_i c_i s_\alpha c_\alpha \tag{9.48}$$

$$\dot{\Omega} = \left[(f_h)_T + (f_h)_{J_2}\right]\frac{s_\alpha}{Vs_i} = \frac{fs_\beta s_\alpha}{Vs_i} - \frac{3J_2R^2}{\mu^3}V^7 c_i s_\alpha^2 \tag{9.49}$$

$$\begin{aligned}\dot{\alpha} &= n - \frac{2(f_r)_{J_2}}{V} - \frac{(f_h)_T s_\alpha}{V\tan i} - \frac{(f_h)_{J_2} s_\alpha}{V\tan i} \\ &= \frac{V^3}{\mu} + \frac{3J_2R^2V^7}{\mu^3}\left[1 + s_\alpha^2(1 - 4s_i^2)\right] - \frac{fs_\beta s_\alpha}{V\tan i}\end{aligned} \tag{9.50}$$

The terms with the subscript T represent the thrust acceleration parts, such that the Hamiltonian of this system is the sum of the thrust parts and the J_2 parts, namely

$$
\begin{aligned}
H &= 1 + \lambda_V \dot{V} + \lambda_i \dot{i} + \lambda_\Omega \dot{\Omega} + \lambda_\alpha \dot{\alpha} = 1 + \lambda_V(-fc_\beta) + \lambda_i \left(\frac{fs_\beta c_\alpha}{V}\right) \\
&+ \lambda_\Omega \left(\frac{fs_\beta s_\alpha}{Vs_i}\right) + \lambda_\alpha \left(\frac{V^3}{\mu} - \frac{fs_\beta s_\alpha}{V \tan i}\right) \\
&+ \lambda_V \left(\frac{3J_2R^2}{\mu^3} V^8 s_i^2 s_\alpha c_\alpha\right) + \lambda_i \left(-\frac{3J_2R^2}{\mu^3} V^7 s_i c_i s_\alpha c_\alpha\right) \\
&+ \lambda_\Omega \left(-\frac{3J_2R^2}{\mu^3} V^7 c_i s_\alpha^2\right) + \lambda_\alpha \left\{\frac{3J_2R^2}{\mu^3} V^7 \left[1 + s_\alpha^2(1 - 4s_i^2)\right]\right\}
\end{aligned}
\tag{9.51}
$$

The Euler–Lagrange equations are, thus, given by

$$
\begin{aligned}
\dot{\lambda}_V &= -\frac{\partial H}{\partial V} = \lambda_i \frac{fs_\beta c_\alpha}{V^2} + \lambda_\Omega \frac{fs_\beta s_\alpha}{V^2 s_i} - \lambda_\alpha \left(\frac{3V^2}{\mu}\right) - \lambda_\alpha \frac{fs_\beta s_\alpha}{V^2 \tan i} \\
&- \lambda_V \left(\frac{3J_2R^2}{\mu^3} 8V^7 s_i^2 s_\alpha c_\alpha\right) - \lambda_i \left(-\frac{3J_2R^2}{\mu^3} 7V^6 s_i c_i s_\alpha c_\alpha\right) \\
&- \lambda_\Omega \left(-\frac{3J_2R^2}{\mu^3} 7V^6 c_i s_\alpha^2\right) - \lambda_\alpha \left\{\frac{3J_2R^2}{\mu^3} 7V^6 \left[1 + s_\alpha^2(1 - 4s_i^2)\right]\right\}
\end{aligned}
\tag{9.52}
$$

$$
\begin{aligned}
\dot{\lambda}_i &= -\frac{\partial H}{\partial i} = \lambda_\Omega \frac{fs_\beta s_\alpha c_i}{Vs_i^2} - \lambda_\alpha \frac{fs_\beta s_\alpha}{Vs_i^2} - \lambda_V \left(\frac{3J_2R^2}{\mu^3} V^8 2s_i c_i s_\alpha c_\alpha\right) \\
&- \lambda_i \left[-\frac{3J_2R^2}{\mu^3} V^7 (c_i^2 - s_i^2) s_\alpha c_\alpha\right] - \lambda_\Omega \left(\frac{3J_2R^2}{\mu^3} V^7 s_i s_\alpha^2\right) \\
&- \lambda_\alpha \left(-\frac{3J_2R^2}{\mu^3} V^7 8 s_\alpha^2 s_i c_i\right)
\end{aligned}
\tag{9.53}
$$

$$
\dot{\lambda}_\Omega = -\frac{\partial H}{\partial \Omega} = 0 \tag{9.54}
$$

$$
\begin{aligned}
\dot{\lambda}_\alpha &= -\frac{\partial H}{\partial \alpha} = \lambda_i \frac{fs_\beta s_\alpha}{V} - \lambda_\Omega \frac{fs_\beta c_\alpha}{Vs_i} + \lambda_\alpha \frac{fs_\beta c_\alpha}{V \tan i} - \lambda_V \left[\frac{3J_2R^2}{\mu^3} V^8 s_i^2 (c_\alpha^2 - s_\alpha^2)\right] \\
&- \lambda_i \left[-\frac{3J_2R^2}{\mu^3} V^7 s_i c_i (c_\alpha^2 - s_\alpha^2)\right] - \lambda_\Omega \left(-\frac{3J_2R^2}{\mu^3} V^7 c_i 2 s_\alpha c_\alpha\right) \\
&- \lambda_\alpha \left[\frac{3J_2R^2}{\mu^3} V^7 2 s_\alpha c_\alpha (1 - 4s_i^2)\right]
\end{aligned}
\tag{9.55}
$$

The optimal firing angle β is once again given by the optimality condition $\partial H/\partial\beta = 0$, leading to

$$\tan\beta = \frac{-\lambda_i \dfrac{c_\alpha}{V} - \lambda_\Omega \dfrac{s_\alpha}{V s_i} + \lambda_\alpha \dfrac{s_\alpha}{V \tan i}}{\lambda_V} = \frac{s_\beta}{c_\beta} \tag{9.56}$$

A search was carried out to solve the same example transfer as discussed in this chapter while simultaneously optimizing both the initial departure location α_0 and the arrival location α_f to determine the overall minimum-time solution. The value for J_2 was set at 1.08263×10^{-3}, and a local minimum, labeled "Minimum 1" in Tables 9.3 and 9.4, was found. Hallman [16] found that, when solving precision-integrated minimum-time transfers with optimized initial and final locations on the initial and final orbits, respectively, the existence of multiple local minima is a distinct possibility. Thus, because of the sensitivity of the problem to both the initial location and, more important, the final location, Hallman ran a series of minimum-time solutions for nearby entry points on the final orbit, meaning that he held the final location α_f fixed each time, optimized the transfer, and found the minimum-time solution while also optimizing the initial departure point. It is essential to fix the final location α_f in cumulative radians measured from the starting point and not an angle between 0 and 360 deg because the minima may occur on different final orbits and not necessarily on the same final one. In Table 9.4, α_f is thus left in radians such that the total number of revolutions is easily obtained by dividing by 2π. Essentially starting each time with guessed values of $(\lambda_V)_0$, $(\lambda_i)_0$, $(\lambda_\Omega)_0$, α_0, and $(\lambda_\alpha)_0 = 0$ and a guessed value of t_f, the five search parameters namely, $(\lambda_V)_0$, $(\lambda_i)_0$, $(\lambda_\Omega)_0$, α_0, and t_f are searched, and the dynamic and adjoint equations (9.47–9.50) and (9.52–9.55) are integrated numerically using the continuously varying β angle given by the optimality condition (9.56) until V_f, i_f, Ω_f, α_f, and $H_f = 0$ are satisfied.

Nonlinear programming software called NLP2, developed at The Aerospace Corporation, was used to carry out the searches with the constraints, denoted c_i, driven to zero to within a small tolerance. Integration controls were set to the 10^{-16} level for both the relative and absolute errors. Thus, at t_f, the constraints $c_1 = V - V_f$, $c_2 = i - i_f$, $c_3 = \Omega - \Omega_f$, $c_4 = \alpha - \alpha_f$, and $c_5 = H - H_f$ were all driven to near zero by NLP2. Spanning a relatively large interval in α_f, Hallman found two other possible minima. Then, a modified set of constraints was used to focus in on the two new minima by replacing $c_4 = \alpha - \alpha_f$ by $c_4 = \lambda_\alpha - (\lambda_\alpha)_f$ with $(\lambda_\alpha)_f = 0$ such that the optimal entry points on the final orbits were finally determined by searching once again on $(\lambda_V)_0$, $(\lambda_i)_0$, $(\lambda_\Omega)_0$, α_0, and t_f with a guessed value of t_f retrieved from the preceding series of runs that corresponded to a local dip in the t_f values. This search was then much easier to carry out because the value of α_f that needed to be optimized was already very close to the optimal value and in its immediate vicinity. Thus, the

TABLE 9.3 SOLUTIONS FROM THE THREE PRECISION-INTEGRATED LOCAL MINIMA (FOUR-STATE-UNAVERAGED) AND THE AVERAGED (THREE-STATE WITH PIECEWISE-CONSTANT YAW) APPROACHES

	$(\lambda_V)_0$ (s·km^{-1}·s^{-1})	$(\lambda_i)_0$ (s/rad)	$(\lambda_\Omega)_0$ (s/rad)	α_0 (deg)	t_f (s)	ΔV (km/s)
Minimum 1	0.689315697×10^5	0.299691308×10^7	-0.178480579×10^6	-13.2309819	3.42012214×10^5	1.1970427
Minimum 2	0.288250837×10^6	0.750433878×10^7	-0.430366566×10^6	-34.388762	3.41673186×10^5	1.1958561
Minimum 3	0.109621864×10^6	0.383795138×10^7	-0.224840404×10^6	-20.6460985	3.41282707×10^5	1.1944895
Averaged	0.546709224×10^6	0.214122398×10^8	-0.547250956×10^6	NA	3.88355734×10^5	1.3592451

TABLE 9.4 INITIAL AND FINAL PARAMETERS OBTAINED FOR THE THREE PRECISION-INTEGRATED LOCAL MINIMA (FOUR-STATE-UNAVERAGED) AND AVERAGED (THREE-STATE WITH PIECEWISE-CONSTANT YAW) APPROACHES

	V_0 (km/s) V_f (km/s)	i_0 (deg) i_f (deg)	Ω_0 (deg) Ω_f (deg)	α_0 (deg) α_f (rad)	$(\lambda_\alpha)_0$ (s/rad) $(\lambda_\alpha)_f$ (s/rad)	H_f
Minimum 1	7.7931587 7.6126915	10.0 4.99999815	20.0 9.99996333	−13.2309819 386.639054	0.0 0.216353×10^{-5}	0.384556×10^{-5}
Minimum 2	7.7931587 7.61269218	10.0 5.00000000	20.0 10.00000000	−34.3887621 380.862035	0.0 -0.206772×10^{-5}	$-0.342145 \times 10^{-12}$
Minimum 3	7.7931587 7.61269218	10.0 5.00000000	20.0 10.00000000	−20.6460985 383.723086	0.0 -0.713620×10^{-6}	$-0.632865 \times 10^{-10}$
Averaged	7.7931587 7.61267551	10.0 4.99999999	20.0 9.99915629	NA NA	NA NA	-0.386454×10^{-4}

other two minima were found as reported in Tables 9.3 and 9.4. As can be seen, the three minima are clustered to within about 5 min of each other in transfer time. Minimum 3 is clearly the global minimum for this example transfer. Note that the Hamiltonian, because it is not an explicit function of time, is constant throughout all three trajectories at its near-zero value. Also, the $(\lambda_\alpha)_f$ values are near-zero for the optimized entry locations, which differ by only about 3 rad in cumulative α values. The final V_f, i_f, and Ω_f parameters are also clearly matched. Note that H is a constant of the motion, being independent of time. Minimum 1 was generated using the algorithm UNCMIN [17], which minimizes the sum of squares of the errors at t_f, whereas minima 2 and 3 were generated by NLP2, resulting in higher resolution, as seen by the values of H_f in Table 9.4.

9.5 THREE-STATE AVERAGED SYSTEM WITH J_2

It is well known that the secular perturbations of the first order do not affect the mean values of a, e, and i. For example

$$\left(\dot{\tilde{i}}\right)_{J_2} = \frac{1}{2\pi}\int_0^{2\pi}(\dot{i})_{J_2}\,\mathrm{d}\alpha = \frac{1}{2\pi}\int_0^{2\pi}\frac{(f_h)_{J_2}c_\alpha\,\mathrm{d}\alpha}{V} = -\frac{1}{2\pi}\frac{3J_2R^2}{\mu^3}V^7 s_i c_i\int_0^{2\pi}s_\alpha c_\alpha\,\mathrm{d}\alpha = 0$$

However

$$\left(\dot{\tilde{\Omega}}\right)_{J_2} = \frac{1}{2\pi}\int_0^{2\pi}(\dot{\Omega})_{J_2}\,\mathrm{d}\alpha = \frac{1}{2\pi}\int_0^{2\pi}\frac{(f_h)_{J_2}s_\alpha\,\mathrm{d}\alpha}{Vs_i} = -\frac{3}{2\pi}\frac{J_2R^2}{\mu^3}V^7c_i\int_0^{2\pi}s_\alpha^2\,\mathrm{d}\alpha$$
$$= -\frac{3}{2}\frac{J_2R^2}{\mu^3}V^7c_i \tag{9.57}$$

which leads to the following system of three equations for V, i, and Ω, while neglecting the $\dot{\alpha}$ and $\dot{M}$ equations,

$$\dot{\tilde{V}} = -fc_\beta \tag{9.58}$$

$$\dot{\tilde{i}} = \frac{2fs_\beta}{\pi V}c_{\theta_{c_n}} \tag{9.59}$$

$$\dot{\tilde{\Omega}} = \frac{2fs_\beta}{\pi Vs_i}s_{\theta_{c_n}} - \frac{3}{2}\frac{J_2R^2}{\mu^3}V^7c_i \tag{9.60}$$

The use of the mean elements effectively neglects the short-period perturbations due to J_2. This assumption becomes increasingly valid as the thrust level decreases.

As in the first section, θ_c is the instantaneous angular position of the relative line of nodes, which is now being perturbed due to the Ω precession induced by J_2. Thus, the thrust averaging holds the β angle piecewise-constant, switching signs at the relative antinodes (i.e., ± 90 deg from θ_c), with the latter represented now by θ_{c_n} in Fig. 9.1. It is thus also necessary to update the value of θ_f represented by θ_{f_n} because it is no longer constant. The instantaneous orbit orientation is now defined by Ω_n and i_n, as these quantities are being integrated numerically. The relative inclination i_n^* of the instantaneous orbit with respect to the final orbit is given by

$$c_{i_n^*} = c_{\Omega_n-\Omega_2} s_{i_n} s_{i_2} + c_{i_n} c_{i_2} \tag{9.61}$$

If $\Omega_n < \Omega_f$ during integration, then, from spherical trigonometry

$$s_{\theta_{f_n}} = s_{i_n} s_{\Omega_2-\Omega_n} / s_{i_n^*} \tag{9.62}$$

$$c_{\theta_{f_n}} = (c_{i_n} - c_{i_2} c_{i_n^*})/(s_{i_2} s_{i_n^*}) \tag{9.63}$$

which allows us to update θ_{c_n} as follows:

$$s_{\theta_{c_n}} = s_{i_2} s_{\Omega_2-\Omega_n} / s_{i_n^*} \tag{9.64}$$

$$c_{\theta_{c_n}} = c_{\Omega_2-\Omega_n} c_{\theta_{f_n}} - s_{\Omega_2-\Omega_n} s_{\theta_{f_n}} c_{i_2} \tag{9.65}$$

If $\Omega_n > \Omega_f$ during integration, then θ_{f_n} and θ_{c_n} are obtained from

$$s_{\theta_{f_n}} = s_{i_n} s_{\Omega_n-\Omega_2} / s_{i_n^*} \tag{9.66}$$

$$c_{\theta_{f_n}} = (c_{i_n^*} c_{i_2} - c_{i_n})/(s_{i_n^*} s_{i_2}) \tag{9.67}$$

$$s_{\theta_{c_n}} = s_{i_2} s_{\Omega_n-\Omega_2} / s_{i_n^*} \tag{9.68}$$

$$c_{\theta_{c_n}} = c_{\theta_{f_n}} c_{\Omega_n-\Omega_2} + s_{\theta_{f_n}} s_{\Omega_n-\Omega_2} c_{i_2} \tag{9.69}$$

This reduced three-state dynamics description now accepts the averaged Hamiltonian given by

$$\begin{aligned} \tilde{H} &= 1 + \tilde{\lambda}_V \dot{\tilde{V}} + \tilde{\lambda}_i \dot{\tilde{i}} + \tilde{\lambda}_\Omega \dot{\tilde{\Omega}} = 1 - \tilde{\lambda}_V f c_\beta + \tilde{\lambda}_i \frac{2 f s_\beta}{\pi \tilde{V}} c_{\theta_{c_n}} \\ &+ \tilde{\lambda}_\Omega \left(\frac{2 f s_\beta}{\pi \tilde{V} s_{\tilde{i}}} s_{\theta_{c_n}} \right) + \tilde{\lambda}_\Omega \left(-\frac{3}{2} \frac{J_2 R^2}{\mu^3} \tilde{V}^7 c_{\tilde{i}} \right) \end{aligned} \tag{9.70}$$

leading to the adjoints

$$\dot{\tilde{\lambda}}_V = -\partial \tilde{H}/\partial \tilde{V} = \frac{2 f s_\beta}{\pi \tilde{V}^2} \left(\tilde{\lambda}_i c_{\theta_{c_n}} + \frac{\tilde{\lambda}_\Omega}{s_{\tilde{i}}} s_{\theta_{c_n}} \right) + \tilde{\lambda}_\Omega \frac{3}{2} \frac{J_2 R^2}{\mu^3} 7 \tilde{V}^6 c_{\tilde{i}} \tag{9.71}$$

$$\dot{\tilde{\lambda}}_i = -\partial \tilde{H}/\partial \tilde{i} = \tilde{\lambda}_\Omega \frac{2 f s_\beta}{\pi \tilde{V}} \frac{c_{\tilde{i}}}{s_{\tilde{i}}^2} s_{\theta_{c_n}} - \tilde{\lambda}_\Omega \left(\frac{3}{2} \frac{J_2 R^2}{\mu^3} \tilde{V}^7 s_{\tilde{i}} \right) \tag{9.72}$$

$$\dot{\tilde{\lambda}}_\Omega = -\partial \tilde{H}/\partial \tilde{\Omega} = 0 \tag{9.73}$$

The firing angle is given by the optimality condition $\partial\tilde{H}/\partial\beta = 0$, leading to

$$\tan\beta = -\left(\tilde{\lambda}_i c_{\theta_{c_n}} + \frac{\tilde{\lambda}_\Omega}{s_{\tilde{i}}} s_{\theta_{c_n}}\right)\frac{2}{\pi\tilde{V}\tilde{\lambda}_V} = \frac{s_\beta}{c_\beta} \tag{9.74}$$

Thus, Eqs. (9.58–9.60) and (9.71–9.73) constitute the set of dynamic and adjoint equations to be integrated simultaneously by using the optimal β angle in Eq. (9.74). The search now consists of $(\tilde{\lambda}_V)_0, (\tilde{\lambda}_i)_0, (\tilde{\lambda}_\Omega)_0$, and t_f, such that, starting from $\tilde{V}_0$, $\tilde{i}_0$, and $\tilde{\Omega}_0$ the final values $\tilde{V}_f$, $\tilde{i}_f$, and $\tilde{\Omega}_f$are matched, along with $\tilde{H}_f = 0$. The solution is listed in Tables 9.3 and 9.4 in the row labelled "Averaged." This optimization problem is much easier to solve than the corresponding precision-integration problem of Sec. 9.4, and it visibly has a single local minimum within the time frame t_f associated with the cases discussed thus far, as the clustered minima vanish because of the averaging. Note that t_f is now some 13.79% larger than the value corresponding to the global precision-integrated minimum 3. Equivalently, the velocity change required for the transfer is equally larger by the same percentage. However, the use of a piecewise-constant β thrust yaw angle is much easier to implement in an operational sense than the continuously varying thrust vector orientation needed to implement the unaveraged transfer. Figure 9.6 shows the variations in the angles θ_{f_n} and θ_{c_n} from $(\theta_{c_n})_0 = 9.7086461$ deg and $(\theta_{f_n})_0 = 19.6329215$ deg

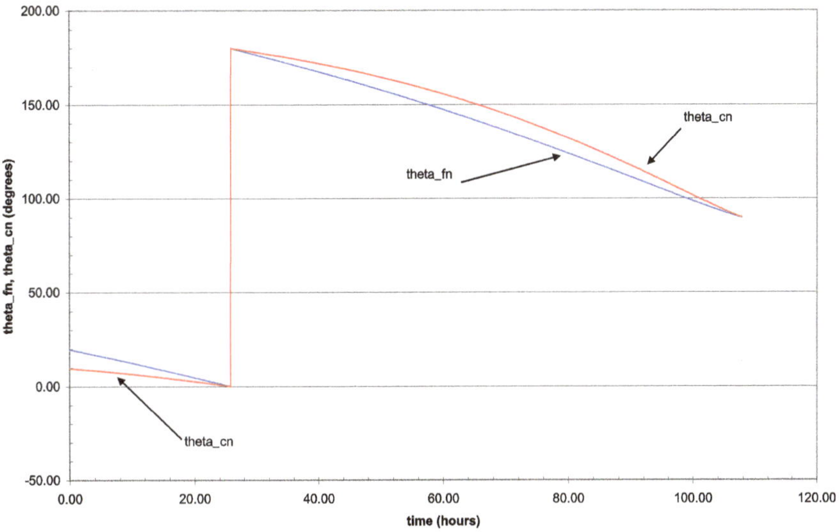

Fig. 9.6 Evolution of the θ_{c_n} and θ_{f_n} angular position angles for the J_2-perturbed averaged theory.

to the essentially common final values of $(\theta_{c_n})_f = 89.9978957$ deg and $(\theta_{f_n})_0 = 89.9970552$ deg. Also i_n^* decreases monotonically from $(i^*)_0 = 5.1489398$ deg to $(i^*)_f = 0.73534 \times 10^{-4}$, or effectively zero, as the current orbit merges with the final target orbit. Figures 9.7 and 9.8 show the variations in inclination for the three exact trajectories, as well as the averaged one, with the latter curve being smooth, without any of the small fluctuations appearing on the exact curves. Note that minimum 2 exhibits the largest decrease in velocity during the actual transfer, meaning that the orbit expands to a larger size before shrinking back toward its target size. A very small similar effect is noted near the end of the trajectory for the averaged transfer.

In Fig. 9.9, the node Ω wanders beyond the final Ω_f value of 10 deg before reversing course and reaching the target 10 deg mark. This drift is very large for the averaged transfer and is responsible for the corresponding longer flight time. This large deviation is due to the strategy employed by the averaged transfer, consisting of rotating the current orbit around the current line of nodes of the current and final orbits. Such an approach is not necessarily very efficient, because unwanted changes in Ω are being generated in addition to the J_2-induced change itself. This is why Ω decreases much more rapidly than desired. However, the rotation strategy will eventually achieve the final orbit because the wedge angle will be driven to zero over time. Note that the inclination curve in Fig. 9.7 stays rather close to the exact curves. A large-scale parametric study involving orbit size and orientation as well as thrust acceleration may

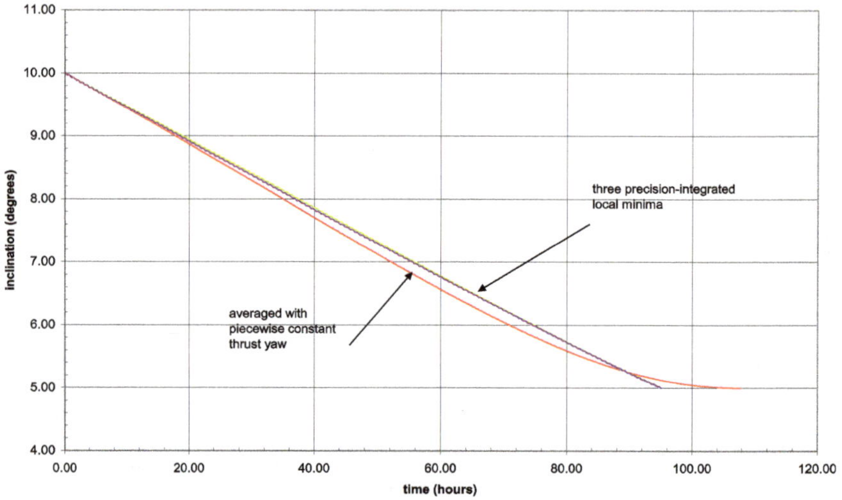

Fig. 9.7 Inclination variation for the three exact local minima and the averaged transfer under the influence of J_2.

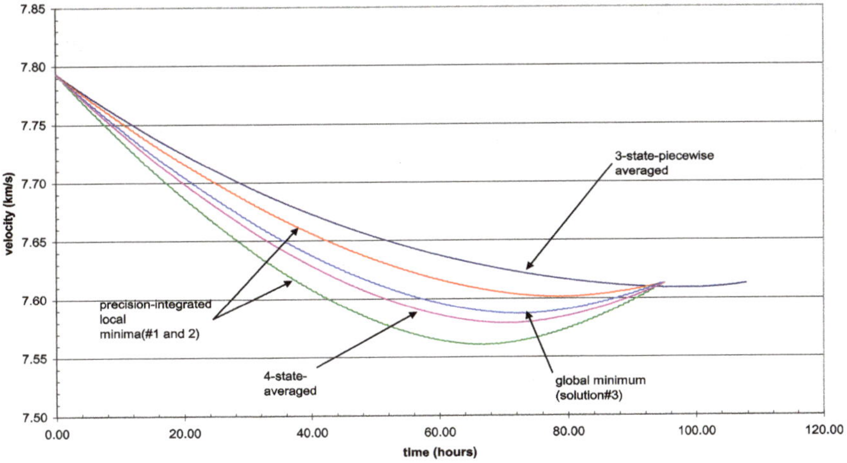

Fig. 9.8 Velocity variation for the three exact local minima and the averaged transfer under the influence of J_2.

reveal the favorable and less favorable regions where it would be more or less to implement this particular averaging strategy without incurring significant penalties in total transfer time.

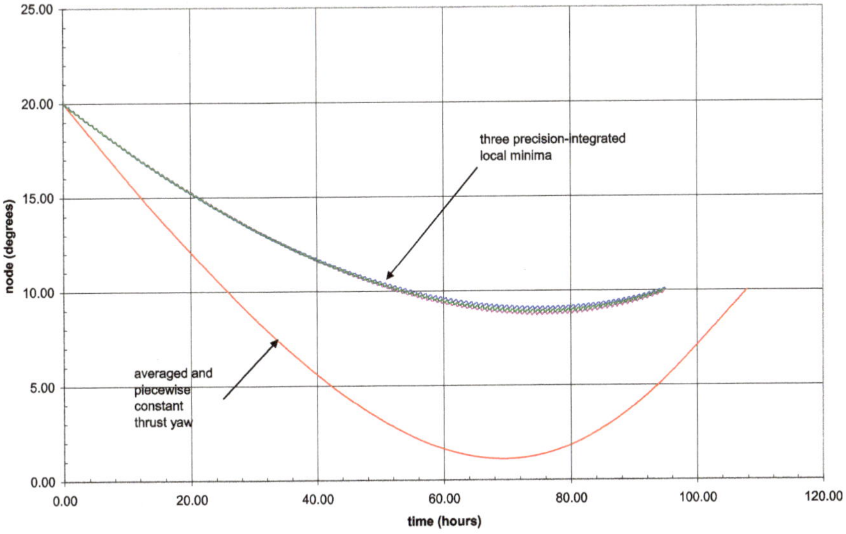

Fig. 9.9 Node variation for the three exact local minima and the averaged transfer under the influence of J_2.

Figure 9.10 shows the evolution of (λ_α) for the last few hours of the exact transfers, with high-frequency oscillations crossing the zero value multiple times, making the enforcement of the constraint $(\lambda_\alpha)_f = 0$ problematic, especially when all three curves end at zero so close to one another. In Fig. 9.11, the continuously varying angle β for all three minima exhibit complex behaviors near the ends of the trajectory, which could make the control of the thrust vector orientation difficult to implement. Figures 9.12 and 9.13 show the small fluctuations in the inclination and node mentioned earlier on a scale that is easily visible near the point where the trajectory ends. The small difference in flight times is clearly visible as i and Ω reach their target values of 5 and 10 deg, respectively. Finally, Fig. 9.14 shows the history of angle β for the global minimum exact solution, with the complex pattern shown near the end as discussed regarding Fig. 9.11, and the much smoother history for the averaged solution, which effectively changes sign at the antinodes within each orbit. As for the thrust-only averaged case, the Hamiltonian $\tilde{H}$ is not constant because of the formulation involving the angle θ_c.

The Hamiltonian $\tilde{H}$ decreases from -4.20434 at time zero to about -4.55239 at $t = 44.4$ h and then increases in a parabolic manner until $\tilde{H} = 0$ at t_f. Several of the plots shown in this chapter can also be presented as a function of running orbit number. For example, Eq. (9.11) for the angular position θ can be integrated using $n = V^3/\mu$ with V given as a function of time t [4]. However, these plots were left as-is for easier interpretation.

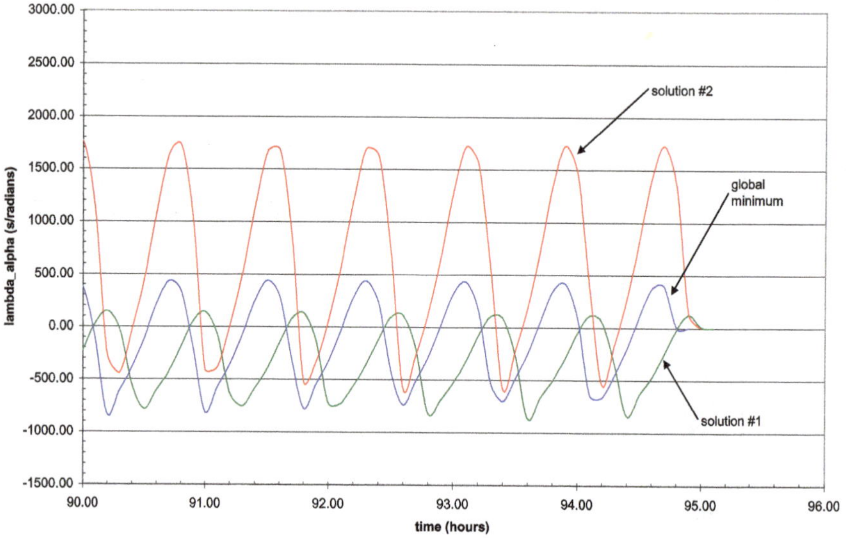

Fig. 9.10 Evolution of λ_α for the three local minima under the influence of J_2 during the last phase of transfers.

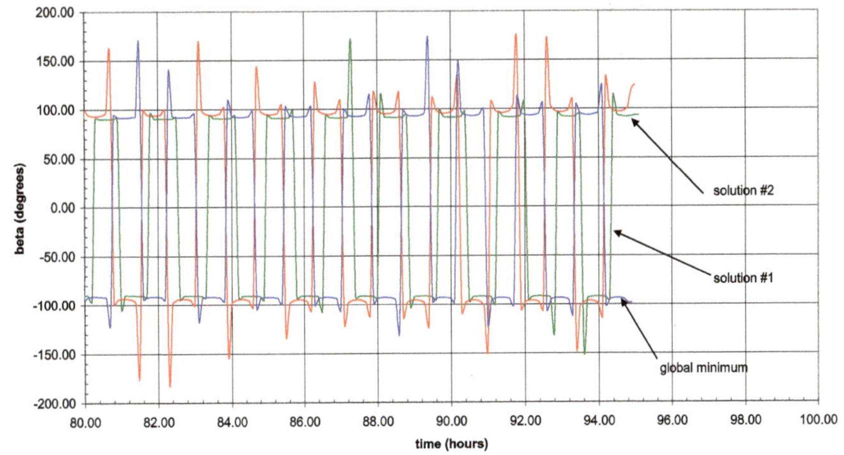

Fig. 9.11 Histories of thrust angle β for the three local minima under the influence of J_2 during the last phase of transfers.

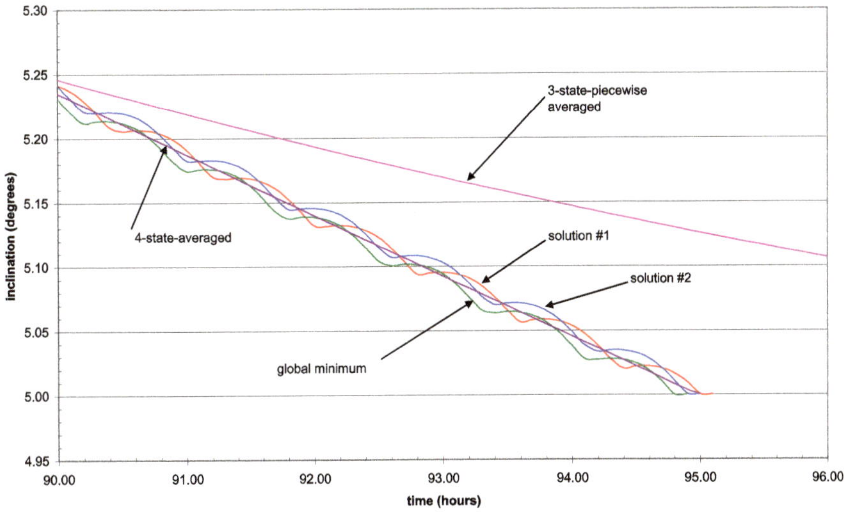

Fig. 9.12 Inclination variation for the three local minima under the influence of J_2 during the last phase of transfers.

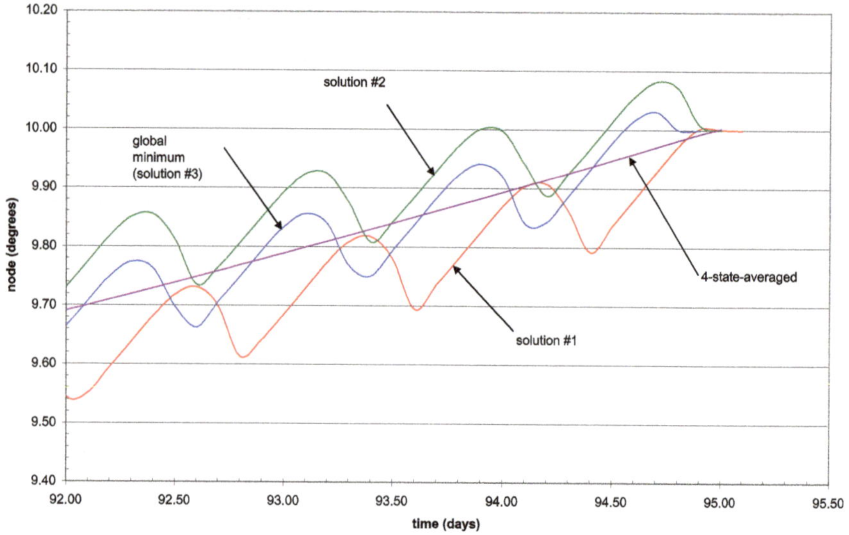

Fig. 9.13 Node variation for the three local minima under the influence of J_2 during the last phase of transfers.

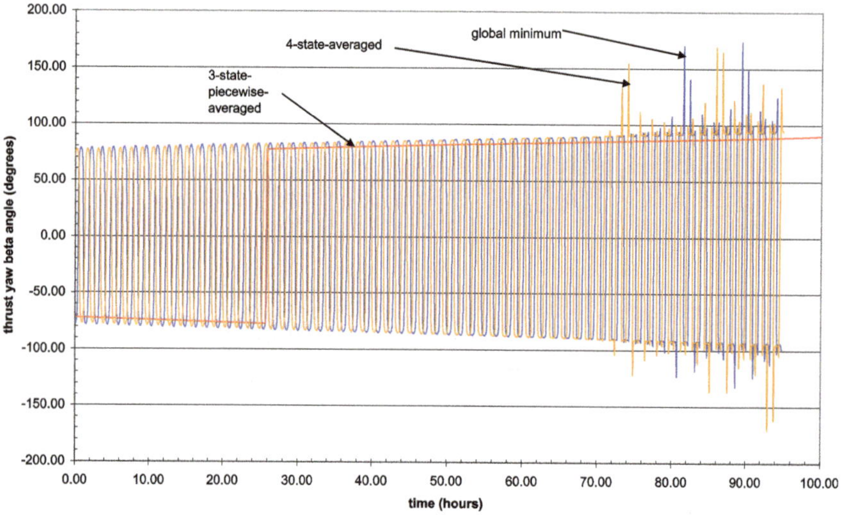

Fig. 9.14 Histories of thrust angle β for the global minimum and averaged transfers under the influence of J_2.

9.6 AVERAGING OF THE J_2-PERTURBED FOUR-STATE DYNAMICS USING NUMERICAL QUADRATURE

Returning to Eqs. (9.47–9.50) for $\dot{V}$, $\dot{i}$, $\dot{\Omega}$, and $\dot{\alpha}$, one can write the averaged Hamiltonian as

$$\tilde{H} = \frac{1}{T_0}\int_0^{T_0} H \, \mathrm{d}t = \frac{1}{T_0}\int_{-\pi}^{\pi} \frac{H \, \mathrm{d}\alpha}{\dot{\alpha}(\tilde{z},\alpha)} \tag{9.75}$$

where H is given by Eq. (9.51); $\tilde{z} = (\tilde{V} \ \tilde{i} \ \tilde{\Omega} \ \tilde{\alpha})^T$ represents the state vector; and T_0 is the orbit period, which is given by $T_0 = 2\pi/\tilde{n}$. From Kepler's equation and the assumption of a circular orbit (i.e., $\tilde{r} = \tilde{a}$), $\tilde{n} = \dot{E} = \dot{\alpha} = \tilde{V}^3/\mu$, where E is the eccentric anomaly in Kepler's equation $\tilde{M} = E - \tilde{e}s_E = \tilde{n}t$. Thus

$$\tilde{H} = \frac{1}{2\pi}\int_{-\pi}^{\pi} H \, \mathrm{d}\alpha \tag{9.76}$$

leading to the state and adjoint equations

$$\dot{\tilde{z}} = \partial\tilde{H}/\partial\tilde{\lambda}_z = \frac{1}{2\pi}\int_{-\pi}^{\pi} \left(\frac{\partial H}{\partial\tilde{\lambda}_z}\right)^T \mathrm{d}\alpha \tag{9.77}$$

$$\dot{\tilde{\lambda}}_z = -(\partial\tilde{H}/\partial\tilde{z})^T = -\frac{1}{2\pi}\int_{-\pi}^{\pi} \left(\frac{\partial H}{\partial\tilde{z}}\right)^T \mathrm{d}\alpha \tag{9.78}$$

such that the averaged state derivatives are obtained by quadrature from the equations

$$\dot{\tilde{V}} = \frac{1}{2\pi}\int_{-\pi}^{\pi} \dot{V} \, \mathrm{d}\alpha = \frac{1}{2\pi}\int_{-\pi}^{\pi} \left(-fc_\beta + \frac{3J_2R^2}{\mu^3} V^8 s_i^2 s_\alpha c_\alpha\right) \mathrm{d}\alpha \tag{9.79}$$

$$\dot{\tilde{i}} = \frac{1}{2\pi}\int_{-\pi}^{\pi} \dot{i} \, \mathrm{d}\alpha = \frac{1}{2\pi}\int_{-\pi}^{\pi} \left(\frac{fs_\beta c_\alpha}{V} - \frac{3J_2R^2}{\mu^3} V^7 s_i c_i s_\alpha c_\alpha\right) \mathrm{d}\alpha \tag{9.80}$$

$$\dot{\tilde{\Omega}} = \frac{1}{2\pi}\int_{-\pi}^{\pi} \dot{\Omega} \, \mathrm{d}\alpha = \frac{1}{2\pi}\int_{-\pi}^{\pi} \left(\frac{fs_\beta s_\alpha}{Vs_i} - \frac{3J_2R^2}{\mu^3} V^7 c_i s_\alpha^2\right) \mathrm{d}\alpha \tag{9.81}$$

$$\begin{aligned}\dot{\tilde{\alpha}} &= \frac{1}{2\pi}\int_{-\pi}^{\pi} \dot{\alpha} \, \mathrm{d}\alpha \\ &= \frac{1}{2\pi}\int_{-\pi}^{\pi} \left\{\frac{V^3}{\mu} - \frac{fs_\beta s_\alpha}{V \tan i} + \frac{3J_2R^2V^7}{\mu^3}\left[1 + s_\alpha^2(1 - 4s_i^2)\right]\right\} \mathrm{d}\alpha\end{aligned} \tag{9.82}$$

In a similar way, the averaged adjoint derivatives are obtained from

$$
\begin{aligned}
\dot{\tilde{\lambda}}_V &= \frac{1}{2\pi}\int_{-\pi}^{\pi} \dot{\lambda}_V \,\mathrm{d}\alpha = \frac{1}{2\pi}\int_{-\pi}^{\pi}\left(\frac{\lambda_i f s_\beta c_\alpha}{V^2} + \frac{\lambda_\Omega f s_\beta s_\alpha}{V^2 s_i} - \lambda_\alpha \frac{3V^2}{\mu} - \lambda_\alpha \frac{f s_\beta s_\alpha}{V^2 \tan i}\right.\\
&\quad - \lambda_V\left(\frac{3J_2R^2}{\mu^3} 8V^7 s_i^2 s_\alpha c_\alpha\right) - \lambda_i\left(-\frac{3J_2R^2}{\mu^3} 7V^6 s_i c_i s_\alpha c_\alpha\right)\\
&\quad \left. -\lambda_\Omega\left(-\frac{3J_2R^2}{\mu^3} 7V^6 c_i s_\alpha^2\right) - \lambda_\alpha\left\{\frac{3J_2R^2}{\mu^3} 7V^6\left[1 + s_\alpha^2(1 - 4s_i^2)\right]\right\}\right) \mathrm{d}\alpha
\end{aligned}
\tag{9.83}
$$

$$
\begin{aligned}
\dot{\tilde{\lambda}}_i &= \frac{1}{2\pi}\int_{-\pi}^{\pi} \dot{\lambda}_i \,\mathrm{d}\alpha = \frac{1}{2\pi}\int_{-\pi}^{\pi}\left[\frac{\lambda_\Omega f s_\beta c_i s_\alpha}{V s_i^2} - \frac{\lambda_\alpha f s_\beta s_\alpha}{V s_i^2} - \lambda_V \frac{3J_2R^2}{\mu^3} V^8 2 s_i c_i s_\alpha c_\alpha\right.\\
&\quad \left. -\lambda_i\left(-\frac{3J_2R^2}{\mu^3} V^7 c_{2i} s_\alpha c_\alpha\right) - \lambda_\Omega\left(\frac{3J_2R^2}{\mu^3} V^7 s_i s_\alpha^2\right) - \lambda_\alpha\left(-\frac{3J_2R^2}{\mu^3} V^7 8 s_i c_i s_\alpha^2\right)\right] \mathrm{d}\alpha
\end{aligned}
\tag{9.84}
$$

$$
\dot{\tilde{\lambda}}_\Omega = \frac{1}{2\pi}\int_{-\pi}^{\pi} \dot{\lambda}_\Omega \,\mathrm{d}\alpha = 0 \tag{9.85}
$$

$$
\begin{aligned}
\dot{\tilde{\lambda}}_\alpha &= \frac{1}{2\pi}\int_{-\pi}^{\pi} \dot{\lambda}_\alpha \,\mathrm{d}\alpha = \frac{1}{2\pi}\int_{-\pi}^{\pi}\left\{\frac{\lambda_i f s_\beta s_\alpha}{V} - \frac{\lambda_\Omega f s_\beta c_\alpha}{V s_i} + \frac{\lambda_\alpha f s_\beta c_\alpha}{V \tan i}\right.\\
&\quad - \lambda_V\left(\frac{3J_2R^2}{\mu^3} V^8 s_i^2 c_{2\alpha}\right) - \lambda_i\left(-\frac{3J_2R^2}{\mu^3} V^7 s_i c_i c_{2\alpha}\right) - \lambda_\Omega\left(-\frac{3J_2R^2}{\mu^3} V^7 c_i 2 s_\alpha c_\alpha\right)\\
&\quad \left. -\lambda_\alpha\left[\frac{3J_2R^2}{\mu^3} V^7 (1 - 4s_i^2) 2 s_\alpha c_\alpha\right]\right\} \mathrm{d}\alpha
\end{aligned}
\tag{9.86}
$$

The optimal firing angle β is given by

$$
\tan\beta = \frac{-\dfrac{\tilde{\lambda}_i c_{\tilde{\alpha}}}{\tilde{V}} - \dfrac{\tilde{\lambda}_\Omega s_{\tilde{\alpha}}}{\tilde{V} s_{\tilde{i}}} + \dfrac{\tilde{\lambda}_\alpha s_{\tilde{\alpha}}}{\tilde{V} \tan \tilde{i}}}{\tilde{\lambda}_V} = \frac{s_\beta}{c_\beta} \tag{9.87}
$$

It is clear that β is continuously varying and that the J_2 terms in the expressions for $\dot{\tilde{V}}$ and $\dot{\tilde{i}}$ involving $s_\alpha c_\alpha$ will not contribute and can be safely neglected. The same is true for the $s_\alpha c_\alpha$ terms in the integrals for $\dot{\tilde{\lambda}}_V$, $\dot{\tilde{\lambda}}_i$, and $\dot{\tilde{\lambda}}_\alpha$. Also, the $c_{2\alpha}$ terms in the expression for $\dot{\tilde{\lambda}}_\alpha$ can be neglected for the same reason.

Because the angle β is associated with the thrust terms and not the J_2 terms in Eqs. (9.79–9.86), the angular position α can be averaged out of the J_2 terms analytically without the need for numerical quadrature. The $s_\alpha c_\alpha$ terms will not contribute, whereas the s_α^2 terms will contribute π from the integrations, such that these equations simplify to the following set:

$$\dot{\tilde{V}} = \frac{1}{2\pi}\int_{-\pi}^{\pi} -fc_\beta \, d\alpha \tag{9.88}$$

$$\dot{\tilde{i}} = \frac{1}{2\pi}\int_{-\pi}^{\pi} \frac{fs_\beta c_\alpha}{V} \, d\alpha \tag{9.89}$$

$$\dot{\tilde{\Omega}} = \frac{1}{2\pi}\int_{-\pi}^{\pi} \frac{fs_\beta s_\alpha}{Vs_i} \, d\alpha - \frac{3}{2}\frac{J_2R^2}{\mu^3}V^7 c_i \tag{9.90}$$

$$\dot{\tilde{\alpha}} = \frac{1}{2\pi}\int_{-\pi}^{\pi} \left(\frac{V^3}{\mu} - \frac{fs_\beta s_\alpha}{V\tan i}\right) d\alpha + \frac{3J_2R^2V^7}{\mu^3}\left(\frac{3}{2} - 2s_i^2\right) \tag{9.91}$$

$$\dot{\tilde{\lambda}}_i = \frac{1}{2\pi}\int_{-\pi}^{\pi} \left(\frac{\lambda_\Omega fs_\beta s_\alpha c_i}{Vs_i^2} - \frac{\lambda_\alpha fs_\beta s_\alpha}{Vs_i^2}\right) d\alpha + \frac{3}{2}\frac{J_2R^2}{\mu^3}V^7(-\lambda_\Omega s_i + 8\lambda_\alpha s_i c_i) \tag{9.92}$$

$$\dot{\tilde{\lambda}}_V = \frac{1}{2\pi}\int_{-\pi}^{\pi} \left(\frac{\lambda_i fs_\beta c_\alpha}{V^2} + \frac{\lambda_\Omega fs_\beta s_\alpha}{V^2 s_i} - \lambda_\alpha\frac{3V^2}{\mu} - \lambda_\alpha\frac{fs_\beta s_\alpha}{V^2\tan i}\right) d\alpha + \frac{3}{2}\frac{J_2R^2}{\mu^3}7V^6\left[\lambda_\Omega c_i - \lambda_\alpha(3 - 4s_i^2)\right] \tag{9.93}$$

$$\dot{\tilde{\lambda}}_\Omega = 0 \tag{9.94}$$

$$\dot{\tilde{\lambda}}_\alpha = \frac{1}{2\pi}\int_{-\pi}^{\pi} \left(\lambda_i\frac{fs_\beta s_\alpha}{V} - \frac{\lambda_\Omega fs_\beta c_\alpha}{Vs_i} + \lambda_\alpha\frac{fs_\beta c_\alpha}{V\tan i}\right) d\alpha \tag{9.95}$$

When J_2 is turned off, because $\tilde{H}$ is not an explicit function of time, it effectively remains constant, and a relatively small number of segments on the order of 16 is needed to carry out the eight-point Gauss–Legendre quadrature in this particular example of relatively high acceleration $f = 3.5 \times 10^{-6}$ km/s^2. However, when J_2 is on, the history of β, which is essentially oscillatory in the beginning of the trajectory, becomes more complicated near the end, as shown in Fig. 9.14 for the exact case. The transition from the purely oscillatory mode to the more complicated one challenges the constancy of $\tilde{H}$ during that period, requiring a large number of segments on the order of 128 and higher for the quadrature to force $\tilde{H}$ to remain constant during the transition period as well, making the numerical integrations more time-consuming. This difficulty may be the result of using relatively high-thrust accelerations combined with J_2 in the LEO environment. The following solution is obtained: $(\lambda_V)_0 = 0.140168691 \times 10^6$ s/km/s, $(\lambda_i)_0 = 0.445773755 \times 10^7$ s/rad, $(\lambda_\Omega)_0 = -0.259177663 \times 10^6$ s/rad, $\alpha_0 = 112.007934$ deg, $(\lambda_\alpha)_f = -0.406767769 \times 10^{-7}$ s/rad, $H_f = -0.148932970 \times 10^{-3}$, and $t_f = 0.341928948 \times 10^6$ s, along with the parameters $V_f = 7.61252604$ km/s,

$i_f = 5.000040$ deg, $\Omega_f = 10.002697$ deg, and $\alpha_f = 385.990903$ deg. Figure 9.8 shows the variation of the velocity, closely following the results corresponding to the exact global minimum, whereas Figs. 9.12 and 9.13 show the corresponding variations in the inclination and node, respectively, without the oscillations that are present in the exact local minima. Finally, Fig. 9.14 shows the β history for the four-state averaged case, with the transition taking place a little earlier than for the precision-integrated global minimum transfer.

9.7 CONCLUSION

The technique of analytic averaging of the dynamic equations needed to solve optimal transfer problems between general circular orbits using low-thrust acceleration is revisited within the context of a reduced set of state variables, including the right ascension of the ascending node Ω, for further extension to the most relevant problem of perturbed motion due to J_2. Edelbaum's results are thus recovered for the thrust-only case by means of numerical integration and compared with an exact solution involving four state variables with unaveraged dynamics that maintains the angular position in the system equations and allows for fully optimized thrust vectoring. The analysis is further extended to the important J_2-perturbed case, by also considering both the analytic averaging of the system dynamics, removing the angular position variable from the system equations, and the full unaveraged system involving all four relevant state variables, as well as continuously optimized thrust vectoring during the transfer. Numerical comparisons on a transfer example show that, for the real-world thrust-and-J_2-perturbation case, the analytic averaging is about 13% less economical to carry out purely in terms of ΔV, but much more easily implemented in actual operations, because the orientation of the thrust vector maintains a constant direction with respect to rotating axes, unlike in the exact case, where this orientation must be continuously steered in time. The existence of multiple minima in the precision-integrated case without the analytic averaging complicates the search even further, as shown by an example. Further investigations spanning significant portions of the parametric space involving orbit size and orientation could reveal the regions where the averaged transfer solutions are truly even more competitive in ΔV with the exact transfers.

REFERENCES

[1] Edelbaum, T. N., "Propulsion requirements for controllable satellites," *ARS Journal*, Vol. 31, No. 8, Aug. 1961, pp. 1079–1089.

[2] Edelbaum, T. N., "Theory of maxima and minima," *Optimization Techniques with Applications to Aerospace Systems*, edited by G. Leitmann, Academic Press, New York, 1962, Chap. 1, pp. 1–32.

[3] Wiesel, W. E., and Alfano, S., "Optimal many-revolution orbit transfer," AAS/AIAA Astrodynamics Specialist Conference, AAS Paper 83-352, Lake Placid, NY, Aug. 1983.

[4] Kéchichian, J. A., "Reformulation of Edelbaum's low-thrust transfer problem using optimal control theory," *Journal of Guidance, Control, and Dynamics*, Vol. 20, No. 5, Sept.-Oct. 1997, pp. 988-994.

[5] Kéchichian, J. A., "Optimal altitude-constrained low-thrust transfer between inclined circular orbits," *Proceedings of the Malcolm D. Shuster Astronautics Symposium*, edited by Crassidis, J. L., Markley, F. L., Junkins, J. L., and K. C., Howell, Advances in the Astronautical Sciences, American Astronautical Society, San Diego, CA, 2005, Vol. 122, pp. 335-354.

[6] Bryson, A. E., Jr., and Ho, Y.-C., *Applied Optimal Control: Optimization, Estimation and Control*, Ginn and Company, Waltham, MA, 1969, pp. 117-127.

[7] Edelbaum, T. N., Sackett, L. L., and Malchow, H. L., "Optimal low-thrust geocentric transfer," 10th AIAA Electric Propulsion Conference, AIAA Paper 73-1074, Lake Tahoe, NV, Oct.-Nov. 1973.

[8] Kéchichian, J. A., "Minimum-time constant acceleration orbit transfer with first-order oblateness effect," *Journal of Guidance, Control, and Dynamics*, Vol. 23, No. 4, July-Aug. 2000, pp. 595-603.

[9] Kéchichian, J. A., "The streamlined and complete set of the nonsingular J_2-perturbed dynamic and adjoint equations for trajectory optimization in terms of eccentric longitude," AAS/AIAA Space Flight Mechanics Meeting, AAS Paper 07-120, Sedona, AZ, Jan.-Feb. 2007.

[10] Kéchichian, J. A., "The inclusion of the higher order J_3 and J_4 zonal harmonics in the modelling of optimal low-thrust orbit transfer," AAS/AIAA Space Flight Mechanics Meeting, AAS Paper 08-198, Galveston, TX, Jan. 2008.

[11] Kéchichian, J. A., "The streamlined and complete set of the nonsingular J_2-perturbed dynamic and adjoint equations for trajectory optimization in terms of eccentric longitude," *Journal of the Astronautical Sciences*, Vol. 55, No. 3, Sept. 2007, pp. 325-348.

[12] Kéchichian, J. A., "Inclusion of higher order harmonics in the modeling of optimal low-thrust orbit transfer," *Journal of the Astronautical Sciences*, Vol. 56, No. 1, March 2008, pp. 41-70.

[13] Kéchichian, J. A., "Optimal low-thrust transfer in general circular orbit using analytic averaging of the system dynamics," F. Landis Markley Astronautics Symposium, AAS Paper 08-272, Cambridge, MD, June-July 2008.

[14] Kéchichian, J. A., "Optimal low-thrust transfer in general circular orbit using analytic averaging of the system dynamics," *Journal of the Astronautical Sciences*, Vol. 57, Nos. 1-2, Jan. 2009, pp. 369-392.

[15] Kéchichian, J. A., *Applied Nonsingular Astrodynamics: Optimal Low-Thrust Orbit Transfer*, Cambridge Aerospace Series No. 45, Cambridge University Press, Cambridge, U.K., 2018.

[16] Hallman, W. P., The Aerospace Corporation, El Segundo, CA, Personal communication, 2008.

[17] Kahaner, D., Moler, C., and Nash, S., *Numerical Methods and Software*, Prentice Hall, Englewood Cliffs, NJ, 1989.

CHAPTER 10

Derivation of the Equations for the Variation of the Orbital Elements in Polar Coordinates for Elliptical and Hyperbolic Trajectories

NOMENCLATURE

a = orbit semimajor axis, m
E_0 = eccentric anomaly
$\boldsymbol{i}_u$ = unit vector along the direction of the ascending node
M_0 = mean anomaly
n_0 = Keplerian mean motion, $(\mu/a_0^3)^{1/2}$
$\boldsymbol{r}$ = position vector
t = time
θ_0^* = true anomaly
ν = "hyperbolic orbit mean motion equivalent", $(\mu/a^3)^{1/2}$
$\boldsymbol{v}$ = velocity vector

10.1 INTRODUCTION

In this chapter, perturbation equations using the Gaussian representation are derived for the hyperbolic case, following the method of Battin [1], which was applied to the elliptical case.

Analytic solutions for the perturbed motion under the influence of J_2 and Sun's gravity can, thus, be obtained for any general approach hyperbola. The errors associated with mapping the error covariances of the orbit determination in the conic B-plane parameters through the region of nonspherical perturbations leading to entry can also be investigated analytically.

The rigorous derivation of the equations for the variation of the orbital elements for hyperbolic trajectories is carried out in this chapter following

Battin's method [1], which was applied to obtain the corresponding equations for the elliptical case. These equations of motion are exact in that they contain the secular, long-period and short-period contributions to the motion. Furthermore, they can be used to express the perturbations affecting the orbit due to a planet's nonsphericity, the effect of third-body gravity, solar pressure, drag, acceleration thrust, and so on. Historically, Lagrange and Gauss developed and applied this special perturbation technique to describe the evolution of certain trajectories, and later, maneuver analysts optimized fuel consumption to carry out a given orbit transfer or rendezvous.

Brouwer and Clemence [2] made use of the method of the variation of arbitrary constants to derive the differential equations for the classical orbit elements, also showing how to derive Gauss's equations directly in the elliptical case. Breakwell [3] used vacant focus theory to derive similar differential equations for the elliptical case. In all of these cases, many of the differential equations are singular and unsuitable for numerical integration for either circular or equatorial orbits because of the use of the classical elements.

These analytic methods can also be of great use to orbit determination analysts who need to calibrate and model the various perturbations affecting the nominal spacecraft trajectory to fit the observational data and extract the error covariance for the state and other parameters.

Another important aspect of these techniques involves the problem of mapping those various error covariances from a certain epoch to a critical event. Such a case arises, for example, in the Galileo navigation area, where a state covariance in the B-plane parameters is produced at epoch and delivered to maneuver analysts for targeting purposes. It is well known that J_2 perturbations during the final hours of the trajectory will swing the B plane so much that the epoch of the B-plane mapping must be backed off such that the orbit used to map the error covariance is essentially a conic.

One can thus carry out an analytic description of this mapping into the region of J_2 and third-body perturbations, of which this analysis constitutes the first step. Here, the equations needed to obtain the perturbed trajectory are derived together with some of the difficult points that arise in the elliptical case analyzed by Battin [1].

10.2 GENERAL ANALYSIS

The second-order differential equation for the position vector of a spacecraft with respect to the center of an attracting body with gravitational constant μ and subject to a vector acceleration $\boldsymbol{a}_d$ is given by

$$\frac{\mathrm{d}^2\boldsymbol{r}}{\mathrm{d}t^2} = -\frac{\mu}{r^3}\boldsymbol{r} + \boldsymbol{a}_d \tag{10.1}$$

In the absence of any such acceleration, the equation of motion reduces to

$$\frac{\mathrm{d}^2\boldsymbol{r}}{\mathrm{d}t^2} = -\frac{\mu}{r^3}\boldsymbol{r} \tag{10.2}$$

Let $\boldsymbol{r}_{\mathrm{osc}}(t)$ be the solution of Eq. (10.2) and $\boldsymbol{r}(t)$ be the solution of Eq. (10.1). Then

$$\boldsymbol{r}(t) = \boldsymbol{r}_{\mathrm{osc}}(t) + \boldsymbol{\delta} \tag{10.3}$$

$$\boldsymbol{v}(t) = \boldsymbol{v}_{\mathrm{osc}}(t) + \boldsymbol{\nu} \tag{10.4}$$

where $\boldsymbol{r}_{\mathrm{osc}}(t)$ and $\boldsymbol{v}_{\mathrm{osc}}(t)$ are the osculating position and velocity vectors at time t and $\boldsymbol{\nu}(t) = \mathrm{d}\boldsymbol{\delta}(t)/\mathrm{d}t$.

Let the two solutions coincide at time t_0; that is, $\boldsymbol{r}(t_0) = \boldsymbol{r}_{\mathrm{osc}}(t_0)$ and $\boldsymbol{v}(t_0) = \boldsymbol{v}_{\mathrm{osc}}(t_0)$. Now, at time $t_1 = t_0 + \Delta t$

$$\boldsymbol{r}(t_1) = \boldsymbol{r}(t_0) + \boldsymbol{v}(t_0)\Delta t + \frac{1}{2}\left(\frac{\mathrm{d}^2\boldsymbol{r}}{\mathrm{d}t^2}\right)_{t_0}\Delta t^2 + \cdots$$

$$\boldsymbol{r}_{\mathrm{osc}}(t_1) = \boldsymbol{r}_{\mathrm{osc}}(t_0) + \boldsymbol{v}_{\mathrm{osc}}(t_0)\Delta t + \frac{1}{2}\left(\frac{\mathrm{d}^2\boldsymbol{r}_{\mathrm{osc}}}{\mathrm{d}t^2}\right)_{t_0}\Delta t^2 + \cdots$$

Equations (10.3) and (10.4) yield

$$\boldsymbol{\delta}(t_1) = \frac{1}{2}\left(\frac{\mathrm{d}^2\boldsymbol{r}}{\mathrm{d}t^2} - \frac{\mathrm{d}^2\boldsymbol{r}_{\mathrm{osc}}}{\mathrm{d}t^2}\right)_{t_0}\Delta t^2 + \cdots = \frac{1}{2}\boldsymbol{a}_d(t_0)\Delta t^2 + \cdots \tag{10.5}$$

$$\boldsymbol{\nu}(t_1) = \boldsymbol{a}_d(t_0)\Delta t + \cdots \tag{10.6}$$

10.2.1 VARIATION OF THE SEMIMAJOR AXIS

From the energy equation, it follows that

$$v^2 = \mu\left(\frac{2}{r} - \frac{1}{a}\right) = \frac{\mu}{ra}(2a - r) \tag{10.7}$$

$$ra\boldsymbol{v}\cdot\boldsymbol{v} = \mu(2a - r) \tag{10.8}$$

Along the osculating orbit at time t_1, with a assumed to be positive throughout this analysis, Eq. (10.8) yields

$$r_{\mathrm{osc}}(t_1)a_0\boldsymbol{v}_{\mathrm{osc}}(t_1)\cdot\boldsymbol{v}_{\mathrm{osc}}(t_1) = \mu[2a_0 - r_{\mathrm{osc}}(t_1)] \tag{10.9}$$

and along the true orbit

$$r(t_1)a_1\boldsymbol{v}(t_1)\cdot\boldsymbol{v}(t_1) = \mu[2a_1 - r(t_1)] \tag{10.10}$$

Using Eqs. (10.3) and (10.4) with $r(t_1) = r_{osc}(t_1) + \delta'$ as the scalar position radius and subtracting Eqs. (10.9) from (10.10) gives

$$\left[r_{osc}(t_1) + \delta'\right] a_1 \left[\boldsymbol{v}_{osc}(t_1) + \boldsymbol{\nu}\right] \cdot \left[\boldsymbol{v}_{osc}(t_1) + \boldsymbol{\nu}\right] - r_{osc}(t_1) a_0 \boldsymbol{v}_{osc}(t_1) \cdot \boldsymbol{v}_{osc}(t_1)$$
$$= \mu\left[2a_1 - r_{osc}(t_1) - \delta'\right] - \mu[2a_0 - r_{osc}(t_1)]$$

Dividing by Δt and rearranging then yields

$$r_{osc}(t_1)\left[\frac{(a_1 - a_0)}{\Delta t} \boldsymbol{v}_{osc}(t_1) \cdot \boldsymbol{v}_{osc}(t_1)\right] + 2a_1 r_{osc}(t_1) \boldsymbol{v}_{osc}(t_1) \cdot \frac{\boldsymbol{\nu}}{\Delta t}$$
$$+ r_{osc}(t_1) a_1 \frac{\boldsymbol{\nu} \cdot \boldsymbol{\nu}}{\Delta t} + \delta' a_1 \frac{\boldsymbol{v}(t_1) \cdot \boldsymbol{v}(t_1)}{\Delta t} = 2\frac{\mu(a_1 - a_0)}{\Delta t} - \mu\frac{\delta'}{\Delta t}$$

Because $t_1 = t_0 + \Delta t$ and $\boldsymbol{\nu} = \boldsymbol{a}_d\, \Delta t$, $\delta' = \frac{1}{2} a_d \Delta t^2$, in the limit as $\Delta t \to 0$, $r_{osc}(t_1) \to r_{osc}(t_0)$, $v_{osc}(t_1) \to v_{osc}(t_0)$, $a_1 = a(t_1) \to a(t_0)$, and so on, such that

$$r\frac{da}{dt} v^2 + 2ar\boldsymbol{v} \cdot \boldsymbol{a}_d = 2\mu\frac{da}{dt}$$

Moreover, from Eq. (10.7), $2\mu - rv^2 = r\mu/a$ such that

$$\frac{da}{dt} = \frac{2a^2}{\mu} \boldsymbol{v} \cdot \boldsymbol{a}_d \tag{10.11}$$

For the case of a hyperbola, $v^2 = \mu(2/r + 1/a)$ with $a > 0$ such that $2\mu - rv^2 = -\mu(r/a)$. Now, Eq. (10.8) must be replaced by

$$ra\boldsymbol{v} \cdot \boldsymbol{v} = \mu(2a + r)$$

However, the same manipulations as carried out for the ellipse lead to the same differential equation

$$r\frac{da}{dt} v^2 + 2ar\boldsymbol{v} \cdot \boldsymbol{a}_d = 2\mu\frac{da}{dt}$$

such that

$$\frac{da}{dt} = \frac{2ar\boldsymbol{v} \cdot \boldsymbol{a}_d}{2\mu - rv^2} = -\frac{2a^2}{\mu} \boldsymbol{v} \cdot \boldsymbol{a}_d \tag{10.11'}$$

Here, a is taken to be a positive quantity. This expression for a hyperbola would be identical to Eq. (10.11) if $a > 0$ for the ellipse and $a < 0$ for the hyperbola. However, because $a > 0$ has been employed throughout this analysis, Eq. (10.11′) should be used for the hyperbola, and Eq. (10.11) should be used for the ellipse. This derivative and all subsequent derivatives defining the variation of the orbital elements are defined at time t_0.

10.2.2 VARIATION OF THE SCALAR ANGULAR MOMENTUM

From the definition of the angular momentum vector

$$\boldsymbol{h} = \boldsymbol{r} \times \boldsymbol{v} \tag{10.12}$$

it follows that

$$h^2 = (\boldsymbol{r} \times \boldsymbol{v}) \cdot (\boldsymbol{r} \times \boldsymbol{v}) \tag{10.13}$$

Applying Eq. (10.13) at time t_1 for the osculating and perturbed orbits and taking the difference gives

$$\begin{aligned}
h_0^2 &= [\boldsymbol{r}_{\text{osc}}(t_1) \times \boldsymbol{v}_{\text{osc}}(t_1)] \cdot [\boldsymbol{r}_{\text{osc}}(t_1) \times \boldsymbol{v}_{\text{osc}}(t_1)] \\
h_1^2 &= [\boldsymbol{r}(t_1) \times \boldsymbol{v}(t_1)] \cdot [\boldsymbol{r}(t_1) \times \boldsymbol{v}(t_1)] \\
h_1^2 - h_0^2 &= [\boldsymbol{r}_{\text{osc}}(t_1) \times \boldsymbol{v}_{\text{osc}}(t_1) + \boldsymbol{r}_{\text{osc}}(t_1) \times \boldsymbol{v} - \boldsymbol{\delta} \times \boldsymbol{v}_{\text{osc}}(t_1) + \boldsymbol{\delta} \times \boldsymbol{v}] \\
&\quad \cdot [\boldsymbol{r}_{\text{osc}}(t_1) \times \boldsymbol{v}_{\text{osc}}(t_1) + \boldsymbol{r}_{\text{osc}}(t_1) \times \boldsymbol{v} + \boldsymbol{\delta} \times \boldsymbol{v}_{\text{osc}}(t_1) + \boldsymbol{\delta} \times \boldsymbol{v}] \\
&\quad - [\boldsymbol{r}_{\text{osc}}(t_1) \times \boldsymbol{v}_{\text{osc}}(t_1)] \cdot [\boldsymbol{r}_{\text{osc}}(t_1) \times \boldsymbol{v}_{\text{osc}}(t_1)] \\
&= [\boldsymbol{r}_{\text{osc}}(t_1) \times \boldsymbol{v}_{\text{osc}}(t_1)] \cdot \{[\boldsymbol{r}_{\text{osc}}(t_1) \times \boldsymbol{v}] + \boldsymbol{\delta} \times \boldsymbol{v}_{\text{osc}}(t_1) + \boldsymbol{\delta} \times \boldsymbol{v}\} \\
&\quad + [\boldsymbol{r}_{\text{osc}}(t_1) \times \boldsymbol{v}] \cdot \{[\boldsymbol{r}_{\text{osc}}(t_1) \times \boldsymbol{v}_{\text{osc}}(t_1)] + \boldsymbol{r}_{\text{osc}}(t_1) \times \boldsymbol{v} \\
&\quad + \boldsymbol{\delta} \times \boldsymbol{v}_{\text{osc}}(t_1) + \boldsymbol{\delta} \times \boldsymbol{v}\} + [\boldsymbol{\delta} \times \boldsymbol{v}_{\text{osc}}(t_1)] \\
&\quad \cdot [\boldsymbol{r}_{\text{osc}}(t_1) \times \boldsymbol{v}_{\text{osc}}(t_1) + \boldsymbol{r}_{\text{osc}}(t_1) \times \boldsymbol{v} + \boldsymbol{\delta} \times \boldsymbol{v}_{\text{osc}}(t_1) + \boldsymbol{\delta} \times \boldsymbol{v}] \\
&\quad + (\boldsymbol{\delta} \times \boldsymbol{v}) \cdot [\boldsymbol{r}_{\text{osc}}(t_1) \times \boldsymbol{v}_{\text{osc}}(t_1) + \boldsymbol{r}_{\text{osc}}(t_1) \times \boldsymbol{v} + \boldsymbol{\delta} \times \boldsymbol{v}_{\text{osc}}(t_1) + \boldsymbol{\delta} \times \boldsymbol{v}] \\
&= 2[\boldsymbol{r}_{\text{osc}}(t_1) \times \boldsymbol{v}(t_1)] \cdot [\boldsymbol{r}_{\text{osc}}(t_1) \times \boldsymbol{v}_{\text{osc}}(t_1)] + \mathcal{O}(\Delta t^2) + \text{HOT}
\end{aligned}$$

where HOT indicates higher-order terms in Δt. This expression divided by Δt reduces to

$$\frac{h_1^2 - h_0^2}{\Delta t} = 2\left\{[\boldsymbol{r}(t_1) - \boldsymbol{\delta}] \times \left(\boldsymbol{a}_d \frac{\Delta t}{\Delta t}\right)\right\} \cdot \{[\boldsymbol{r}(t_1) - \boldsymbol{\delta}] \times [\boldsymbol{v}(t_1) - \boldsymbol{a}_d \Delta t]\}$$

with $\boldsymbol{\delta} = (1/2)\boldsymbol{a}_d \Delta t^2$. Taking the limit as $\Delta t \to 0$, all of the higher-order terms in Δt will yield zero, so that we obtain

$$\begin{aligned}
\frac{\mathrm{d}h^2}{\mathrm{d}t} = 2h\frac{\mathrm{d}h}{\mathrm{d}t} &= 2[\boldsymbol{r}(t_0) \times \boldsymbol{a}_d] \cdot [\boldsymbol{r}(t_0) \times \boldsymbol{v}(t_0)] \\
\frac{\mathrm{d}h}{\mathrm{d}t} &= \frac{1}{h}(\boldsymbol{r} \times \boldsymbol{a}_d) \cdot (\boldsymbol{r} \times \boldsymbol{v})
\end{aligned} \tag{10.14}$$

which is equivalent to

$$\frac{\mathrm{d}h}{\mathrm{d}t} = \frac{1}{h}\left[r^2(\boldsymbol{v} \cdot \boldsymbol{a}_d) - (\boldsymbol{r} \cdot \boldsymbol{v})(\boldsymbol{r} \cdot \boldsymbol{a}_d)\right] \tag{10.15}$$

To show this, let $\boldsymbol{h} = \boldsymbol{r} \times \boldsymbol{v}$ in Eq. (10.14); then

$$\frac{dh}{dt} = \frac{1}{h}(\boldsymbol{r} \times \boldsymbol{a}_d) \cdot \boldsymbol{h} = \frac{1}{h}(\boldsymbol{h} \times \boldsymbol{r}) \cdot \boldsymbol{a}_d = \frac{1}{h}(\boldsymbol{r} \times \boldsymbol{v} \times \boldsymbol{r}) \cdot \boldsymbol{a}_d$$
$$= \frac{1}{h}[(\boldsymbol{r} \cdot \boldsymbol{r})\boldsymbol{v} - (\boldsymbol{r} \cdot \boldsymbol{v})\boldsymbol{r}] \cdot \boldsymbol{a}_d = \frac{1}{h}\left[r^2(\boldsymbol{v} \cdot \boldsymbol{a}_d) - (\boldsymbol{r} \cdot \boldsymbol{v})(\boldsymbol{r} \cdot \boldsymbol{a}_d)\right]$$

which is the expression in Eq. (10.15). This expression involves only dot products, unlike Eq. (10.14), which involves cross products, so this expression may be easier to use.

10.2.3 VARIATION OF THE ECCENTRICITY

The variation of the eccentricity can be obtained in terms of the variations of the semimajor axis and the scalar angular momentum just developed. From Eq. (10.16), which is valid for the ellipse

$$h^2 = \mu a(1 - e^2) \tag{10.16}$$

$$2h\frac{dh}{dt} = \mu(1 - e^2)\frac{da}{dt} + \mu a(-2e)\frac{de}{dt}$$

Making use of Eqs. (10.16), (10.11), and (10.15) and letting $p = h^2/\mu$, we obtain

$$\frac{de}{dt} = \frac{1}{\mu ae}\left[(pa - r^2)(\boldsymbol{v} \cdot \boldsymbol{a}_d) + (\boldsymbol{r} \cdot \boldsymbol{v})(\boldsymbol{v} \cdot \boldsymbol{a}_d)\right] \tag{10.17}$$

which is valid for the ellipse. For the hyperbola, let a be positive

$$h^2 = \mu a(e^2 - 1)$$

$$2h\frac{dh}{dt} = \mu(e^2 - 1)\frac{da}{dt} + \mu a(2e)\frac{de}{dt}$$

$$\frac{de}{dt} = \frac{1}{\mu ae}\left[r^2(\boldsymbol{v} \cdot \boldsymbol{a}_d) - (\boldsymbol{r} \cdot \boldsymbol{v})(\boldsymbol{r} \cdot \boldsymbol{a}_d)\right] + \frac{pa}{\mu ae}(\boldsymbol{v} \cdot \boldsymbol{a}_d)$$

$$\frac{de}{dt} = \frac{1}{\mu ae}\left[(pa + r^2)(\boldsymbol{v} \cdot \boldsymbol{a}_d) - (\boldsymbol{r} \cdot \boldsymbol{v})(\boldsymbol{r} \cdot \boldsymbol{a}_d)\right] \tag{10.17'}$$

This expression is identical to Eq. (10.17) because a is considered to be positive here. If Eq. (10.17) were used instead, with $a < 0$ for the hyperbola, then that expression would be valid for both the ellipse and the hyperbola provided that $a > 0$ is used in the first case and $a < 0$ in the second.

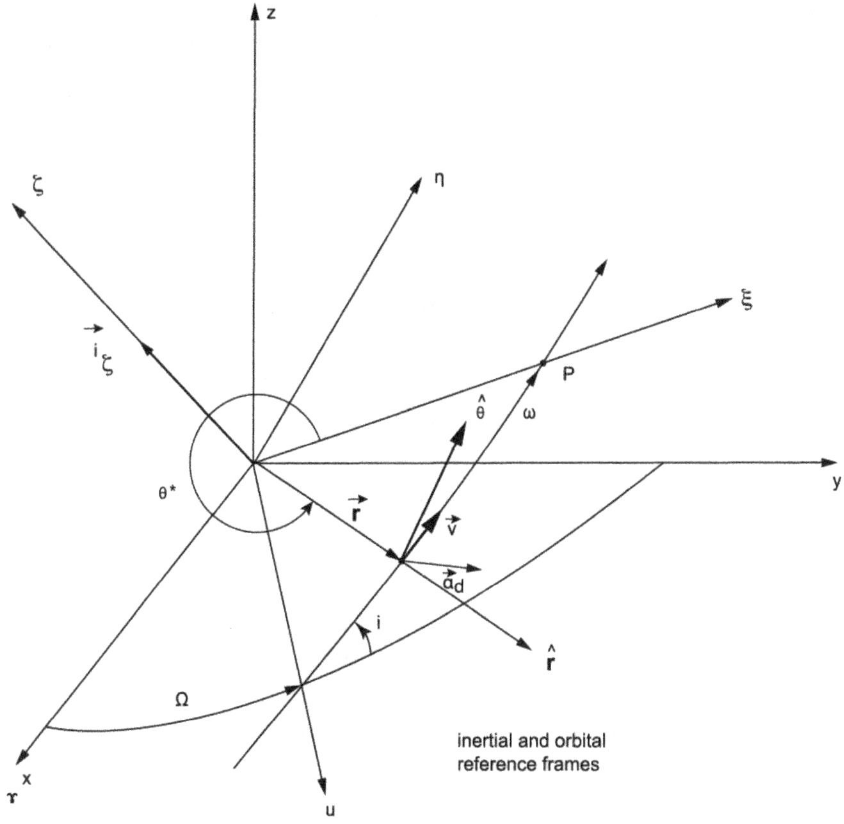

Fig. 10.1 Inertial and orbital reference frames.

10.2.4 VARIATIONS OF THE INCLINATION AND NODE

From Fig. 10.1, the angular momentum vector expressed in the inertial *xyz* system has the form

$$\boldsymbol{h} = (\boldsymbol{r} \times \boldsymbol{v}) = h\boldsymbol{i}_\zeta = h\begin{pmatrix} l_3 \\ m_3 \\ n_3 \end{pmatrix} = h\begin{pmatrix} s_\Omega s_i \\ -c_\Omega s_i \\ c_i \end{pmatrix} \tag{10.18}$$

This can be seen by rotation of the $\boldsymbol{i}_\zeta$ unit vector to the *xyz* system by means of the rotation matrix $\boldsymbol{R}$

$$\boldsymbol{R} = \begin{pmatrix} \boldsymbol{i}_x \cdot \boldsymbol{i}_\xi & \boldsymbol{i}_x \cdot \boldsymbol{i}_\eta & \boldsymbol{i}_x \cdot \boldsymbol{i}_\zeta \\ \boldsymbol{i}_y \cdot \boldsymbol{i}_\xi & \boldsymbol{i}_y \cdot \boldsymbol{i}_\eta & \boldsymbol{i}_y \cdot \boldsymbol{i}_\zeta \\ \boldsymbol{i}_z \cdot \boldsymbol{i}_\xi & \boldsymbol{i}_z \cdot \boldsymbol{i}_\eta & \boldsymbol{i}_z \cdot \boldsymbol{i}_\zeta \end{pmatrix} = \begin{pmatrix} l_1 & l_2 & l_3 \\ m_1 & m_2 & m_3 \\ n_1 & n_2 & n_3 \end{pmatrix}$$

where $\boldsymbol{\xi}$ is along perigee, $\boldsymbol{\eta}$ is in the orbit plane along the direction of the velocity, and the direction cosines are given in terms of the Euler angles i, ω, and Ω by

$$
\begin{aligned}
l_1 &= c_\Omega c_\omega - s_\Omega s_\omega c_i \\
l_2 &= -c_\Omega s_\omega - s_\Omega c_\omega c_i \\
l_3 &= s_\Omega s_i \\
m_1 &= s_\Omega c_\omega + c_\Omega s_\omega c_i \\
m_2 &= -s_\Omega s_\omega + c_\Omega c_\omega c_i \\
m_3 &= -c_\Omega s_i \\
n_1 &= s_\omega s_i \\
n_2 &= c_\omega s_i \\
n_3 &= c_i
\end{aligned}
$$

Now

$$
\frac{\mathrm{d}\boldsymbol{h}}{\mathrm{d}t} = \frac{\mathrm{d}\boldsymbol{r}}{\mathrm{d}t} \times \boldsymbol{v} + \boldsymbol{r} \times \frac{\mathrm{d}\boldsymbol{v}}{\mathrm{d}t} = \boldsymbol{r} \times \boldsymbol{a}_d = \begin{pmatrix} \frac{\mathrm{d}}{\mathrm{d}t}(h s_\Omega s_i) \\ \frac{\mathrm{d}}{\mathrm{d}t}(-h c_\Omega s_i) \\ \frac{\mathrm{d}}{\mathrm{d}t}(h c_i) \end{pmatrix}
$$

$$
\boldsymbol{r} \times \boldsymbol{a}_d = \begin{pmatrix} c_\Omega & -s_\Omega c_i & s_\Omega s_i \\ s_\Omega & c_\Omega c_i & -c_\Omega s_i \\ 0 & s_i & c_i \end{pmatrix} \begin{pmatrix} h s_i \dot{\Omega} \\ -h \dot{i} \\ \dot{h} \end{pmatrix} \tag{10.19}
$$

where $\boldsymbol{r} \times \boldsymbol{a}_d$ is a column vector. This matrix is an orthonormal transformation, $M^{-1} = M^T$, such that, in the inertial xyz frame

$$
\dot{\Omega} = \frac{1}{h s_i} \begin{pmatrix} c_\Omega & s_\Omega & 0 \end{pmatrix} (\boldsymbol{r} \times \boldsymbol{a}_d) \tag{10.20}
$$

$$
\dot{i} = -\frac{1}{h} \begin{pmatrix} -s_\Omega c_i & c_\Omega c_i & s_i \end{pmatrix} (\boldsymbol{r} \times \boldsymbol{a}_d) \tag{10.21}
$$

$$
\dot{h} = \begin{pmatrix} s_\Omega s_i & -c_\Omega s_i & c_i \end{pmatrix} (\boldsymbol{r} \times \boldsymbol{a}_d) \tag{10.22}
$$

These equations hold for both the elliptical and hyperbolic cases.

10.2.5 VARIATION OF THE TRUE ANOMALY

Let $\boldsymbol{r}(t_0)$ be the position vector of the spacecraft at time t_0, and let θ_0^*, E_0, and M_0 be the true, eccentric, and mean anomalies, respectively, associated with the osculating elliptical orbit. At time $t_1 = t_0 + \Delta t$, the position vector is $\boldsymbol{r}(t_1)$, and the corresponding osculating orbit has the anomalies θ_1^*, E_1, and M_1. Then

$$
\eta(t_1) = \theta_1^*(t_1) - \theta_0^*(t_1) \tag{10.23}
$$

$$
\beta(t_1) = E_1(t_1) - E_0(t_1) \tag{10.24}
$$

$$
\gamma(t_1) = M_1(t_1) - M_0(t_1) \tag{10.25}
$$

For the hyperbolic case, Eq. (10.24) is replaced by

$$\beta(t_1) = H_1(t_1) - H_0(t_1) \tag{10.26}$$

with the other two relations remaining unchanged except that M is now replaced by the quantity $N = \nu(t - \tau)$ analogous to the mean anomoly $M = n(t - \tau)$ of elliptic motion, τ being time of pericenter passage.

At $t = t_0$, $\eta(t_0) = \beta(t_0) = \gamma(t_0) = 0$, and because the polar equation of the orbit is the same for both the elliptical and the hyperbolic cases, the differential equation for the instantaneous rate of η at time t_0 will be valid for both cases as well. This is true because, from

$$r(1 + e\cos\theta^*) = h^2/\mu \tag{10.27}$$

for time t_1, we can write

$$r_{osc}(t_1)\left[1 + e_0\cos\theta_0^*(t_1)\right] = h_0^2/\mu \tag{10.28}$$

$$r(t_1)\left[1 + e_1\cos\theta_1^*(t_1)\right] = h_1^2/\mu \tag{10.29}$$

However, we know that $r(t_1) = r_{osc}(t_1) + \delta'$, and using this expression and Eq. (10.23), we can transform Eq. (10.29) into

$$\left[r_{osc}(t_1) + \delta'\right]\left\{1 + e_1\cos\left[\theta_0^*(t_1) + \eta(t_1)\right]\right\} = h_1^2/\mu \tag{10.30}$$

Subtracting Eq. (10.28) from Eq. (10.30) gives

$$\begin{aligned} &r_{osc}(t_1)\left\{1 + e_1\left[\cos\theta_0^*(t_1)\cos\eta(t_1) - \sin\theta_0^*(t_1)\sin\eta(t_1)\right]\right\} \\ &\quad + \delta'\left\{1 + e_1\left[\cos\theta_0^*(t_1)\cos\eta(t_1) - \sin\theta_0^*(t_1)\sin\eta(t_1)\right]\right\} \\ &\quad - r_{osc}(t_1)\left[1 + e_0\cos\theta_0^*(t_1)\right] = (h_1^2 - h_0^2)/\mu \end{aligned}$$

Observing that $\cos\eta(t_1) \simeq 1$ and $\sin\eta(t_1) \simeq \eta(t_1)$, dividing the preceding equation by Δt, and neglecting resulting terms in $\mathcal{O}(\Delta t)$ or higher because these terms will vanish as $\Delta t \rightarrow 0$, we obtain

$$\begin{aligned} &r_{osc}(t_1)\cos\theta_0^*(t_1)\left[\frac{e_1\cos\eta(t_1) - e_0}{\Delta t}\right] \\ &\quad - r_{osc}(t_1)\left[\frac{e_1\sin\theta_0^*(t_1)\sin\eta(t_1)}{\Delta t}\right] = \frac{(h_1^2 - h_0^2)}{\mu\Delta t} \end{aligned}$$

As $\Delta t \rightarrow 0$, $t_1 = t_0 + \Delta t \rightarrow t_0$, and because $\eta(t_0) = 0$, $\eta(t_1) = \eta(t_1) - \eta(t_0) = \mathrm{d}\eta$, the preceding equation reduces to

$$re\sin\theta^*\frac{\mathrm{d}\eta}{\mathrm{d}t} = r\cos\theta^*\frac{\mathrm{d}e}{\mathrm{d}t} - \frac{2h}{\mu}\frac{\mathrm{d}h}{\mathrm{d}t} \tag{10.31}$$

This is the contribution to the change in the true anomaly due solely to the changes in the orbital elements caused by the perturbative action of the acceleration $\boldsymbol{a}_d$ and not to the motion of the spacecraft along the osculating orbit. All of the terms in Eq. (10.31) are defined at time t_0. This expression is valid for both the elliptical and the hyperbolic cases provided that de/dt in Eq. (10.17) is used with the correct sign for the semimajor axis.

10.2.6 VARIATION OF THE ECCENTRIC ANOMALY

In the case of the ellipse, the identity

$$\cos E = \frac{e \cos \theta^*}{1 + e \cos \theta^*} \tag{10.32}$$

is used at t_1 for both osculating orbits as before, yielding

$$\cos E_0(t_1)\left[1 + e_0 \cos \theta_0^*(t_1)\right] = e_0 + \cos \theta_0^*(t_1) \tag{10.33}$$

$$\cos E_1(t_1)\left[1 + e_1 \cos \theta_1^*(t_1)\right] = e_1 + \cos \theta_1^*(t_1) \tag{10.34}$$

Using $E_1(t_1) = E_0(t_1) + \beta(t_1)$ and $\theta_1^*(t_1) = \theta_0^*(t_1) + \eta(t_1)$, we subtract Eq. (10.33) from Eq. (10.34) to obtain

$$\begin{aligned}
&\cos E_0(t_1)[\cos\beta(t_1) - 1] \\
&\quad + \cos E_0(t_1)\left[\cos\beta(t_1) e_1 \cos\theta_0^*(t_1)\cos\eta(t_1) - e_0\cos\theta_0^*(t_1)\right] \\
&\quad + \cos E_0(t_1)\cos\beta(t_1)\left[-e_1\sin\theta_0^*(t_1)\sin\eta(t_1)\right] \\
&\quad - \sin E_0(t_1)\sin\beta(t_1)\left\{1 + e_1\left[\cos\theta_0^*(t_1)\cos\eta(t_1) - \sin\theta_0^*(t_1)\sin\eta(t_1)\right]\right\} \\
&\quad = e_1 - e_0 + \cos\theta_0^*(t_1)[\cos\eta(t_1) - 1] - \sin\theta_0^*(t_1)\sin\eta(t_1)
\end{aligned}$$

which, after division by Δt and elimination of the $\mathcal{O}(\Delta t)$ and higher-order terms, is reduced to the following:

$$\begin{aligned}
&\cos E_0(t_1)\left[\cos\theta_0^*(t_1)\frac{(e_1 - e_0)}{\Delta t}\right] + \cos E_0(t_1)\left\{-e_1\sin\theta_0^*(t_1)\left[\frac{\eta(t_1) - \eta(t_0)}{\Delta t}\right]\right\} \\
&\quad - \sin E_0(t_1)\frac{\beta(t_1) - \beta(t_0)}{\Delta t}\left(1 + e_1\left\{\cos\theta_0^*(t_1) - \sin\theta_0^*(t_1)[\eta(t_1) - \eta(t_0)]\right\}\right) \\
&\quad = \frac{(e_1 - e_0)}{\Delta t} - \sin\theta_0^*(t_1)\left[\frac{\eta(t_1) - \eta(t_0)}{\Delta t}\right]
\end{aligned}$$

Neglecting the term in $d\eta\, d\beta$ and taking the limit as $\Delta t \to 0$ gives

$$(1 - e_1 \cos E)\sin\theta^* \frac{d\eta}{dt} - \sin E \frac{d\beta}{dt}(1 + e_1\cos\theta^*) = \frac{de}{dt}(1 - \cos E\cos\theta^*)$$

Because

$$\sin E = \frac{\sqrt{1-e^2}\sin\theta^*}{1+e\cos\theta^*}$$

the preceding equation reduces to

$$\sqrt{1-e^2}\frac{\mathrm{d}\beta}{\mathrm{d}t} = \frac{r}{a}\frac{\mathrm{d}\eta}{\mathrm{d}t} - \frac{\mathrm{de}}{\mathrm{d}t}\left(\frac{1}{\sin\theta^*} - \frac{\cos E\cos\theta^*}{\sin\theta^*}\right)$$

and with Eq. (10.32), this reduces to the final form

$$\sqrt{1-e^2}\frac{\mathrm{d}\beta}{\mathrm{d}t} = \frac{r}{a}\frac{\mathrm{d}\eta}{\mathrm{d}t} - \frac{r\sin\theta^*}{p}\frac{\mathrm{de}}{\mathrm{d}t} \tag{10.35}$$

In the case of the hyperbola, the identity

$$\cosh H = \frac{e+\cos\theta^*}{1+e\cos\theta^*} \tag{10.36}$$

is used for the two osculating orbits at time t_1 as before, yielding

$$\cosh H_0(t_1)\left[1+e_0\cos\theta_0^*(t_1)\right] = e_0 + \cos\theta_0^*(t_1) \tag{10.37}$$

$$\cosh H_1(t_1)\left[1+e_1\cos\theta_1^*(t_1)\right] = e_1 + \cos\theta_1^*(t_1) \tag{10.38}$$

Using the relations $H_1(t_1) = H_0(t_1) + \beta(t_1)$ and $\theta_1^*(t_1) = \theta_0^*(t_1) + \eta(t_1)$, taking the difference between Eqs. (10.37) and (10.38), eliminating the higher-order terms, and expanding the hyperbolic functions yields

$$\begin{aligned}\cosh H_0(t_1)&\left[\cos\theta_0^*(t_1)(e_1-e_0)\right] - \cosh H_0(t_1)\left[e_1\cos\theta_0^*(t_1)\eta(t_1)\right]\\ &+ \sinh H_0(t_1)\left\{1+e_1\left[\cos\theta_0^*(t_1) - \sin\theta_0^*(t_1)\eta(t_1)\right]\right\}\\ &= (e_1-e_0) - \sin\theta_0^*(t_1)\eta(t_1)\end{aligned}$$

Here

$$\cosh\beta(t_1) \simeq 1 + \frac{\beta^2}{2}(t_1) \simeq 1$$

$$\sinh\beta(t_1) \simeq \beta(t_1) + \frac{\beta^3}{3!}(t_1) \simeq \beta(t_1)$$

Observing that $\eta(t_1) = \eta(t_1) - \eta(t_0) = \mathrm{d}\eta$, dividing by Δt, and taking the limit as $\Delta t \to 0$, we obtain

$$\sinh H(1+e\cos\theta^*)\frac{\mathrm{d}\beta}{\mathrm{d}t} = -\sin\theta^*(1-e\cosh H)\frac{\mathrm{d}\eta}{\mathrm{d}t} + (1-e\cos\theta^*\cosh H)\frac{\mathrm{de}}{\mathrm{d}t}$$

However

$$\sinh H = \frac{\sqrt{e^2 - 1}\sin\theta^*}{(1 + e\cos\theta^*)}$$

and $r = a(e\cosh H - 1)$ with $a > 0$. Making use of Eq. (10.36), the final result is obtained as

$$\sqrt{e^2 - 1}\frac{\mathrm{d}\beta}{\mathrm{d}t} = \frac{r}{a}\frac{\mathrm{d}\eta}{\mathrm{d}t} + \frac{r}{p}\sin\theta^*\frac{\mathrm{d}e}{\mathrm{d}t} \tag{10.39}$$

where $p = a(e^2 - 1)$ with $a > 0$. In this equation $\mathrm{d}e/\mathrm{d}t$ as given in Eq. (10.17′) should be used to have $a > 0$ throughout in a consistent manner.

10.2.7 VARIATION OF THE MEAN ANOMALY

In the case of the ellipse, Kepler's equation for the mean anomaly in terms of the eccentric anomaly is used at time t_1 for both osculating orbits as follows:

$$E_0(t_1) - e_0\sin E_0(t_1) = M_0(t_1) \tag{10.40}$$

$$E_1(t_1) - e_1\sin E_1(t_1) = M_1(t_1) = M_0(t_1) + \gamma(t_1) \tag{10.41}$$

Taking the difference, eliminating higher-order terms, and dividing by Δt with the observation that $E_1(t_1) = E_0(t_1) + \beta(t_1)$ yields

$$\frac{\beta(t_1) - \beta(t_0)}{\Delta t} - \frac{(e_1 - e_0)}{\Delta t}\sin E_0(t_1) - e_1\frac{[\beta(t_1) - \beta(t_0)]}{\Delta t}\cos E_0(t_1) = \frac{\gamma(t_1) - \gamma(t_0)}{\Delta t}$$

Taking the limit as $\Delta t \to 0$ and using $t_1 = t_0 + \Delta t \to t_0$, we obtain

$$\frac{\mathrm{d}\beta}{\mathrm{d}t}(1 - e\cos E) - \sin E\frac{\mathrm{d}e}{\mathrm{d}t} = \frac{\mathrm{d}\gamma}{\mathrm{d}t}$$

However

$$\sin E = \frac{\sqrt{1 - e^2}\sin\theta^*}{1 + e\cos\theta^*}$$

$r/a = (1 - e\cos E)$, and $(1 + e\cos\theta^*) = p/r$ with $p = a(1 - e^2)$, so

$$\frac{\mathrm{d}\gamma}{\mathrm{d}t} = \frac{r}{a}\frac{\mathrm{d}\beta}{\mathrm{d}t} - \frac{r}{p}\sqrt{1 - e^2}\sin\theta^*\frac{\mathrm{d}e}{\mathrm{d}t} \tag{10.42}$$

For the case of the hyperbola

$$N = e\sinh H - H \tag{10.43}$$

$$e_0\sinh H_0(t_1) - H_0(t_1) = N_0(t_1) \tag{10.44}$$

$$e_1 \sinh H_1(t_1) - H_1(t_1) = N_1(t_1) = N_0(t_1) + \gamma(t_1) \tag{10.45}$$

Because $H_1(t_1) = H_0(t_1) + \beta(t_1)$, taking the difference between Eqs. (10.44) and (10.45), expanding the hyperbolic functions, and using the expressions

$$\cosh \beta(t_1) \simeq 1 + \frac{\beta^2}{2}(t_1) \simeq 1$$

$$\sinh \beta(t_1) \simeq \beta(t_1) + \frac{\beta^3}{3!}(t_1) \simeq \beta(t_1)$$

we obtain

$$\begin{aligned} &\sinh H_0(t_1)(e_1 - e_0) + e_1 \cosh H_0(t_1)[\beta(t_1) - \beta(t_0)] - [\beta(t_1) - \beta(t_0)] \\ &\quad = \gamma(t_1) - \gamma(t_0) \end{aligned}$$

where terms such as $\beta(t_0)$ and $\gamma(t_0)$ have been added for convenience as before to produce $\mathrm{d}\beta$ and $\mathrm{d}\gamma$ because $\beta(t_0) = \gamma(t_0) = 0$. Now, dividing by Δt, taking the limit as $\Delta t \to 0$, and observing that

$$\sinh H = \frac{\sqrt{e^2 - 1}\sin\theta^*}{(1 + e\cos\theta^*)}$$

$p/r = (1 + e\cos\theta^*)$, $r = a(e\cosh H - 1)$, and $p = a(e^2 - 1)$ with $a > 0$ throughout, the final result valid for hyperbolic orbits is obtained as

$$\frac{\mathrm{d}\gamma}{\mathrm{d}t} = \frac{r}{p}\sqrt{e^2 - 1}\sin\theta^* \frac{\mathrm{d}e}{\mathrm{d}t} + \frac{r}{a}\frac{\mathrm{d}\beta}{\mathrm{d}t} \tag{10.46}$$

Because $a > 0$, $\mathrm{d}\beta/\mathrm{d}t$ is given by Eq. (10.39), and $\mathrm{d}e/\mathrm{d}t$ is given by Eq. (10.17′).

10.2.8 TOTAL VARIATIONS OF THE TRUE, MEAN, AND ECCENTRIC ANOMALIES

The total variation of the true anomaly is obtained by adding the contributions due to the Keplerian motion and to the changes in the orbital elements resulting from the effect of the perturbative acceleration; that is

$$\dot{\theta}^* = \dot{\theta}_0^* + \frac{\mathrm{d}\eta}{\mathrm{d}t} \tag{10.47}$$

where $\dot{\theta}_0^* = h/r^2$ is due to the Keplerian motion, $\mathrm{d}\eta/\mathrm{d}t$ is given by Eq. (10.31) as a function of $\mathrm{d}e/\mathrm{d}t$ and $\mathrm{d}h/\mathrm{d}t$, and $\mathrm{d}e/\mathrm{d}t$ is given in Eq. (10.17) for the ellipse and in Eq. (10.17′) for the hyperbola with $a > 0$.

The total variation of the eccentric anomaly in the case of the ellipse is obtained from

$$\lim_{\Delta t\to 0}\left[\frac{E_1(t_1)-E_0(t_0)}{\Delta t}-\frac{E_0(t_1)-E_0(t_0)}{\Delta t}\right]=\lim_{\Delta t\to 0}\left[\frac{\beta(t_1)}{\Delta t}-\frac{\beta(t_0)}{\Delta t}\right]=\frac{\mathrm{d}\beta}{\mathrm{d}t}$$

Given that $\beta(t_1)=E_1(t_1)-E_0(t_1)$ and $E_0(t_0)=E_1(t_0)$ because $\beta(t_0)=0$

$$\frac{\mathrm{d}E}{\mathrm{d}t}=\frac{\mathrm{d}E_0}{\mathrm{d}t}+\frac{\mathrm{d}\beta}{\mathrm{d}t}$$

However, $r=a_0(1-e_0\cos E_0)$, so

$$\dot r=a_0e_0\sin E_0\dot E_0=\frac{\mu e_0}{h_0}\sin\theta_0^*$$

Replacing $\sin E_0$ by $(r\mu/h_0^2)\sqrt{1-e_0^2}\sin\theta_0^*$ and observing that $h_0=\sqrt{\mu a_0(1-e_0^2)}$, we obtain

$$\dot E_0=\frac{\mathrm{d}E_0}{\mathrm{d}t}=\frac{n_0a_0}{r}$$

where $n_0=\sqrt{\mu/a_0^3}$ is the Keplerian mean motion. Therefore

$$\dot E=\dot E_0+\frac{\mathrm{d}\beta}{\mathrm{d}t} \tag{10.48}$$

with $\mathrm{d}\beta/\mathrm{d}t$ given by Eq. (10.35) and because it is expressed as a function of $\mathrm{d}e/\mathrm{d}t$, the expression for $\mathrm{d}e/\mathrm{d}t$ in Eq. (10.17) should be used.

In the case of the hyperbola, $\beta(t_1)=H_1(t_1)-H_0(t_1)$ with $\beta(t_0)=0$, as before. Then

$$\lim_{\Delta t\to 0}\left[\frac{H_1(t_1)-H_1(t_0)}{\Delta t}-\frac{H_0(t_1)-H_0(t_0)}{\Delta t}\right]=\lim_{\Delta t\to 0}\left[\frac{\beta(t_1)}{\Delta t}-\frac{\beta(t_0)}{\Delta t}\right]$$

$$\frac{\mathrm{d}H}{\mathrm{d}t}=\frac{\mathrm{d}H_0}{\mathrm{d}t}+\frac{\mathrm{d}\beta}{\mathrm{d}t}$$

However, $r=a_0(e_0\cosh H_0-1)$, so

$$\dot r=a_0e_0\sinh H_0\dot H_0=\frac{\mu e_0}{h_0}\sin\theta_0^*$$

Observing that

$$\sinh H_0=\frac{\mu r}{h_0^2}\sqrt{e_0^2-1}\sin\theta_0^*$$

and $h_0 = \sqrt{\mu a_0(e_0^2 - 1)}$with $a_0 > 0$, we can write $\dot{H}_0 = (a_0/r)\nu$, where $\nu = \sqrt{\mu/a_0^3}$ is the orbit mean motion. Therefore

$$\frac{dH}{dt} = \frac{\nu a}{r} + \frac{d\beta}{dt} \tag{10.49}$$

with $d\beta/dt$ given by Eq. (10.39), in which de/dt is now given by Eq. (10.17′).

The mean anomaly calculations were carried out as follows. For the ellipse, with $\gamma(t_1) = M_1(t_1) - M_0(t_1)$

$$\lim_{\Delta t \to 0}\left[\frac{M_1(t_1) - M_1(t_0)}{\Delta t} - \frac{M_0(t_1) - M_0(t_0)}{\Delta t}\right] = \lim_{\Delta t \to 0}\left[\frac{\gamma(t_1)}{\Delta t} - \frac{\gamma(t_0)}{\Delta t}\right]$$

Of course, $M_1(t_0) = M_0(t_0)$ because $\gamma(t_0) = 0$. In addition

$$\frac{dM}{dt} = \frac{dM_0}{dt} + \frac{d\gamma}{dt}$$

However, $M_0 = n_0(t - \tau)$ yields $\dot{M}_0 = n_0$, such that

$$\frac{dM}{dt} = n + \frac{d\gamma}{dt} \tag{10.50}$$

where all of the variables and constants are defined at time t_0 and where $d\gamma/dt$ is given by Eq. (10.42), de/dt by Eq. (10.17), and $d\beta/dt$ by Eq. (10.35). In the case of the hyperbola, $N_0 = \nu_0\,(t - \tau)$ such that $\dot{N}_0 = \nu_0$; therefore

$$\frac{dN}{dt} = \nu + \frac{d\gamma}{dt} \tag{10.51}$$

with $d\gamma/dt$ defined in terms of de/dt and $d\beta/dt$ as given in Eqs. (10.46), (10.17′), and (10.39), respectively, all with $a > 0$.

The expression for $d\eta/dt$ in Eq. (10.31) is defined in terms of the time rate of change of the eccentricity and the scalar angular momentum. However, an expression independent of de/dt can be obtained as follows. From

$$\mu e r \sin\theta^* = h(\boldsymbol{r}\cdot\boldsymbol{v})$$

$$\mu e_0 r_{osc}(t_1)\sin\theta_0^*(t_1) = h_0[\boldsymbol{r}_{osc}(t_1)\cdot\boldsymbol{v}_{osc}(t_1)] \tag{10.52}$$

$$\mu e_1 r(t_1)\sin\theta_1^*(t_1) = h_1[\boldsymbol{r}(t_1)\cdot\boldsymbol{v}(t_1)] \tag{10.53}$$

Subtracting Eq. (10.52) from Eq. (10.53) and using the relations $r(t_1) = r_{osc}(t_1) + \delta'$, $\boldsymbol{r}(t_1) = \boldsymbol{r}_{osc}(t_1) + \boldsymbol{\delta}$, $\boldsymbol{v}(t_1) = \boldsymbol{v}_{osc}(t_1) + \boldsymbol{\nu}$, and $\theta_1^*(t_1) = \theta_0^*(t_1) + \eta(t_1)$, as well as $\cos\eta(t_1) \simeq 1 - [\eta^2(t_1)]/2 \simeq 1$ and

$\sin\eta(t_1) \simeq \eta(t_1)$, we obtain

$$\begin{aligned}&\mu r_{\rm osc}(t_1)\sin\theta_0^*(t_1)(e_1-e_0)+\mu e_1 r_{\rm osc}(t_1)\cos\theta_0^*(t_1)\eta(t_1)\\&\quad+\mu e_1\delta'\left[\sin\theta_0^*(t_1)+\eta(t_1)\cos\theta_0^*(t_1)\right]\\&\quad=(h_1-h_0)[\boldsymbol{r}_{\rm osc}(t_1)\cdot\boldsymbol{v}_{\rm osc}(t_1)]+h_1[\boldsymbol{r}_{\rm osc}(t_1)\cdot\boldsymbol{\nu}+\boldsymbol{\delta}\cdot\boldsymbol{v}_{\rm osc}(t_1)+\boldsymbol{\delta}\cdot\boldsymbol{\nu}]\end{aligned}$$

Dividing by Δt, neglecting terms of $\mathcal{O}(\Delta t)$ and higher, and observing that

$$\boldsymbol{r}_{\rm osc}(t_1)\cdot\boldsymbol{v}_{\rm osc}(t_1)=\boldsymbol{r}(t_1)\cdot\boldsymbol{v}(t_1)+\text{HOT}=\frac{\mu r(t_1)}{h_1}e_1\sin\theta_1^*(t_1)$$

the preceding difference expression reduces to

$$\begin{aligned}&\mu\left[\sin\theta_1^*(t_1)-d\eta\cos\theta_1^*(t_1)\right]r(t_1)\frac{\Delta e}{\Delta t}\\&\quad+\mu e_1 r(t_1)\left[\cos\theta_1^*(t_1)+d\eta\sin\theta_1^*(t_1)\right]\frac{\Delta\eta}{\Delta t}\\&\qquad=\frac{\mu r(t_1)}{h_1}e_1\sin\theta_1^*(t_1)\frac{\Delta h}{\Delta t}+h_1\boldsymbol{r}(t_1)\cdot\boldsymbol{a}_d\end{aligned}$$

Now, because $h/\mu=p/h$ and $\mu re\sin\theta^*=h(\boldsymbol{r}\cdot\boldsymbol{v})$, in the limit as $\Delta t\to 0$, we have

$$re\cos\theta^*\frac{\mathrm{d}\eta}{\mathrm{d}t}=-r\sin\theta^*\frac{\mathrm{d}e}{\mathrm{d}t}+\frac{p}{h}\left(\boldsymbol{r}\cdot\boldsymbol{a}_d+\boldsymbol{r}\cdot\boldsymbol{v}\frac{1}{h}\frac{\mathrm{d}h}{\mathrm{d}t}\right)\tag{10.54}$$

This expression is valid for both the ellipse and the hyperbola. Now for the ellipse, if $\mathrm{d}e/\mathrm{d}t$ is eliminated between Eqs. (10.54) and (10.31), the following result is obtained:

$$\begin{aligned}re\frac{\mathrm{d}\eta}{\mathrm{d}t}&=-2\frac{h}{\mu}\sin\theta^*\frac{\mathrm{d}h}{\mathrm{d}t}+\frac{p}{h}\cos\theta^*\left(\boldsymbol{r}\cdot\boldsymbol{a}_d+\boldsymbol{r}\cdot\boldsymbol{v}\frac{1}{h}\frac{\mathrm{d}h}{\mathrm{d}t}\right)\\\frac{\mathrm{d}\eta}{\mathrm{d}t}&=\frac{1}{reh}\left[p\cos\theta^*(\boldsymbol{r}\cdot\boldsymbol{a}_d)-2p\sin\theta^*\frac{\mathrm{d}h}{\mathrm{d}t}+re\cos\theta^*\sin\theta^*\frac{\mathrm{d}h}{\mathrm{d}t}\right]\end{aligned}$$

Moreover, because $re\cos\theta^*=p-r$, we can write

$$\frac{\mathrm{d}\eta}{\mathrm{d}t}=\frac{1}{reh}\left[p\cos\theta^*(\boldsymbol{r}\cdot\boldsymbol{a}_d)-(p+r)\sin\theta^*\frac{\mathrm{d}h}{\mathrm{d}t}\right]\tag{10.55}$$

Because the form of $\mathrm{d}e/\mathrm{d}t$ is immaterial to this discussion, its elimination makes Eq. (10.55) valid for both the ellipse and the hyperbola, as the term in $\mathrm{d}h/\mathrm{d}t$ given in Eq. (10.15) is equally valid for both types of trajectories.

10.2.9 VARIATION OF THE ARGUMENT OF PERIGEE

Let $\hat{\boldsymbol{i}}_u$ be a unit vector along the direction of the ascending node. Then, it is clear from Fig. 10.1 that

$$\hat{\boldsymbol{i}}_u \cdot \boldsymbol{r} = r\cos(\omega + \theta^*) = r\cos\theta$$

However, $\hat{\boldsymbol{i}}_u = (\,c_\Omega \quad s_\Omega \quad 0\,)$ in the inertial xyz frame, so that taking derivatives with respect to time gives

$$(-s_\Omega \quad c_\Omega \quad 0\,)\boldsymbol{r}\dot{\Omega} = -r\sin(\omega + \theta^*)(\dot{\omega} + \dot{\theta}^*) \tag{10.56}$$

where $\dot{\theta}^* = \dot{\theta}_0^* + \mathrm{d}\eta/\mathrm{d}t$. Moreover, because changes in ω are investigated by holding r fixed, the Keplerian contribution can be ignored.

Now, from Fig. 10.1, we have

$$\boldsymbol{r} = r\cos\theta^*\hat{\boldsymbol{i}}_\xi + r\sin\theta^*\hat{\boldsymbol{i}}_\eta \tag{10.57}$$

and by differentiation, with $\dot{r} = (\mu e/h)\sin\theta^*$ and $\dot{\theta}^* = h/r^2$, we obtain

$$\boldsymbol{v} = \sqrt{\frac{\mu}{p}}\left[-\sin\theta^*\hat{\boldsymbol{i}}_\xi + (e + \cos\theta^*)\hat{\boldsymbol{i}}_\eta\right] \tag{10.58}$$

However, in the $\xi\eta\zeta$ frame

$$\hat{\boldsymbol{i}}_\xi = \begin{pmatrix} 1 \\ 0 \\ 0 \end{pmatrix} \quad \text{and} \quad \hat{\boldsymbol{i}}_\eta = \begin{pmatrix} 0 \\ 1 \\ 0 \end{pmatrix}$$

and from Sec. 10.2.4 of this chapter, these two unitized vectors transform to the following forms in the inertial xyz frame:

$$\hat{\boldsymbol{i}}_\xi = \begin{pmatrix} l_1 & l_2 & l_3 \\ m_1 & m_2 & m_3 \\ n_1 & n_2 & n_3 \end{pmatrix}\begin{pmatrix} 1 \\ 0 \\ 0 \end{pmatrix} = \begin{pmatrix} l_1 \\ m_1 \\ n_1 \end{pmatrix} = \begin{pmatrix} c_\Omega c_\omega - s_\Omega s_\omega c_i \\ s_\Omega c_\omega + c_\Omega s_\omega c_i \\ s_\omega s_i \end{pmatrix}$$
$$\hat{\boldsymbol{i}}_\eta = \begin{pmatrix} l_1 & l_2 & l_3 \\ m_1 & m_2 & m_3 \\ n_1 & n_2 & n_3 \end{pmatrix}\begin{pmatrix} 0 \\ 1 \\ 0 \end{pmatrix} = \begin{pmatrix} l_2 \\ m_2 \\ n_2 \end{pmatrix} = \begin{pmatrix} -c_\Omega s_\omega - s_\Omega c_\omega c_i \\ -s_\Omega s_\omega + c_\Omega c_\omega c_i \\ c_\omega s_i \end{pmatrix}$$

Consequently, $\boldsymbol{r}$ in Eq. (10.57) is reduced to

$$\boldsymbol{r} = r\begin{pmatrix} c_\Omega c_\theta - s_\Omega s_\theta c_i \\ s_\Omega c_\theta + c_\Omega s_\theta c_i \\ s_\theta s_i \end{pmatrix} \tag{10.59}$$

with $\theta = \theta^* + \omega$. In a similar manner, Eq. (10.58) for the velocity vector reduces to

$$\boldsymbol{v} = \frac{\mu}{h}\begin{bmatrix} -c_\Omega(s_\theta + es_\omega) - s_\Omega c_i(c_\theta + ec_\omega) \\ -s_\Omega(s_\theta + es_\omega) + c_\Omega c_i(c_\theta + ec_\omega) \\ s_i(c_\theta + ec_\omega) \end{bmatrix} \tag{10.60}$$

Now, returning to Eq. (10.56) and replacing $\boldsymbol{r}$ by its expression in Eq. (10.59) gives

$$c_i\frac{\mathrm{d}\Omega}{\mathrm{d}t} = -\frac{\mathrm{d}\omega}{\mathrm{d}t} - \frac{\mathrm{d}\eta}{\mathrm{d}t}$$

$$\dot{\omega} = -c_i\dot{\Omega} - \frac{\mathrm{d}\eta}{\mathrm{d}t} \tag{10.61}$$

which is valid for both the ellipse and the hyperbola.

To convert these equations for the variation of the orbital elements into polar or Gaussian form, it is necessary to decompose the acceleration $\boldsymbol{a}_d$ defined so far in the nonrotating frame, along the radial, perpendicular, and out-of-plane directions, namely, $\hat{\boldsymbol{r}}$, $\hat{\boldsymbol{\theta}}$, and $\hat{\boldsymbol{i}}_\zeta$, where $\hat{\boldsymbol{r}}$ is a unit vector along the instantaneous radial direction, $\hat{\boldsymbol{\theta}}$ is a unit vector in the orbit plane 90 deg ahead of $\hat{\boldsymbol{r}}$ in the direction of motion, and $\hat{\boldsymbol{i}}_\zeta$ is a unit vector along the angular momentum vector $\boldsymbol{h}$ or the out-of-plane direction. Now

$$\boldsymbol{a}_d = a_{d_r}\hat{\boldsymbol{r}} + a_{d_\theta}\hat{\boldsymbol{\theta}} + a_{d_\zeta}\hat{\boldsymbol{i}}_\zeta$$

and with

$$\boldsymbol{v} = \dot{r}\hat{\boldsymbol{r}} + r\dot{\theta}\hat{\boldsymbol{\theta}} = \frac{\mu e}{h}\sin\theta^*\hat{\boldsymbol{r}} + \frac{h}{r}\hat{\boldsymbol{\theta}}$$

$$\boldsymbol{r}\cdot\boldsymbol{a}_d = ra_{d_r} \tag{10.62}$$

$$\boldsymbol{v}\cdot\boldsymbol{a}_d = \frac{\mu e}{h}\sin\theta^* a_{d_r} + \frac{h}{r}a_{d_\theta} \tag{10.63}$$

The cross product $\boldsymbol{r}\times\boldsymbol{a}_d$ appearing in some of the equations developed so far is defined in the inertial xyz reference system of Fig. 10.1. The transformation that must be applied to this vector to convert it to the local polar coordinate system $\hat{\boldsymbol{r}}$, $\hat{\boldsymbol{\theta}}$, $\hat{\boldsymbol{i}}_\zeta$ consists of the three rotations Ω, i, and $\omega + \theta^*$ or θ applied in reverse order such that the orthogonal matrix $\boldsymbol{S}^{-1}$ is given by

$$\boldsymbol{S}^{-1} = \begin{pmatrix} c_\Omega c_\theta - s_\Omega s_\theta c_i & -c_\Omega s_\theta - s_\Omega c_\theta c_i & s_\Omega s_i \\ s_\Omega c_\theta + c_\Omega s_\theta c_i & -s_\Omega s_\theta + c_\Omega c_\theta c_i & -c_\Omega s_i \\ s_\theta s_i & c_\theta s_i & c_i \end{pmatrix}^{-1}$$

with $S^{-1} = S^T$. As a result, from Eq. (10.19), we obtain

$$(\boldsymbol{r} \times \boldsymbol{a}_d)_{\text{polar}} = \boldsymbol{S}^{-1}(\boldsymbol{r} \times \boldsymbol{a}_d)_{\text{inertial}}$$
$$= \boldsymbol{S}^{-1}\begin{pmatrix} c_\Omega & -s_\Omega c_i & s_\Omega s_i \\ s_\Omega & c_\Omega c_i & -c_\Omega s_i \\ 0 & s_i & c_i \end{pmatrix}\begin{pmatrix} hs_i\dot{\Omega} \\ -h\dot{i} \\ \dot{h} \end{pmatrix}$$
$$= \begin{pmatrix} c_\theta & s_\theta & 0 \\ -s_\theta & c_\theta & 0 \\ 0 & 0 & 1 \end{pmatrix}\begin{pmatrix} hs_i\dot{\Omega} \\ -h\dot{i} \\ \dot{h} \end{pmatrix} \tag{10.64}$$

However, in the local polar coordinate system

$$\boldsymbol{r} \times \boldsymbol{a}_d = \begin{pmatrix} \hat{\boldsymbol{r}} & \hat{\boldsymbol{\theta}} & \hat{\boldsymbol{i}}_\zeta \\ r & 0 & 0 \\ a_{dr} & a_{d\theta} & a_{d\zeta} \end{pmatrix} = (0)\hat{\boldsymbol{r}} + (-ra_{d\zeta})\hat{\boldsymbol{\theta}} + (ra_{d\theta})\hat{\boldsymbol{i}}_\zeta = r\begin{pmatrix} 0 \\ -a_{d\zeta} \\ a_{d\theta} \end{pmatrix}$$

This expression is now used in Eq. (10.64), in which the matrix is also orthogonal, to yield the rates of change of Ω, i, and h in terms of the polar component of the acceleration vector $\boldsymbol{a}_d$

$$\dot{\Omega} = \frac{r}{hs_i} s_\theta a_{d\zeta} \tag{10.65}$$

$$\dot{i} = \frac{r}{h} c_\theta a_{d\zeta} \tag{10.66}$$

$$\dot{h} = ra_{d\theta} \tag{10.67}$$

These equations are valid for both the ellipse and the hyperbola, and Eq. (10.67) can be derived directly from Eq. (10.15) using Eqs. (10.62) and (10.63) and the cross product

$$\boldsymbol{r} \times \boldsymbol{v} = \frac{\mu re}{h} \sin\theta^*$$

such that

$$\dot{h} = \frac{1}{h}\left[r^2(\boldsymbol{v}\cdot\boldsymbol{a}_d) - (\boldsymbol{r}\cdot\boldsymbol{v})(\boldsymbol{r}\cdot\boldsymbol{a}_d)\right]$$
$$= \frac{1}{h}\left[r^2\left(\frac{\mu e\sin\theta^*}{h}a_{dr} + \frac{h}{r}a_{d\theta}\right) - \frac{\mu re}{h}\sin\theta^* ra_{dr}\right] = ra_{d\theta}$$

The expression for $\mathrm{d}\eta/\mathrm{d}t$ given in Eq. (10.55), which is common to both the ellipse and the hyperbola, will be transformed into

$$\begin{aligned}\frac{\mathrm{d}\eta}{\mathrm{d}t} &= \frac{1}{reh}\left[pc_{\theta^*}(\boldsymbol{r}\cdot\boldsymbol{a}_d) - (p+r)s_{\theta^*}\dot{h}\right]\\ \dot{\eta} &= \frac{1}{reh}\left[pc_{\theta^*}ra_{dr} - (p+r)s_{\theta^*}ra_{d\theta}\right]\\ \dot{\eta} &= \frac{1}{eh}\left[pc_{\theta^*}a_{dr} - (p+r)s_{\theta^*}a_{d\theta}\right]\end{aligned} \tag{10.68}$$

The rate of change of the argument of perigee given in Eq. (10.61) is also valid for both conics, such that use of Eqs. (10.65) and (10.68) gives

$$\begin{aligned}\dot{\omega} &= -c_i\dot{\Omega} - \dot{\eta}\\ \dot{\omega} &= -\frac{r}{h}s_\theta \cot i a_{d\zeta} - \frac{1}{eh}\left[pc_{\theta^*}a_{dr} - (p+r)s_{\theta^*}a_{d\theta}\right]\end{aligned} \tag{10.69}$$

For the case of the ellipse, the rate of change of the semimajor axis is obtained from Eq. (10.11) along with Eq. (10.63) as follows:

$$\begin{aligned}\frac{\mathrm{d}a}{\mathrm{d}t} &= \frac{2a^2}{\mu}\boldsymbol{v}\cdot\boldsymbol{a}_d = \frac{2a^2}{\mu}\left(\frac{\mu e\sin\theta^*}{h}a_{dr} + \frac{h}{r}a_{d\theta}\right)\\ \frac{\mathrm{d}a}{\mathrm{d}t} &= \frac{2a^2}{\mu}\left(e\sin\theta^* a_{dr} + \frac{p}{r}a_{d\theta}\right)\end{aligned} \tag{10.70}$$

For the hyperbola, with $a > 0$, Eq. (10.17′) is used instead of Eq. (10.17) such that

$$\frac{\mathrm{d}a}{\mathrm{d}t} = \frac{-2a^2}{\mu}\boldsymbol{v}\cdot\boldsymbol{a}_d = \frac{-2a^2}{\mu}\left(e\sin\theta^* a_{dr} + \frac{p}{r}a_{d\theta}\right) \tag{10.71}$$

The rate of change of the eccentricity for the ellipse is obtained from Eq. (10.17) as

$$\begin{aligned}\frac{\mathrm{d}e}{\mathrm{d}t} &= \frac{1}{\mu ae}\left[(pa - r^2)(\boldsymbol{v}\cdot\boldsymbol{a}_d) + (\boldsymbol{r}\cdot\boldsymbol{v})(\boldsymbol{r}\cdot\boldsymbol{a}_d)\right]\\ &= \frac{1}{\mu ae}\left\{(pa - r^2)\left[\frac{\mu e\sin\theta^*}{h}a_{dr} + \frac{h}{r}a_{d\theta}\right] + \frac{\mu re\sin\theta^*}{h}ra_{dr}\right\}\\ &= \frac{p}{h}\sin\theta^* a_{dr} + \frac{1}{\mu ae}(pa - r^2)\frac{h}{r}a_{d\theta}\\ &= \frac{p}{h}\sin\theta^* a_{dr} + \frac{p}{h}\left[\frac{(1+e\cos\theta^*)}{e} - \frac{r}{ae}\right]a_{d\theta}\\ &= \frac{p}{h}\sin\theta^* a_{dr} + \frac{p}{h}\left[\cos\theta^* + \frac{r}{p}(e+\cos\theta^*)\right]a_{d\theta}\end{aligned} \tag{10.72}$$

This expression is also valid for the hyperbola with $a > 0$ and $p = a(e^2 - 1)$ because use of Eq. (10.17′) results in

$$\begin{aligned}
\dot{e} &= \frac{1}{\mu a e}\left[(pa + r^2)(\boldsymbol{v} \cdot \boldsymbol{a}_d) - (\boldsymbol{r} \cdot \boldsymbol{v})(\boldsymbol{r} \cdot \boldsymbol{a}_d)\right] \\
&= \frac{1}{\mu a e}\left\{(pa + r^2)\left[\frac{\mu e \sin\theta^*}{h} a_{dr} + \frac{h}{r} a_{d\theta}\right] - \frac{\mu r e \sin\theta^*}{h} r a_{dr}\right\} \\
&= \frac{p}{h}\sin\theta^* a_{dr} + \frac{1}{\mu a e}(pa + r^2)\frac{h}{r} a_{d\theta} \\
&= \frac{p}{h}\sin\theta^* a_{dr} + \frac{p}{h}\left[\frac{(1 + e\cos\theta^*)}{e} + \frac{r}{ae}\right] a_{d\theta} \\
&= \frac{p}{h}\sin\theta^* a_{dr} + \frac{p}{h}\left[\frac{(a+p)}{(1 + e\cos\theta^*)ae} + \frac{\cos\theta^*}{(1 + e\cos\theta^*)} + \cos\theta^*\right] a_{d\theta} \\
&= \frac{p}{h}\sin\theta^* a_{dr} + \frac{p}{h}\left[\cos\theta^* + \frac{r}{p}(e + \cos\theta^*)\right] a_{d\theta}
\end{aligned} \tag{10.73}$$

which is identical to Eq. (10.72)

In the case of the eccentric anomalies, the rates can be found as follows. For the ellipse, according to Eq. (10.35)

$$\sqrt{1 - e^2}\frac{\mathrm{d}\beta}{\mathrm{d}t} = \frac{r}{a}\frac{\mathrm{d}\eta}{\mathrm{d}t} - \frac{r\sin\theta^*}{p}\frac{\mathrm{d}e}{\mathrm{d}t}$$

Using Eq. (10.68) for $\dot{\eta}$ and Eq. (10.72) for $\dot{e}$, we can transform this equation into

$$\begin{aligned}
\sqrt{1 - e^2}\dot{\beta} &= \left(\frac{rpc_{\theta^*}}{aeh} - \frac{r}{h}s_{\theta^*}^2\right) a_{dr} - \frac{rs_{\theta^*}}{h}\left[\frac{(p+r)}{ae} + c_{\theta^*} + \frac{r}{p}(e + c_{\theta^*})\right] a_{d\theta} \\
&= \frac{p}{eh}(c_{\theta^*} - e) a_{dr} - \frac{r}{h}s_{\theta^*}\left[\frac{(p+r)}{ae} + c_{\theta^*} + \frac{r}{p}(e + c_{\theta^*})\right] a_{d\theta} \\
\dot{\beta} &= \frac{(c_{\theta^*} - e)}{aen} a_{dr} - \frac{s_{\theta^*} r}{anp}\left[\frac{(p+r)}{ae} + \frac{r(e + c_{\theta^*})}{p} + c_{\theta^*}\right] a_{d\theta}
\end{aligned} \tag{10.74}$$

For the hyperbola, Eq. (10.39) is transformed into

$$\begin{aligned}
\sqrt{e^2 - 1}\dot{\beta} &= \frac{r}{a}\dot{\eta} + \frac{rs_{\theta^*}}{p}\dot{e} \\
&= \frac{r}{h}\left(\frac{pc_{\theta^*}}{ae} + s_{\theta^*}^2\right) a_{dr} - \frac{rs_{\theta^*}}{h}\left[\frac{(p+r)}{ae} - c_{\theta^*} - \frac{r}{p}(e + c_{\theta^*})\right] a_{d\theta} \\
\dot{\beta} &= \frac{-(c_{\theta^*} - e)}{aev} a_{dr} - \frac{s_{\theta^*} r}{avp}\left[\frac{(p+r)}{ae} - \frac{r(e + c_{\theta^*})}{p} - c_{\theta^*}\right] a_{d\theta}
\end{aligned} \tag{10.75}$$

where ν is the orbital mean motion. This equation is, of course, valid for $a > 0$ and $p = a(e^2 - 1)$.

Finally, we can obtain the variations of the mean motion from Eq. (10.42), using Eq. (10.74) for $\dot{\beta}$ and Eq. (10.72) for $\dot{e}$, along with

$$\begin{aligned}
h &= \sqrt{\mu a(1-e^2)} \\
\dot{\gamma} &= \frac{r}{a}\dot{\beta} - \frac{r}{p}\sqrt{(1-e^2)}s_{\theta^*}\dot{e} \\
&= \left[\frac{r(c_{\theta^*}-e)}{a^2en} - \frac{rs_{\theta^*}^2}{h}\right]a_{dr} \\
&\quad - \left\{\frac{r^2(p+r)}{a^2nepa}s_{\theta^*} + \frac{r(p+r)}{a^2np}s_{\theta^*}\left[c_{\theta^*} + \frac{r}{p}(e+c_{\theta^*})\right]\right\}a_{d\theta} \\
&= \frac{(pc_{\theta^*}-2re)}{a^2en}a_{dr} - \frac{s_{\theta^*}(p+r)}{a^2ne}\left\{\frac{r^2}{pa} + \frac{re}{p}\left[c_{\theta^*} + \frac{r}{p}(e+c_{\theta^*})\right]\right\}a_{d\theta} \\
\dot{\gamma} &= \frac{(pc_{\theta^*}-2re)}{a^2en}a_{dr} - \frac{s_{\theta^*}(p+r)}{a^2ne}a_{d\theta}
\end{aligned} \tag{10.76}$$

For the hyperbola, with $a > 0$, Eq. (10.46), rewritten in the form

$$\dot{\gamma} = \frac{r}{a}\dot{\beta} + \frac{r}{p}\sqrt{e^2-1}s_{\theta^*}\dot{e}$$

is used for $\dot{\gamma}$, with $\dot{\beta}$ given by Eq. (10.75) and $\dot{e}$ given by Eq. (10.72). With $h = \sqrt{\mu a(e^2-1)}$, $p = a(e^2-1)$, $\nu = \sqrt{\mu/a^3}$, and $r = a(\cosh H - 1)$, consistent with $a > 0$, after some manipulation, we obtain

$$\begin{aligned}
\dot{\gamma} &= \left(\frac{r\sqrt{e^2-1}}{h}s_{\theta^*}^2 - \frac{r(c_{\theta^*}-e)}{a^2e\nu}\right)a_{dr} \\
&\quad + \left\{\frac{rs_{\theta^*}}{a^2\nu}\left(1+\frac{r}{p}\right)\left[c_{\theta^*} + \frac{r}{p}(e+c_{\theta^*})\right] - \frac{r^2s_{\theta^*}(p+r)}{a^2\nu\, pae}\right\}a_{d\theta} \\
\dot{\gamma} &= \frac{1}{a^2\nu e}(2re - pc_{\theta^*})a_{dr} + \frac{(p+r)}{a^2\nu e}s_{\theta^*}a_{d\theta}
\end{aligned} \tag{10.77}$$

The complete set of equations for both the ellipse and the hyperbola are then recapitulated as follows:

For the ellipse with $h = \sqrt{\mu a(1-e^2)}$, $p = a(1-e^2)$, $n = \sqrt{\mu/a^3}$, $\theta = \theta^* + \omega$, and $r = a(1 - e\cos E)$

$$\dot{a} = \frac{2a^2}{h}\left(es_{\theta^*}a_{dr} + \frac{p}{r}a_{d\theta}\right) \tag{10.78}$$

$$\dot{e} = \frac{p}{h} s_{\theta^*} a_{dr} + \frac{p}{h}\left[c_{\theta^*} + \frac{r}{p}(e + c_{\theta^*})\right] a_{d\theta} \tag{10.79}$$

$$\dot{h} = r a_{d\theta} \tag{10.80}$$

$$\dot{\Omega} = \frac{r}{h s_i} s_\theta a_{d\zeta} \tag{10.81}$$

$$\dot{i} = \frac{r}{h} c_\theta a_{d\zeta} \tag{10.82}$$

$$\dot{\omega} = -\frac{r}{h} s_\theta \cot i a_{d\zeta} - \frac{1}{eh}[p c_{\theta^*} a_{dr} - (p + r) s_{\theta^*} a_{d\theta}] \tag{10.83}$$

$$\dot{\theta}^* = \frac{h}{r^2} + \frac{1}{eh}[p c_{\theta^*} a_{dr} - (p + r) s_{\theta^*} a_{d\theta}] \tag{10.84}$$

$$\dot{E} = \frac{na}{r} + \frac{(c_{\theta^*} - e)}{aen} a_{dr} - \frac{s_{\theta^*} r}{anp}\left[\frac{(p + r)}{ae} + \frac{r}{p}(e + c_{\theta^*}) + c_{\theta^*}\right] a_{d\theta} \tag{10.85}$$

A slightly more compact expression for $\dot{E}$ can be written as

$$\dot{E} = \frac{na}{r} + \frac{1}{nae}\left[(c_{\theta^*} - e) a_{dr} - \left(1 + \frac{r}{a}\right) s_{\theta^*} a_{d_\theta}\right] \tag{10.86}$$

It is easy to show that both Eqs. (10.85) and (10.86) are equivalent.

$$\dot{M} = n + \frac{(p c_{\theta^*} - 2re)}{a^2 en} a_{dr} - \frac{s_{\theta^*}(p + r)}{a^2 ne} a_{d\theta} \tag{10.87}$$

For the hyperbola, with $a > 0$, $p = a(e^2 - 1)$, $h = \sqrt{\mu a(e^2 - 1)}$, $\nu = \sqrt{\mu / a^3}$, and $r = a(e\cosh H - 1)$

$$\dot{a} = -\frac{2a^2}{h}\left(e s_{\theta^*} a_{dr} + \frac{p}{r} a_{d\theta}\right) \tag{10.88}$$

$$\dot{e} = \frac{p}{h} s_{\theta^*} a_{dr} + \frac{p}{h}\left[c_{\theta^*} + \frac{r}{p}(e + c_{\theta^*})\right] a_{d\theta} \tag{10.89}$$

$$\dot{h} = r a_{d\theta} \tag{10.90}$$

$$\dot{\Omega} = \frac{r}{h s_i} s_\theta a_{d\zeta} \tag{10.91}$$

$$\dot{i} = \frac{r}{h} c_\theta a_{d\zeta} \tag{10.92}$$

$$\dot{\omega} = -\frac{r}{h} s_\theta \cot i a_{d\zeta} - \frac{1}{eh}[p c_{\theta^*} a_{dr} - (p + r) s_{\theta^*} a_{d\theta}] \tag{10.93}$$

$$\dot{\theta}^* = \frac{h}{r^2} + \frac{1}{eh}[p c_{\theta^*} a_{dr} - (p + r) s_{\theta^*} a_{d\theta}] \tag{10.94}$$

$$\dot{H} = \frac{a\nu}{r} - \frac{(c_{\theta^*} - e)}{ae\nu} a_{dr} - \frac{s_{\theta^*} r}{a\nu p}\left[\frac{(p + r)}{ae} - \frac{r}{p}(e + c_{\theta^*}) - c_{\theta^*}\right] a_{d\theta} \tag{10.95}$$

A slightly more compact form for $\dot{H}$ can be written as

$$\dot{H} = \frac{av}{r} - \frac{(c_{\theta^*} - e)}{aev} a_{dr} + \frac{s_{\theta^*}}{aev}\left(1 - \frac{r}{a}\right) a_{d\theta} \tag{10.96}$$

It is easy to show that Eqs. (10.95) and (10.96) are indeed equivalent.

$$\dot{N} = v + \frac{1}{a^2 ve}(2re - pc_{\theta^*}) a_{dr} + \frac{(p + r)}{a^2 ve} s_{\theta^*} a_{d\theta} \tag{10.97}$$

10.3 CONCLUSION

In this chapter, the perturbation equations for the variations of the parameters in the Gaussian representation have been derived for both the elliptical and hyperbolic orbits. The method used is based on Battin's analysis for the elliptical case [1], which was thus extended to the hyperbolic case. Both the elliptical case and hyperbolic case are included here for completeness, and the full sets of the variation-of-parameters equations are shown for the classical orbit elements. For convenience and commonality, these derivations assume that the semimajor axis a for the hyperbola is positive.

REFERENCES

[1] Battin, R. N. *Astronautical Guidance*, McGraw-Hill, New York, 1964.

[2] Brouwer, D., and Clemence, G. M., *Methods of Celestial Mechanics*, Academic Press, New York, 1961.

[3] Breakwell, J. V., "Lecture Notes: Space Mechanics," Course AA279A, Department of Aeronautics and Astronautics, Stanford University, Palo Alto, CA, 1974.

CHAPTER 11

Analysis of the Effects of Drag and Zonal Harmonics on a Near-Circular Orbit

NOMENCLATURE

a = orbit semimajor axis
e = orbit eccentricity
E = eccentric anomaly
n = orbit mean motion $\sqrt{\frac{\mu}{a^3}}$, rad/s
ρ = air density
C_D = drag coefficient
$\hat{\boldsymbol{x}}, \hat{\boldsymbol{y}}, \hat{\boldsymbol{z}}$ = Earth-centered inertial frame
$\hat{\boldsymbol{r}}, \hat{\boldsymbol{\theta}}, \hat{\boldsymbol{h}}$ = Euler–Hill rotating frame
J_2, J_3 = second and third zonal harmonics of the Earth potential function
$\boldsymbol{\omega}_F$ = angular velocity of the $(\hat{\boldsymbol{r}}, \hat{\boldsymbol{\theta}}, \hat{\boldsymbol{h}})$ frame

11.1 INTRODUCTION

In the first attempt to analyze these perturbations, the zonal harmonics are considered (J_2 and J_3 but easily generalized to include the higher-order zonals) together with the drag perturbation. An exponential atmospheric density model is assumed, and a Fourier–Bessel series approximation is used to describe the exponential term.

Rates of change for the semimajor axis and the eccentricity are obtained using the Euler–Hill formulation for near-circular orbits.

11.2 ANALYSIS

We assume that the drag is acting along a direction 90 deg away from the radial direction for a near-circular orbit and that the flight path angle is very small, so that the velocity vector and the transverse direction $\hat{\boldsymbol{\theta}}$ shown in Fig. 11.1 are almost coincident; the drag, in reality, is opposite to the velocity vector.

If the atmosphere of the planet is exponential, then the density can be described by

$$\rho = \rho_p \exp\left(-\frac{r - r_p}{H}\right) \tag{11.1}$$

where ρ_p is the density at the reference radius r_p (periapse) and H is the scale height. The orbit equation as a function of the eccentric anomaly yields

$$r = a(1 - e\cos E) \tag{11.2}$$

$$r_p = a(1 - e) \tag{11.3}$$

$$\rho = \rho_p \exp\left[\frac{-ae}{H}(1 - \cos E)\right] = \rho_p \exp[-k(1 - \cos E)] \tag{11.4}$$

where

$$k = \frac{ae}{H} \tag{11.5}$$

Kepler's equation

$$E - e\sin E = M = nt \tag{11.6}$$

can be approximated for small e (near-circular orbits) as

$$E = nt \tag{11.7}$$

where n is the mean motion of the spacecraft, given by

$$n = \sqrt{\frac{\mu}{a^3}} \tag{11.8}$$

The drag perturbation can be expressed as

$$D = \frac{1}{2}\rho C_D A V^2 = B\rho V^2 \tag{11.9}$$

where C_D is the drag coefficient; A is the cross section of the spacecraft normal to the velocity vector; and V is the velocity, which can be expressed as a function of eccentric anomaly to yield

$$D \simeq B\rho_p \frac{\mu}{a} \exp(-k)\exp(k\cos nt) \tag{11.10}$$

Appendix A11.1 describes how to express the exponential in terms of Bessel functions in their asymptotic expansion forms for large powers k. Using Eq. (11.1), the drag is now written as

$$D = B\rho_p \frac{\mu}{a} \mathrm{e}^{-k}\left[I_0(k) + 2\sum_{j=1}^{\infty} I_j(k)\cos jnt\right] \tag{11.11}$$

Next, the Euler–Hill equations are used with x along the radial direction and y 90 deg ahead in the direction of motion, with z being the out-of-plane component. This frame is attached to the spacecraft. Therefore

$$\ddot{x} - 2n\dot{y} - 3n^2x = f_r = 0 \tag{11.12}$$

$$\ddot{y} + 2n\dot{x} = f_\theta = -D \tag{11.13}$$

$$\ddot{z} + n^2z = f_h = 0 \tag{11.14}$$

Equation (11.13) can be integrated once to give

$$\dot{y} + 2nx = -B\rho_p\frac{\mu}{a}e^{-k} \times \left[I_0(k) + 2I_1(k)\frac{1}{n}s_{nt} + 2I_2(k)\frac{1}{2n}s_{2nt} + 2I_3(k)\frac{1}{3n}s_{3nt} + \cdots\right] \tag{11.15}$$

Substituting for $\dot{y}$ in Eq. (11.12) gives the following second-order differential equation in x:

$$\ddot{x} + n^2x = -2nB\rho_p\frac{\mu}{a}e^{-k}\left[I_0(k) + 2I_1(k)\frac{1}{n}s_{nt} + 2I_2(k)\frac{1}{2n}s_{2nt} + \cdots\right] = f(t) \tag{11.16}$$

The right-hand side of Eq. (11.16) is a function of time $f(t)$ and can be regarded as the forcing function.

The homogenous equation has the solution

$$x = C_1c_{nt} + C_2s_{nt} \tag{11.17}$$

with the general solution

$$x = C_1c_{nt} + C_2s_{nt} + \frac{1}{n}\int_0^t f(u)\sin n(t-u)du \tag{11.18}$$

By carrying out the integration and ignoring the homogenous solution, we obtain

$$x_{\text{drg}} = -\frac{2}{n}B\rho_p\frac{\mu}{a}e^{-k}\left[I_0(k)t - I_1(k)tc_{nt} + s_{nt}\left(-\frac{I_0}{n} + \frac{3}{4n}I_+\frac{2}{3n}I_2\right) - \frac{I_1}{4n}s_{3nt} - \frac{I_2}{3n}s_{2nt} + \frac{I_1}{2n}c_{nt}s_{2nt} + \text{HOT}\right] \tag{11.19}$$

We can now use Eq. (11.13) to solve for y as

$$y_{\rm drg} = B\rho_p \frac{\mu}{a} e^{-k}\left[\frac{3}{2} I_0 t^2 - \frac{4I_1}{n} t s_{nt} - \frac{4}{n} c_{nt}\left(-\frac{I_0}{n} + \frac{3}{2}\frac{I_1}{n} + \frac{2}{3n} I_2\right) + \frac{7}{6}\frac{I_2}{n^2} c_{2nt}\right] \tag{11.20}$$

From Eq. (11.19), we have

$$\begin{aligned}\dot{x}_{\rm drg} = -\frac{2}{n} B\rho_p \frac{\mu}{a} e^{-k}\Bigg[& I_0 - I_1 c_{nt} + nI_1 t s_{nt} + c_{nt}\left(-I_0 + \frac{3}{4} I_1 + \frac{2}{3} I_2\right) \\ & - \frac{3}{4} I_1 c_{3nt} - \frac{2}{3} I_2 c_{2nt} + \frac{I_1}{2}(-s_{nt} s_{2nt} + 2c_{nt} c_{2nt})\Bigg]\end{aligned} \tag{11.21}$$

If we keep only the terms in x that involve the linear time t, and follow Breakwell's analysis [1], then the leading part of x can be written as

$$x = -\frac{2}{n} B\rho_p \frac{\mu}{a} {\rm e}^{-k}[I_0(k)t - I_1(k)t c_{nt}] \tag{11.22}$$

The main contribution to $\dot{x}$ can therefore be written as

$$\dot{x} = -\frac{2}{n} B\rho_p \frac{\mu}{a} {\rm e}^{-k}[I_0(k) - I_1(k) c_{nt}] \tag{11.23}$$

Taking the derivative with respect to time of the orbit equation in Eq. (11.2) gives

$$\dot{r} = \dot{a} - c_{nt}(\dot{a}e + a\dot{e}) + aens_{nt} \tag{11.24}$$

Because x is along r, $\dot{x} = \dot{r}$, and therefore, from Eq. (11.23) and for $e = 0$ in Eq. (11.24), which reduces to $\dot{r} = \dot{a} - a\dot{e}c_{nt}$, we have Breakwell's equations for $\dot{a}$ and $\dot{e}$ as

$$\dot{a} = -2a^2 nB\rho_p I_0(k)e^{-k} \tag{11.25}$$

$$\dot{e} = -2nB\rho_p a e^{-k} I_1(k) \tag{11.26}$$

These equations describe how the semimajor axis and the eccentricity are affected by the drag alone. If we include the second zonal J_2 and zonal J_3, for example, then we need to express the perturbation components of these zonals in the Euler–Hill frame, and this is shown in Appendix 11.2.

For the J_2 zonal harmonic, the differential equations for the x and y coordinates can be written using the perturbation components f_r and f_θ in Eqs. (A11.34) of Appendix 11.2 as follows:

$$\frac{\ddot{x}}{a} - 2n\frac{\dot{y}}{a} - 3n^2\frac{x}{a} = -3n^2 J_2\left(\frac{R}{a}\right)^2\left(\frac{1}{2} - \frac{3}{4}s_i^2 + \frac{3}{4}s_i^2 c_{2nt}\right) \tag{11.27}$$

$$\frac{\ddot{y}}{a} + 2n\frac{\dot{x}}{a} = -\frac{3}{2}n^2 J_2 \left(\frac{R}{a}\right)^2 s_i^2 s_{2nt} \tag{11.28}$$

Integrating Eq. (11.28) once, ignoring the initial conditions and the constant of integration for simplicity, and carrying only the J_2 terms gives

$$\frac{\dot{y}}{a} + 2n\frac{x}{a} = \frac{3}{2}\frac{n^2}{2n} J_2 \left(\frac{R}{a}\right)^2 s_i^2 c_{2nt} \tag{11.29}$$

Next, replacing the $\dot{y}/a$ term in Eq. (11.27) will yield the forced-harmonic-oscillator form

$$\frac{\ddot{x}}{a} + n^2\frac{x}{a} = -\frac{3}{2}n^2 J_2 \left(\frac{R}{a}\right)^2 (1 - 3s_i^2) - \frac{3}{4}n^2 J_2 \left(\frac{R}{a}\right)^2 s_i^2 c_{2nt} \tag{11.30}$$

whose solution is obtained from

$$\begin{aligned} \frac{x}{a} = c_1 c_{nt} + c_2 s_{nt} + \frac{1}{n}\int_0^t \Bigg[& -\frac{3}{2}n^2 J_2 \left(\frac{R}{a}\right)^2 \left(1 - \frac{3}{2}s_i^2\right) \\ & - \frac{3}{4}n^2 J_2 \left(\frac{R}{a}\right)^2 s_i^2 c_{2nu} \Bigg] \sin n(t-u) du \end{aligned} \tag{11.31}$$

After carrying out the integrations and keeping only the J_2 terms while ignoring the initial conditions, we obtain the solution for x as

$$x_{J_2} = -3aJ_2 \left(\frac{R}{a}\right)^2 \left[\left(\frac{1}{2} - \frac{3}{4}s_i^2\right) - \frac{s_i^2 c_{2nt}}{12} - \left(\frac{1}{2} - \frac{5}{6}s_i^2\right) c_{nt} \right] \tag{11.32}$$

This expression is then used in Eq. (11.29) for $\dot{y}$, which is then integrated to yield the expression for y as

$$y_{J_2} = 3naJ_2 \left(\frac{R}{a}\right)^2 \left[\left(1 - \frac{3}{2}s_i^2\right) t - \frac{1}{n}\left(1 - \frac{5}{3}s_i^2\right) s_{nt} \right] + \frac{1}{8}aJ_2 \left(\frac{R}{a}\right)^2 s_i^2 s_{2nt} \tag{11.33}$$

The velocities and accelerations, namely, $\dot{x}$, $\dot{y}$, $\ddot{x}$, and $\ddot{y}$ are obtained as

$$\dot{x}_{J_2} = -3aJ_2 \left(\frac{R}{a}\right)^2 \left[\frac{s_i^2}{6} n s_{2nt} + \left(\frac{1}{2} - \frac{5}{6}s_i^2\right) n s_{nt} \right] \tag{11.34}$$

$$\ddot{x}_{J_2} = -3aJ_2 \left(\frac{R}{a}\right)^2 \left[\frac{s_i^2}{3} n^2 c_{2nt} + \left(\frac{1}{2} - \frac{5}{6}s_i^2\right) n^2 c_{nt} \right] \tag{11.35}$$

$$\dot{y}_{J_2} = 3naJ_2\left(\frac{R}{a}\right)^2\left[\left(1-\frac{3}{2}s_i^2\right)-\left(1-\frac{5}{3}s_i^2\right)c_{nt}\right]+\frac{1}{4}aJ_2\left(\frac{R}{a}\right)^2 s_i^2 n c_{2nt} \quad (11.36)$$

$$\ddot{y}_{J_2} = 3naJ_2\left(\frac{R}{a}\right)^2\left[\left(1-\frac{5}{3}s_i^2\right)ns_{nt}\right]-\frac{1}{2}aJ_2\left(\frac{R}{a}\right)^2 s_i^2 n^2 s_{2nt} \quad (11.37)$$

These expressions are then used in the original second-order differential equations, namely, Eqs. (11.27) and (11.28), to verify that these differential equations are satisfied, thereby ensuring that the derivations are error-free.

For the J_3 zonal harmonic, the second-order differential equations for the rotating x and y coordinates are given by

$$\ddot{x} - 2n\dot{y} - 3n^2x = f_r = \mu J_3\left(\frac{R^3}{r^5}\right)s_i\left[3s_\theta\left(\frac{5}{2}s_i^2-2\right)-\frac{5}{2}s_i^2 s_{3\theta}\right] \quad (11.38)$$

$$\ddot{y} + 2n\dot{x} = f_\theta = \mu J_3\frac{R^3}{r^5}s_i\left[\frac{3}{2}c_\theta\left(1-\frac{5}{4}s_i^2\right)+\frac{15}{8}s_i^2 c_{3\theta}\right] \quad (11.39)$$

Using $\theta = nt$ and integrating Eq. (11.39) yields the $\dot{y}$ expression as

$$\dot{y} = -2nx + n^2a\left(\frac{R}{a}\right)^3 J_3 s_i\left[\frac{3}{2n}\left(1-\frac{5}{4}s_i^2\right)s_{nt}+\frac{5}{8n}s_i^2 s_{3nt}\right] \quad (11.40)$$

which is then used in Eq. (11.38) to yield the forced-harmonic-oscillator form

$$\ddot{x} + n^2x = n^2aJ_3\left(\frac{R}{a}\right)^3 s_i\left[3\left(\frac{5}{4}s_i^2-1\right)s_{nt}-\frac{5}{4}s_i^2 s_{3nt}\right] \quad (11.41)$$

Ignoring once again the homogenous solution and the constant of integration and keeping only the J_3 contribution, we obtain

$$x_{J_3} = \frac{1}{n}\int_0^t n^2aJ_3\left(\frac{R}{a}\right)^3 s_i\left[3\left(\frac{5}{4}s_i^2-1\right)s_{nu}-\frac{5}{4}s_i^2 s_{3nu}\right]\sin n(t-u)du \quad (11.42)$$

After the integrations have been carried out, x_{J_3} can be cast in the form

$$x_{J_3} = n^2aJ_3\left(\frac{R}{a}\right)^3 s_i\left[\left(\frac{5}{2}s_i^2-3\right)\frac{(3s_{nt}-s_{3nt})}{8n^2} + 3\left(\frac{5}{4}s_i^2-1\right)\left(\frac{-tc_{nt}}{2n}+\frac{s_{3nt}}{8n^2}+\frac{s_{nt}}{8n^2}\right)\right] \quad (11.43)$$

An expression for $\dot{x}_{J_3}$ is now readily obtained by differentiating the x_{J_3} solution in Eq. (11.43) to give

$$\begin{aligned}\dot{x}_{J_3} &= n^2 aJ_3\left(\frac{R}{a}\right)^3 s_i\left[\left(\frac{5}{2}s_i^2-3\right)\frac{3c_{nt}}{8n}+3\left(\frac{5}{4}s_i^2-1\right)\left(\frac{-c_{nt}}{2n}+\frac{ts_{nt}}{2}\right)\right.\\ &\left.+\frac{5}{4}s_i^2 3\frac{c_{3nt}}{8n}+3\left(\frac{5}{4}s_i^2-1\right)\frac{c_{nt}}{8n}\right]\end{aligned} \tag{11.44}$$

Inserting the expression for x_{J_3} in Eq. (11.43) into Eq. (11.40) for $\dot{y}$ results in

$$\begin{aligned}\dot{y}_{J_3} &= a\left(\frac{R}{a}\right)^3 J_3 s_i\left[\frac{15n}{4}(1-s_i^2)s_{nt}-\frac{3n}{2}\left(\frac{5}{4}s_i^2-1\right)c_{nt}s_{2nt}\right.\\ &\left.+3n^2\left(\frac{5}{4}s_i^2-1\right)tc_{nt}+\frac{n}{4}(5s_i^2-3)s_{3nt}\right]\end{aligned} \tag{11.45}$$

A final integration yields the expression for y_{J_3} as

$$y_{J_3} = a\left(\frac{R}{a}\right)^3 J_3 s_i\left[\frac{15}{2}\left(\frac{9}{8}s_i^2-1\right)c_{nt}+\frac{3}{n}\left(\frac{5}{4}s_i^2-1\right)ts_{nt}-\frac{5}{48}s_i^2 c_{3nt}\right] \tag{11.46}$$

The $\ddot{x}_{J_3}$ and $\ddot{y}_{J_3}$ accelerations are obtained by direct differentiation of the expressions for $\dot{x}_{J_3}$ and $\dot{y}_{J_3}$, such that

$$\begin{aligned}\ddot{x}_{J_3} &= n^2 aJ_3\left(\frac{R}{a}\right)^3 s_i\left[-\left(\frac{5}{2}s_i^2-3\right)\frac{3}{8}s_{nt}+3\left(\frac{5}{4}s_i^2-1\right)\left(s_{nt}+\frac{nt}{2}c_{nt}\right)\right.\\ &\left.-\frac{5}{4}s_i^2\frac{9}{8}s_{3nt}-3\left(\frac{5}{4}s_i^2-1\right)\frac{s_{nt}}{8}\right]\end{aligned} \tag{11.47}$$

$$\begin{aligned}\ddot{y}_{J_3} &= aJ_3\left(\frac{R}{a}\right)^3 s_i\left[\frac{15}{4}n^2(1-s_i^2)c_{nt}-\frac{3n}{2}\left(\frac{5}{4}s_i^2-1\right)(-ns_{nt}s_{2nt}+2nc_{nt}s_{2nt})\right.\\ &\left.+3n^2\left(\frac{5}{4}s_i^2-1\right)(c_{nt}-nts_{nt})+\frac{n}{4}(5s_i^2-3)3nc_{3nt}\right]\end{aligned} \tag{11.48}$$

A check is made by verifying that the original second-order differential equations (11.38) and (11.39) are satisfied by the solutions just derived to ensure that these solutions are indeed error-free.

Considering drag and J_3 first, Eq. (A11.35) can now be used as before to obtain the following differential equations:

$$\ddot{x}-2n\dot{y}-3n^2x = f_r = \mu J_3\frac{R^3}{r^5}s_i\left[3s_\theta\left(\frac{5}{2}s_i^2-2\right)-\frac{5}{2}s_i^2 s_{3\theta}\right] \tag{11.49}$$

$$\ddot{y} + 2n\dot{x} = f_\theta = -B\rho_p \frac{\mu}{a} e^{-k}\left[I_0(k) + 2\sum_{j=1}^{\infty} I_j(k)\cos jnt \right] + \mu J_3 \frac{R^3}{r^5} s_i \left[\frac{3}{2} c_\theta \left(1 - \frac{5}{4} s_i^2\right) + \frac{15}{8} s_i^2 c_{3\theta} \right] \tag{11.50}$$

Once again using the expressions

$$r \simeq a \tag{11.51}$$

$$n^2 a^3 = \mu \tag{11.52}$$

$$\theta = nt \tag{11.53}$$

we can add the solutions for drag and J_3 linearly to provide their combined effects.

For $e = 0$, we can write, as before, $\dot{r} = \dot{a} - a\dot{e}c_{nt}$. Moreover, from $\dot{x} = \dot{r}$, retaining only the leading terms in $\dot{x} = \dot{x}_{\text{drg}} + \dot{x}_{J_3}$, we can extract the expressions for $\dot{a}$ and $\dot{e}$ as

$$\dot{a} = -\frac{2}{n} B\rho_p \frac{\mu}{a} e^{-k} I_0(k) \tag{11.54}$$

$$\dot{e} = -2naB\rho_p e^{-k} I_1(k) + \frac{3}{2} n J_3 \left(\frac{R}{a}\right)^3 s_i \left(\frac{5}{4} s_i^2 - 1\right) \tag{11.55}$$

The J_2 zonal harmonic does not affect eccentricity, so Eqs. (11.54) and (11.55) are unchanged upon the introduction of the J_2 perturbation.

Finally, the combined effects of drag, J_2, and J_3 yield the following general solution for x

$$\begin{aligned} x = {} & c_1 c_{nt} + c_2 s_{nt} - 3aJ_2 \left(\frac{R}{a}\right)^3 \left[\left(\frac{1}{2} - \frac{3}{4} s_i^2\right) - \frac{s_i^2 c_{2nt}}{12} - c_{nt}\left(\frac{1}{2} - \frac{5}{6} s_i^2\right) \right] \\ & - \frac{2}{n} B\rho_p \frac{\mu}{a} e^{-k} I_0(k) t + \left[\frac{2}{n} B\rho_p \frac{\mu}{a} e^{-k} I_1(k) - \frac{3n}{2} aJ_3 \left(\frac{R}{a}\right)^3 s_i \left(\frac{5}{4} s_i^2 - 1\right) \right] t c_{nt} \\ & - \left\{ \frac{2}{n} B\rho_p \frac{\mu}{a} e^{-k} \left[\frac{-I_0(k)}{n} + \frac{3}{4n} I_1(k) + \frac{2}{3n} I_2(k) \right] \right. \\ & \left. - \frac{3}{8} aJ_3 \left(\frac{R}{a}\right)^3 s_i \left(\frac{5}{2} s_i^2 - 3\right) \right\} s_{nt} + \frac{2}{3n^2} B\rho_p \frac{\mu}{a} e^{-k} I_2(k) s_{2nt} \\ & + \left[\frac{1}{2n^2} B\rho_p \frac{\mu}{a} e^{-k} I_1(k) - \frac{aJ_3}{8} \left(\frac{R}{a}\right)^3 s_i \left(\frac{5}{2} s_i^2 - 3\right) \right] s_{3nt} \\ & - \left[\frac{B}{n^2} \rho_p \frac{\mu}{a} e^{-k} I_1(k) - \frac{3}{4} aJ_3 \left(\frac{R}{a}\right)^3 s_i \left(\frac{5}{4} s_i^2 - 1\right) \right] c_{nt} s_{2nt} \end{aligned} \tag{11.56}$$

with

$$\dot{x} = -3aJ_2\left(\frac{R}{a}\right)^2\left[\frac{s_i^2}{12}2ns_{2nt} + ns_{nt}\left(\frac{1}{2} - \frac{5}{6}s_i^2\right)\right] - \frac{2}{n}B\rho_p\frac{\mu}{a}e^{-k}I_0(k)$$
$$+\left[\frac{2}{n}B\rho_p\frac{\mu}{a}e^{-k}I_1(k) - \frac{3n}{2}aJ_3\left(\frac{R}{a}\right)^3 s_i\left(\frac{5}{4}s_i^2 - 1\right)\right](c_{nt} - tns_{nt})$$
$$-\left\{\frac{2}{n}B\rho_p\frac{\mu}{a}e^{-k}\left[\frac{-I_0(k)}{n} + \frac{3}{4n}I_1(k) + \frac{2}{3n}I_2(k)\right] - \frac{3}{8}aJ_3\left(\frac{R}{a}\right)^3 s_i\left(\frac{5}{2}s_i^2 - 3\right)\right.$$
$$\left.-\frac{3}{8}aJ_3\left(\frac{R}{a}\right)^3 s_i\left(\frac{5}{4}s_i^2 - 1\right)\right\}nc_{nt} + \frac{3}{8}naJ_3\left(\frac{R}{a}\right)^3\frac{5}{4}s_i^3 c_{3nt} \tag{11.57}$$

The different Bessel functions appearing in these equations are evaluated in Eqs. (A11.20–A11.34) of Appendix 11.1. The solutions described can also be generalized further to include more zonal harmonic terms, namely, J_4 and beyond, by the method shown in Appendix 11.2. Each term can be evaluated independently because of the linear structure of the potential function in the J_n terms. A further generalization would require consideration of the tesseral and sectoral harmonics representation of the potential function. The analytic solutions are then dependent on the longitude λ, as well as the declination δ, which has so far appeared by itself.

The author has provided a more elaborate analysis of this problem in elsewhere [2] for the case of a near-circular orbit, that is, for orbits with small eccentricity. In the same book [2], the author then shows how to use the main perturbation due to J_2, as well the perturbations due to air drag and luni-solar gravity, to accurately predict the motion of a spacecraft at future times and, for the case of the J_2-perturbed trajectories, how to use the relative coordinates with respect to rotating reference frames to solve for terminal rendezvous using impulsive velocity changes.

11.3 BREAKWELL'S ANALYSIS OF THE J_2 PERTURBATION

Breakwell [1] has provided remarkable analytic forms and useful expressions for several orbit parameters affected by the J_2 perturbation, for both quick analysis and mathematical insight. In this section, we present some of the most useful elements of his analysis, starting from some general astrodynamics equations.

Denoting the general perturbation vector by $\boldsymbol{f}$, we can write the equations of motion in a straightforward manner following Newton's law as

$$\ddot{\boldsymbol{r}} = -\frac{\mu\boldsymbol{r}}{r^3} + \boldsymbol{f} \tag{11.58}$$

From the definition of the angular momentum

$$\boldsymbol{h} = \boldsymbol{r} \times \dot{\boldsymbol{r}} \tag{11.59}$$

and in view of Eq. (11.58), we have

$$\dot{\boldsymbol{h}} = \frac{\mathrm{d}}{\mathrm{d}t}(\boldsymbol{r} \times \dot{\boldsymbol{r}}) = \boldsymbol{r} \times \boldsymbol{f} \tag{11.60}$$

Furthermore

$$\boldsymbol{h} \times \boldsymbol{r} = (\boldsymbol{r} \times \dot{\boldsymbol{r}}) \times \boldsymbol{r} = (\boldsymbol{r} \cdot \boldsymbol{r})\dot{\boldsymbol{r}} - (\boldsymbol{r} \cdot \dot{\boldsymbol{r}})\boldsymbol{r} = r^2\dot{\boldsymbol{r}} - (\boldsymbol{r} \cdot \dot{\boldsymbol{r}})\boldsymbol{r} \tag{11.61}$$

From $r^2 = \boldsymbol{r} \cdot \boldsymbol{r}$, we have $2r\dot{r} = 2(\boldsymbol{r} \cdot \dot{\boldsymbol{r}})$ such that

$$\dot{r} = \frac{\dot{\boldsymbol{r}} \cdot \boldsymbol{r}}{r} = \dot{\boldsymbol{r}} \cdot \hat{\boldsymbol{r}} \tag{11.62}$$

The rate of the unit vector $\dot{\hat{\boldsymbol{r}}}$ is given by

$$\dot{\hat{\boldsymbol{r}}} = \frac{\mathrm{d}}{\mathrm{d}t}\left(\frac{\boldsymbol{r}}{r}\right) = \frac{r\dot{\boldsymbol{r}} - \boldsymbol{r}\dot{r}}{r_2} \tag{11.63}$$

Using Eq. (11.62), this equation can be rewritten as

$$\dot{\hat{\boldsymbol{r}}} = \frac{r^2\dot{\boldsymbol{r}} - (\boldsymbol{r} \cdot \dot{\boldsymbol{r}})\boldsymbol{r}}{r^3} \tag{11.64}$$

$$\dot{\hat{\boldsymbol{r}}} = \frac{\boldsymbol{h} \times \boldsymbol{r}}{r^3} = \frac{1}{\mu}\boldsymbol{h} \times (\boldsymbol{f} - \ddot{\boldsymbol{r}}) \tag{11.65}$$

where use is made of Eqs. (11.61) and (11.58). If $\boldsymbol{f} = 0$, there are no perturbations, so from Eq. (11.65), we obtain

$$\frac{\mathrm{d}}{\mathrm{d}t}\left(\frac{\boldsymbol{r}}{r}\right) = \frac{1}{\mu}\ddot{\boldsymbol{r}} \times \boldsymbol{h} \tag{11.66}$$

such that, with $\dot{\boldsymbol{h}} = 0$, we have

$$\hat{\boldsymbol{r}} = \frac{\boldsymbol{r}}{r} = \frac{1}{\mu}\dot{\boldsymbol{r}} \times \boldsymbol{h} - \boldsymbol{e} \tag{11.67}$$

where $\boldsymbol{e}$ is a constant vector. If $\boldsymbol{f} \neq 0$, then from

$$\boldsymbol{e} = \frac{1}{\mu}\dot{\boldsymbol{r}} \times \boldsymbol{h} - \hat{\boldsymbol{r}}$$

we have

$$\dot{\boldsymbol{e}} = \frac{1}{\mu}(\ddot{\boldsymbol{r}} \times \boldsymbol{h} + \dot{\boldsymbol{r}} \times \dot{\boldsymbol{h}}) - \dot{\hat{\boldsymbol{r}}}$$

and with the use of Eqs. (11.65) and (11.60), we can write

$$\dot{\boldsymbol{e}} = \frac{1}{\mu}[\boldsymbol{f} \times \boldsymbol{h} + \dot{\boldsymbol{r}} \times (\boldsymbol{r} \times \boldsymbol{f})] \tag{11.68}$$

The orbit equation is readily obtained from

$$\boldsymbol{e} \cdot \boldsymbol{r} + r = \frac{1}{\mu}(\boldsymbol{r} \cdot \dot{\boldsymbol{r}} \times \dot{\boldsymbol{h}}) = \frac{1}{\mu}(\boldsymbol{r} \cdot \dot{\boldsymbol{r}}) \cdot \dot{\boldsymbol{h}} = \frac{1}{\mu}\boldsymbol{h} \cdot \boldsymbol{h} = \frac{h^2}{\mu} = \boldsymbol{r}(1 + ec_{\theta^*}) \tag{11.69}$$

because the eccentricity vector is pointed toward perigee and $\boldsymbol{r}$ is inclined toward perigee by an angle θ^*, which is the true anomaly. Of course, from Eq. (11.64), it is easy to show that $\dot{\hat{\boldsymbol{r}}} \cdot \hat{\boldsymbol{r}} = 0$ such that $\dot{\hat{\boldsymbol{r}}}$ is orthogonal to $\hat{\boldsymbol{r}}$.

If we denote the angular velocity of the $(\hat{\boldsymbol{r}}, \hat{\boldsymbol{\theta}}, \hat{\boldsymbol{h}})$ frame by $\boldsymbol{\omega}_F$, then from $\boldsymbol{h} = h\hat{\boldsymbol{h}}$

$$\dot{\boldsymbol{h}} = \dot{h}\hat{\boldsymbol{h}} + h\boldsymbol{\omega}_F \times \hat{\boldsymbol{h}} = \hat{\boldsymbol{r}} \times \hat{\boldsymbol{f}} = r(f_\theta\hat{\boldsymbol{h}} - f_h\hat{\boldsymbol{\theta}}) \tag{11.70}$$

because $\dot{\hat{\boldsymbol{h}}} = \boldsymbol{\omega}_F \times \hat{\boldsymbol{h}}$. The angular velocity $\boldsymbol{\omega}_F$ can be written as

$$\boldsymbol{\omega}_F = \dot{\Omega}\hat{\boldsymbol{z}} + \dot{i}\hat{\boldsymbol{A}} + \dot{\theta}\hat{\boldsymbol{h}} \tag{11.71}$$

Upon substitution of $c_\Omega\hat{\boldsymbol{x}} + s_\Omega\hat{\boldsymbol{y}}$ for $\hat{\boldsymbol{A}}$, $\dot{i}\hat{\boldsymbol{A}}$ reduces to $\dot{i}(c_\theta\hat{\boldsymbol{r}} + s_\theta\hat{\boldsymbol{\theta}})$ in view of the transformation matrix in Eq. (A11.30), which is also used to write $\boldsymbol{\omega}_F$ directly in terms of the $\hat{\boldsymbol{r}}$, $\hat{\boldsymbol{\theta}}$, and $\hat{\boldsymbol{h}}$ such that

$$\boldsymbol{\omega}_F = (\dot{\Omega}s_is_\theta + \dot{i}c_\theta)\hat{\boldsymbol{r}} + (\dot{\Omega}s_ic_\theta - \dot{i}s_\theta)\hat{\boldsymbol{\theta}} + (\dot{\Omega}c_i + \dot{\theta})\hat{\boldsymbol{h}} = \omega_r\hat{\boldsymbol{r}} + \omega_\theta\hat{\boldsymbol{\theta}} + \omega_h\hat{\boldsymbol{h}} \tag{11.72}$$

From Eq. (11.70), we have

$$\dot{\boldsymbol{h}} = rf_\theta \tag{11.73}$$

and

$$\boldsymbol{\omega}_F \times \hat{\boldsymbol{h}} = -\frac{rf_h\hat{\boldsymbol{\theta}}}{h} = \omega_\theta\hat{\boldsymbol{r}} - \omega_r\hat{\boldsymbol{\theta}} \tag{11.74}$$

Therefore, $\omega_r = rf_h/h$ and $\omega_\theta = 0$ such that

$$\omega_r = \boldsymbol{\omega}_F \cdot \hat{\boldsymbol{r}} = \dot{\Omega}s_is_\theta + \dot{i}c_\theta = \frac{rf_h}{h} = \frac{f_h}{v_\theta} \tag{11.75}$$

$$\omega_\theta = \boldsymbol{\omega}_F \cdot \hat{\boldsymbol{\theta}} = \dot{\Omega}s_ic_\theta - \dot{i}s_\theta = 0 \tag{11.76}$$

$$\omega_h = \boldsymbol{\omega}_F \cdot \hat{\boldsymbol{h}} = \dot{\theta} + \dot{\Omega}c_i = \frac{h}{r^2} \tag{11.77}$$

The relation $\omega_h = h/r^2$ in Eq. (11.77) can be shown to be true from the Lagrange–Gauss equations, namely

$$\dot{\theta}^* = \frac{h}{r^2} + \frac{1}{eh}[pc_{\theta^*}f_r - (p+r)s_{\theta^*}f_\theta] \tag{11.78}$$

$$\dot{\omega} = \frac{1}{eh}[-pc_{\theta^*}f_r + (p+r)s_{\theta^*}f_\theta] - \frac{rs_\theta c_i}{hs_i}f_h \tag{11.79}$$

$$\dot{\Omega} = \frac{rs_\theta}{hs_i}f_h \tag{11.80}$$

such that

$$\dot{\theta} = \dot{\theta}^* + \dot{\omega} = \frac{h}{r^2} - \frac{rs_\theta c_i}{hs_i}f_h \tag{11.81}$$

leading to

$$\dot{\theta} + \dot{\Omega}c_i = \frac{h}{r^2} \tag{11.82}$$

The symbol v_θ in Eq. (11.75) is the velocity component along the $\hat{\boldsymbol{\theta}}$ direction, and p is the orbit parameter, given by $p = a(1-e^2)$. Therefore, $\boldsymbol{\omega}_F$ consists of an in-plane rotation about $\hat{\boldsymbol{h}}$ plus a rotation of the orbit plane about $\hat{\boldsymbol{r}}$. One can easily obtain $\dot{i}$, $\dot{\Omega}$, and $\dot{\theta}$ in a more direct manner from Eqs. (11.75) and (11.76) by multiplying the first equation by c_θ and the second equation by s_θ and adding, yielding

$$\dot{i} = \frac{rf_h}{h}c_\theta \tag{11.83}$$

Similarly, multiplying the first equation by s_θ and the second by c_θ and adding gives

$$\dot{\Omega} = \frac{rf_h}{hs_i}s_\theta \tag{11.84}$$

In turn, Eq. (11.77) provides

$$\dot{\theta} = \frac{h}{r^2} - \dot{\Omega}c_i \tag{11.85}$$

Using $\boldsymbol{e} = ec_{\theta^*}\hat{\boldsymbol{r}} - es_{\theta^*}\hat{\boldsymbol{\theta}}$, Breakwell provided a further expression [1] for $\dot{\boldsymbol{e}}$ in the $(\hat{\boldsymbol{r}}, \hat{\boldsymbol{\theta}}, \hat{\boldsymbol{h}})$ system as

$$\dot{\boldsymbol{e}} = \left[\dot{e}c_{\theta^*} + \left(\frac{h}{r^2} - \dot{\theta}^*\right)es_{\theta^*}\right]\hat{\boldsymbol{r}} + \left[-\dot{e}s_{\theta^*} + \left(\frac{h}{r^2} - \dot{\theta}^*\right)ec_{\theta^*}\right]\hat{\boldsymbol{\theta}} - \frac{rf_h}{h}es_{\theta^*}\hat{\boldsymbol{h}} \tag{11.86}$$

with the following Lagrange equation for the classical element e:

$$\dot{e} = \frac{r}{h}\left[s_{\theta^*}(1 + ec_{\theta^*})f_r + (e + 2c_{\theta^*} + ec_{\theta^*}^2)f_\theta\right] \tag{11.87}$$

Returning to the J_2 perturbation equations (11.27) and (11.28) and including the third differential equation from Eq. (A11.34), we have once again the full system of differential equations for the x, y, and z coordinates as

$$\frac{\ddot{x}}{a} - 2n\frac{\dot{y}}{a} - 3n^2\frac{x}{a} = -3n^2 J_2\left(\frac{R}{a}\right)^2\left(\frac{1}{2} - \frac{3}{4}s_i^2 + \frac{3}{4}s_i^2 c_{2nt}\right) \tag{11.88}$$

$$\frac{\ddot{y}}{a} + 2n\frac{\dot{x}}{a} = -\frac{3}{2}n^2 J_2\left(\frac{R}{a}\right)^2 s_i^2 s_{2nt} \tag{11.89}$$

$$\frac{\ddot{z}}{a} + n^2\frac{z}{a} = -3n^2 J_2\left(\frac{R}{a}\right)^2 s_i c_i s_{nt} \tag{11.90}$$

Breakwell used a particular solution of this system that corresponds to a particular choice of the fictitious rotating orbit such as for x and y

$$\frac{x}{a} = \frac{1}{2}J_2\left(\frac{R}{a}\right)^2\left(1 - \frac{3}{2}s_i^2\right) + \frac{1}{4}J_2\left(\frac{R}{a}\right)^2 s_i^2 c_{2nt} \tag{11.91}$$

$$\frac{y}{a} = \frac{1}{8}J_2\left(\frac{R}{a}\right)^2 s_i^2 s_{2nt} \tag{11.92}$$

It is easy to obtain the expressions for $\dot{x}, \ddot{x}, \dot{y}$, and $\ddot{y}$, and indeed, the second-order differential equations (11.88 – 11.90) are satisfied, as was done with the solution in the previous section.

For $\dot{x}, \ddot{x}, \dot{y}$, and $\ddot{y}$ of Breakwell's particular solution, we have the following forms:

$$\frac{\dot{x}}{a} = -\frac{1}{2}nJ_2\left(\frac{R}{a}\right)^2 s_i^2 s_{2nt}$$

$$\frac{\ddot{x}}{a} = -n^2 J_2\left(\frac{R}{a}\right)^2 s_i^2 c_{2nt}$$

$$\frac{\dot{y}}{a} = \frac{1}{4}nJ_2\left(\frac{R}{a}\right)^2 s_i^2 c_{2nt}$$

$$\frac{\ddot{y}}{a} = -\frac{1}{2}n^2 J_2\left(\frac{R}{a}\right)^2 s_i^2 s_{2nt}$$

For the next step, we start from $\dot{\Omega} = (-3\mu J_2 R^2/r^3 h)c_i s_\theta$ and $\dot{\theta} = (h/r^2) - \dot{\Omega}c_i$, where $f_h = (-3\mu J_2 R^2/r^4)s_i c_i s_\theta$ from Eqs. (A11.34) is used for $\dot{\Omega} = (rf_h/hs_i)s_\theta$

in Eq. (11.84) and $\mathrm{d}\Omega/\mathrm{d}\theta \simeq (r^2/h)\dot{\Omega}$ is cast as

$$\frac{\mathrm{d}\Omega}{\mathrm{d}\theta} = \frac{-3\mu J_2 R^2}{h^2} c_i \frac{s_\theta^2}{p}(1 + e c_{\theta^*})$$

Further replacing θ^* by $\theta - \omega$ and h^2 by μp; observing that the average of $s_\theta^2(1 + e c_{\theta-\omega})$ is equal to 1/2, and keeping only the J_2 term while neglecting the J_2^2 term, because of the approximation used in $\mathrm{d}\Omega/\mathrm{d}\theta \simeq (r^2/h)\dot{\Omega}$, which amounts to neglecting the $\dot{\Omega} c_i$ term in $\dot{\theta} = (h/r^2) - \dot{\Omega} c_i$ and using only $\dot{\theta} = h/r^2$ over the single-orbit integration to obtain the averaged effect due to J_2, which amounts to neglecting the J_2^2 term, we obtain the following well-known expression:

$$\left(\frac{\mathrm{d}\Omega}{\mathrm{d}\theta}\right)_{\mathrm{av}} = \frac{-3}{2} J_2 \left(\frac{R}{p}\right)^2 c_i \tag{11.93}$$

It can also be shown that $(\mathrm{d}i/\mathrm{d}\theta)_{\mathrm{av}} = 0$, $(\mathrm{d}p/\mathrm{d}\theta)_{\mathrm{av}} = 0$, $(\mathrm{d}e/\mathrm{d}\theta)_{\mathrm{av}} = 0$, and $(\mathrm{d}a/\mathrm{d}\theta)_{\mathrm{av}} = 0$, as well as $(\mathrm{d}h/\mathrm{d}\theta)_{\mathrm{av}} = 0$ from

$$\frac{1}{h}\frac{\mathrm{d}h}{\mathrm{d}\theta} \simeq \frac{r^2}{h^2} r f_\theta = -3 J_2 \left(\frac{R}{p}\right)^2 s_i^2 s_\theta c_\theta (1 + e c_{\theta-\omega})$$

such that $(\mathrm{d}h/\mathrm{d}\theta)_{\mathrm{av}} = 0$.

Returning to Eq. (11.85), which can be written as

$$\frac{h}{r^2} - \dot{\theta}^* = \dot{\Omega} c_i + \dot{\omega} \tag{11.94}$$

and using the form for $\dot{\theta}^*$ given by Eq. (11.78), we can write

$$\dot{\omega} + \dot{\Omega} c_i = \frac{r}{he}[-c_{\theta^*}(1 + e c_{\theta^*}) f_r + s_{\theta^*}(2 + e c_{\theta^*}) f_\theta] \tag{11.95}$$

Next, replacing f_r and f_θ by the expressions in Eqs. (A11.34) and taking the average over one revolution, yields, after some algebra

$$\left(\frac{\mathrm{d}\omega}{\mathrm{d}\theta}\right)_{\mathrm{av}} + c_i \left(\frac{\mathrm{d}\Omega}{\mathrm{d}\theta}\right)_{\mathrm{av}} = \frac{3}{2} J_2 \left(\frac{R}{p}\right)^2 \left(1 - \frac{3}{2} s_i^2\right) \tag{11.96}$$

which leads to the following form for $(\mathrm{d}\omega/\mathrm{d}\theta)_{\mathrm{av}}$ once $(\mathrm{d}\Omega/\mathrm{d}\theta)_{\mathrm{av}}$ is replaced by its form in Eq. (11.93):

$$\left(\frac{\mathrm{d}\omega}{\mathrm{d}\theta}\right)_{\mathrm{av}} = \frac{3}{2} J_2 \left(\frac{R}{p}\right)^2 \left(2 - \frac{5}{2} s_i^2\right) \tag{11.97}$$

In summary, the orbit plane precesses according to Eq. (11.93), and perigee moves according to Eq. (11.97), traveling backward between the angles $i = 63.4$ deg and $i = 116.6$ deg, and forward otherwise.

For $i = 0$

$$\left[\frac{\mathrm{d}}{\mathrm{d}\theta}(\Omega + \omega)\right]_{\mathrm{av}} = \frac{3}{2}J_2\left(\frac{R}{p}\right)^2$$

from Eqs. (11.93) and (11.97), or in other words, $\Omega + \omega$ increases by $(3/2)J_2(R/p)^2 \cdot 2\pi = 3\pi J_2(R/p)^2$ radians per revolution. It is easy to show that from

$$\boldsymbol{r} = rc_\theta\hat{\boldsymbol{x}} + rs_\theta\hat{\boldsymbol{y}} \tag{11.98}$$

$\dot{\boldsymbol{r}}$ can be cast as

$$\dot{\boldsymbol{r}} = -\frac{\mu}{h}(s_\theta + es_\omega)\hat{\boldsymbol{x}} + \frac{\mu}{h}(c_\theta + ec_\omega)\hat{\boldsymbol{y}} \tag{11.99}$$

$$\dot{\boldsymbol{r}} = \frac{\mu}{h}\left[es_{\theta^*}\hat{\boldsymbol{r}} + (1 + ec_{\theta^*})\hat{\boldsymbol{\theta}}\right] \tag{11.100}$$

along the inertial and rotating directions, respectively. This is true because, from the orbit equation, $r = (h^2/\mu)/(1 + ec_{\theta^*})$, we have $\dot{r} = (\mu/h)es_{\theta^*}$ with $\dot{\theta} = h/r^2$.

Finally, Breakwell's analytic expressions for r and θ are obtained for orbits with small eccentricities (i.e., $e < 0.1$) for the combined effect of the J_2 oblateness and the air drag.

For $e = 0$, $r = a + x$ leads to

$$r = a\left[1 + \frac{1}{2}J_2\left(\frac{R}{a}\right)^2\left(1 - \frac{3}{2}s_i^2\right) + \frac{1}{4}J_2\left(\frac{R}{a}\right)^2 s_i^2 c_{2\theta}\right] \tag{11.101}$$

where a is the semimajor axis of the fictitious orbit. Thus, from

$$(\dot{\theta} + \dot{\Omega}c_i)_{\mathrm{av}} = \left(\frac{h}{r^2}\right)_{\mathrm{av}} = n = \sqrt{\frac{\mu}{a^3}} \tag{11.102}$$

we have, without accounting for the y motion

$$\frac{\mathrm{d}\theta}{\mathrm{d}t} + \frac{\mathrm{d}\Omega}{\mathrm{d}\theta}\frac{\mathrm{d}\theta}{\mathrm{d}t}c_i = \left[1 - \frac{3}{2}J_2\left(\frac{R}{a}\right)^2 c_i^2\right]\dot{\theta} = n \tag{11.103}$$

such that

$$\dot{\theta} = n\left[1 - \frac{3}{2}J_2\left(\frac{R}{a}\right)^2 c_i^2\right]^{-1}$$

$$\theta \simeq n(t - t_A)\left[1 + \frac{3}{2}J_2\left(\frac{R}{a}\right)^2 c_i^2\right]$$

where t_A is the time at the ascending mode. To account for the y motion, we add the angle y/a due to y to obtain

$$\theta = n(t - t_A)\left[1 + \frac{3}{2}J_2\left(\frac{R}{a}\right)^2 c_i^2\right] + \frac{1}{8}J_2\left(\frac{R}{a}\right)^2 s_i^2 s_{2\theta} \tag{11.104}$$

We can then use the rescaling

$$a_1 = a\left[1 + \frac{1}{2}J_2\left(\frac{R}{a}\right)^2\left(1 - \frac{3}{2}s_i^2\right)\right] \tag{11.105}$$

such that $n_1 = \sqrt{\mu/a_1^3}$ is given by

$$n_1 = \mu^{1/2}a^{-3/2}\left[1 + \frac{1}{2}J_2\left(\frac{R}{a}\right)^2\left(1 - \frac{3}{2}s_i^2\right)\right]^{-3/2}$$

leading to

$$n = n_1\left[1 + \frac{3}{4}J_2\left(\frac{R}{a}\right)^2\left(1 - \frac{3}{2}s_i^2\right)\right] \tag{11.106}$$

where the J_2^2 term has been neglected.

The expression for θ can now be written as

$$\theta = n(t - t_A)\left\{1 + \frac{3}{2}J_2\left(\frac{R}{a_1}\right)^2\left[1 + \frac{1}{2}J_2\left(\frac{R}{a}\right)^2\left(1 - \frac{3}{2}s_i^2\right)\right]c_i^2\right\} + \frac{y}{a}$$

Neglecting J_2^2 and substituting for n in terms of n_1 according to Eq. (11.106) leads to

$$\theta = n_1(t - t_A)\left[1 + \frac{3}{2}J_2\left(\frac{R}{a_1}\right)^2 c_i^2\right]\left[1 + \frac{3}{4}J_2\left(\frac{R}{a}\right)^2\left(1 - \frac{3}{2}s_i^2\right)\right] + \frac{y}{a}$$

where, once again, the J_2^2 terms have been neglected, including the one from replacing a by a_1 in the last bracket of the preceding equation and in y/a, such that

$$\theta = n_1(t - t_A)\left[1 + J_2\left(\frac{R}{a_1}\right)^2\left(\frac{9}{4} - \frac{21}{8}\right)s_i^2\right] + \frac{1}{8}J_2\left(\frac{R}{a_1}\right)^2 s_i^2 s_{2\theta} \tag{11.107}$$

In a similar manner, r can be written in terms of a_1 as

$$r = a_1\left[1 + \frac{1}{4}J_2\left(\frac{R}{a_1}\right)^2 s_i^2 c_{2\theta}\right] \tag{11.108}$$

Also, from Eq. (11.97)

$$(\dot{\omega})_{\text{av}} = n_1 \frac{3}{2} J_2 \left(\frac{R}{a_1}\right)^2 \left(2 - \frac{5}{2} s_i^2\right) \tag{11.109}$$

such that $\dot{\theta}^* = \dot{\theta} - \dot{\omega}$ leads to

$$(\dot{\theta}^*)_{\text{av}} = n_1 \left[1 + J_2 \left(\frac{R}{a_1}\right)^2 \left(-\frac{3}{4} + \frac{9}{8} s_i^2\right)\right] = n^* \tag{11.110}$$

When eccentricity and drag are included, with $e < 0.1$, Breakwell neglects terms involving eJ_2 and J_2^2, and after denoting a_1 as a_0, e as e_0, and $a_0(1 - e_0^2)$ as p_0, he obtained the useful expressions for r, ω, Ω, and θ^* such as

$$r = \frac{p_0(1 - k\theta)}{1 + \left[e_0 - k\theta \dfrac{I_1(\beta)}{I_0(\beta)}\right] c_{\theta^*} - \dfrac{1}{4} J_2 \left(\dfrac{R}{p_0}\right)^2 s_i^2 c_{2\theta}} \tag{11.111}$$

where $\beta = a_0 e_0 / H$ and $k = -(1/a)(\mathrm{d}a/\mathrm{d}\theta)_{\text{av}}$, with H being the scale height. In the denominator, the term $k\theta\{[I_1(\beta)]/[I_0(\beta)]\}$ represents the drop in eccentricity due to drag, and both of the k terms in Eq. (11.111) indicate that the ellipse is gradually shrinking. The last term in the denominator is due to the oscillation in r. The ω and Ω expressions are given by

$$\omega = \omega_0 + \frac{3}{2} J_2 \left(\frac{R}{p_0}\right)^2 \left(2 - \frac{5}{2} s_i^2\right) \theta \tag{11.112}$$

$$\Omega = \Omega_0 - \frac{3}{2} J_2 \left(\frac{R}{p_0}\right)^2 c_i \theta \tag{11.113}$$

and from the time equation $E - es_E = n(t - t_p) = M$, we can write

$$E - E_{A_0} = n_0^*(t - t_{A_0}) + es_E - e_0 s_{E_{A_0}} + \frac{1}{8} J_2 \left(\frac{R}{p_0}\right)^2 s_i^2 s_{2\theta} + \frac{3}{4} k\theta^2 \tag{11.114}$$

The last term is due to the drag acceleration term. From

$$\tan \frac{\theta^*}{2} = \sqrt{\frac{1 + e}{1 - e}} \tan \frac{E}{2}$$

we have

$$\theta^* = E - 2 \tan^{-1} \frac{es_E}{1 + \sqrt{1 - e^2} - ec_E} \tag{11.115}$$

REFERENCES

[1] Breakwell, J. V., "Lecture Notes: Space Mechanics," Course AA279A, Department of Aeronautics and Astronautics, Stanford University, Palo Alto, CA, 1974.

[2] Kéchichian, J. A., *Orbital Relative Motion and Terminal Rendezvous*, Springer Nature, Cham, Switzerland, 2021.

APPENDIX 11.1: FOURIER–BESSEL SERIES EXPANSION OF THE DRAG EXPONENTIAL FOR NEAR-CIRCULAR ORBITS

The assumption that the atmosphere can be modeled by an exponential leads to the evaluation of the following term:

$$\exp(k\cos nt) = I_0(k) + 2\sum_{j=1}^{\infty} I_j(k)\cos jnt \tag{A11.1}$$

where $I_j(k)$ is the modified Bessel function of the first kind and of order j in the variable k. From the definition of the exponential, we obtain

$$\exp(k\cos nt) = 1 + k\cos nt + \frac{k^2}{2!}\cos^2 nt + \frac{k^3}{3!}\cos^3 nt + \cdots \tag{A11.2}$$

Use of the trigonometric multiple-angle relations

$$\cos^2 nt = \frac{\cos 2nt + 1}{2} \tag{A11.3}$$

$$\cos^3 nt = \frac{\cos 3nt + 3\cos nt}{4} \tag{A11.4}$$

and so forth allows Eq. (A11.2) to be reduced to

$$\begin{aligned}\exp(k\cos nt) &= \left(1 + \frac{k^2}{4} + \cdots\right) + \left(k + \frac{k^3}{8} + \cdots\right)\cos nt \\ &\quad + \left(\frac{k^2}{4} + \cdots\right)\cos^2 2nt + \cdots\end{aligned} \tag{A11.5}$$

which can be written as

$$\begin{aligned}\exp(k\cos nt) &= \left(1 + \frac{k^2}{4} + \cdots\right) + 2\left\{\frac{k}{2}\left[1 - \frac{(ik/2)^2}{2} - \frac{(ik/2)^4}{12} + \cdots\right]\right\}\cos nt \\ &\quad - 2\left[\frac{(ik/2)^2}{2} - \frac{(ik/2)^4}{6} + \cdots\right]\cos 2nt + \cdots\end{aligned} \tag{A11.6}$$

However, the Bessel functions of the first kind and of order 0 and 1 have the following expanded forms

$$J_0(ik) = 1 - (ik/2)^2 + \frac{(ik/2)^4}{4} - \frac{(ik/2)^6}{36} + \cdots \tag{A11.7}$$

$$J_1(ik) = \frac{ik}{2}\left[1 - \frac{(ik/2)^2}{2} + \frac{(ik/2)^4}{12} - \frac{(ik/2)^6}{144} + \cdots\right] \tag{A11.8}$$

and in general, the following recursive relation generates the higher-order series

$$J_{v+1}(x) + J_{v-1}(x) = \frac{2v}{x} J_v(x) \tag{A11.9}$$

which yields, for example

$$J_2(ik) = \frac{2}{ik} J_1(ik) - J_0(ik) = \frac{(ik/2)^2}{2} - \frac{(ik/2)^4}{6} + \cdots \tag{A11.10}$$

Returning to Eq. (A11.6) and using the relations in Eqs. (A11.7–A11.10), we obtain

$$\exp(k\cos nt) = J_0(ik) + 2\left[\frac{J_1(ik)}{i}\cos nt - J_2(ik)\cos 2nt + \cdots\right] \tag{A11.11}$$

Using now the modified Bessel functions of the first kind (always a real function) and the relation that links them to the original Bessel functions, namely

$$I_v(x) = \frac{1}{i^v} J_v(ix) \tag{A11.12}$$

one has

$$I_1(k) = -iJ_1(ik) \tag{A11.13}$$

$$I_2(k) = -J_2(ik) \tag{A11.14}$$

and so on.

Finally, Eq. (A11.11) becomes

$$\begin{aligned}\exp(k\cos nt) &= I_0(k) + 2[I_1(k)c_{nt} + I_2(k)c_{2nt} + I_3(k)c_{3nt}\cdots] \\ &= I_0(k) + 2\sum_{j=1}^{\infty} I_j(k)\cos jnt\end{aligned} \tag{A11.15}$$

which is the desired result.

For large k, the modified Bessel functions accept the following asymptotic form

$$I_\nu(k) = \frac{e^k}{\sqrt{2\pi k}}\left[\sum_{\mu=0}^{n}\frac{(-1)^\mu(\nu,\mu)}{(2k)^\mu} + O(k^{-n-1})\right] \tag{A11.16}$$

where the following notation due to Lebedev is used

$$(\nu,\mu) = \frac{(-1)^\mu}{\mu!}\left(\frac{1}{2}-\nu\right)^\mu\left(\frac{1}{2}+\nu\right)^\mu = \frac{(4\nu^2-1)(4\nu^2-3^2)\ldots\left[4\nu^2-(2\mu-1)^2\right]}{2^{2\mu}\,\mu!} \tag{A11.17}$$

with the definition

$$(\nu,0) = 1 \tag{A11.18}$$

For example

$$\left.\begin{array}{lll}
(0{,}0) = 1 & (1{,}0) = 1 & (2{,}0) = 1 \\
(0{,}1) = -\dfrac{1}{4} & (1{,}1) = \dfrac{3}{4} & (2{,}1) = \dfrac{15}{4} \\
(0{,}2) = \dfrac{9}{32} & (1{,}2) = \dfrac{-15}{32} & (2{,}2) = \dfrac{105}{32} \\
(0{,}3) = -\dfrac{225}{384} & (1{,}3) = -\dfrac{285}{384} & (2{,}3) = -\dfrac{945}{384} \\
(3{,}0) = 1 & (4{,}0) = 1 & \\
(3{,}1) = \dfrac{35}{4} & (4{,}1) = \dfrac{63}{4} & \\
(3{,}2) = \dfrac{945}{32} & (4{,}2) = \dfrac{3465}{32} & \\
(3{,}3) = \dfrac{10395}{384} & (4{,}3) = \dfrac{135135}{384} &
\end{array}\right\} \tag{A11.19}$$

Therefore, Eq. (A11.16) gives

$$I_0(k) = \frac{e^k}{\sqrt{2\pi k}}\left(1+\frac{1}{8k}+\frac{9}{128k^2}+\frac{225}{3072k^3}+\cdots\right) \tag{A11.20}$$

$$I_1(k) = \frac{e^k}{\sqrt{2\pi k}}\left(1-\frac{3}{8k}-\frac{15}{128k^2}+\frac{285}{3072k^3}+\cdots\right) \tag{A11.21}$$

$$I_2(k) = \frac{e^k}{\sqrt{2\pi k}}\left(1-\frac{15}{8k}+\frac{105}{128k^2}+\frac{945}{3072k^3}+\cdots\right) \tag{A11.22}$$

$$I_3(k) = \frac{e^k}{\sqrt{2\pi k}}\left(1-\frac{35}{8k}+\frac{945}{128k^2}-\frac{10395}{3072k^3}+\cdots\right) \tag{A11.23}$$

$$I_4(k) = \frac{e^k}{\sqrt{2\pi k}}\left(1 - \frac{63}{8k} + \frac{3465}{128k^2} - \frac{135135}{3072k^3} + \cdots\right) \tag{A11.24}$$

APPENDIX 11.2: PERTURBATION COMPONENTS OF THE ZONAL HARMONICS IN THE EULER–HILL FRAME

Assuming a symmetrical planet about its polar axis, the potential at a point a distance of r from the center of mass and with declination δ (Fig. 11.1) is

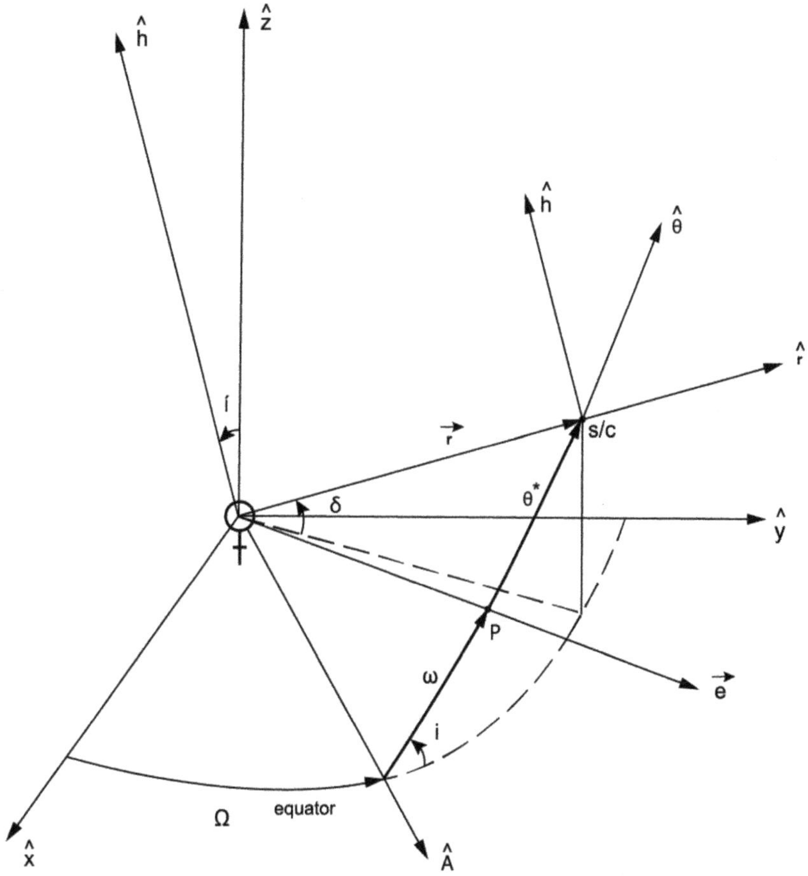

Fig. 11.1 Orbit geometry.

given by

$$U = \frac{\mu}{r}\left[1 - \sum_{n=2}^{\infty} J_n \left(\frac{R}{r}\right)^n P_n(s_\delta)\right] \tag{A11.25}$$

where J_n is the nth zonal harmonic, $P_n(s_\delta)$ is the Legendre polynomial of order n in $\sin\,\delta$, and R is the equatorial radius of the planet. The disturbing potential F is then

$$F = U - U_0 = U - \frac{\mu}{r} = -\frac{\mu}{r}\sum_{n=2}^{\infty} J_n \left(\frac{R}{r}\right)^n P_n(s_\delta) \tag{A11.26}$$

Restricting ourselves to the J_2 and J_3 zonals, we can write

$$F = -\frac{\mu}{r}\left[J_2\left(\frac{R}{r}\right)^2\left(\frac{3}{2}s_\delta^2 - \frac{1}{2}\right) + J_3\left(\frac{R}{r}\right)^3\left(\frac{5}{2}s_\delta^2 - \frac{3}{2}\right)s_\delta\right] \tag{A11.27}$$

We can also replace s_δ using

$$s_\delta = \frac{z}{r} \tag{A11.28}$$

where z is along the north pole of the planet, with $\hat{\boldsymbol{x}}$ being the unit vector pointing toward the vernal equinox and $\hat{\boldsymbol{y}}$ completing the frame.

The perturbation vector $\boldsymbol{f} = (f_r, f_\theta, f_h)$, with components along the radial direction, the direction 90 deg ahead in the orbit plane, and the direction parallel to the angular momentum vector $(\hat{\boldsymbol{r}}, \hat{\boldsymbol{\theta}}, \hat{\boldsymbol{h}})$ describing the Euler–Hill frame (Fig. 11.1), is now given by

$$\begin{aligned}\boldsymbol{f} &= \nabla\left[U(\boldsymbol{r}) - \frac{\mu}{r}\right] = \nabla F = \nabla\left[-\frac{\mu}{r}J_2\left(\frac{R}{r}\right)^2\left(\frac{3z^2}{2r^2} - \frac{1}{2}\right) - \frac{\mu}{r}J_3\left(\frac{R}{r}\right)^3\left(\frac{5z^2}{2r^2} - \frac{3}{2}\right)\frac{z}{r}\right] \\ &= -\mu J_2 R^2\left[\frac{3}{r^5}z\hat{\boldsymbol{z}} + \left(\frac{3}{2r^4} - \frac{15z^2}{2r^6}\right)\hat{\boldsymbol{r}}\right] - \mu J_3 R^3\left[\left(\frac{15z^2}{2r^7} - \frac{3}{2r^5}\right)\hat{\boldsymbol{z}} + \left(\frac{15}{2r^6}z - \frac{35z^3}{2r^8}\right)\hat{\boldsymbol{r}}\right]\end{aligned} \tag{A11.29}$$

The $(\hat{\boldsymbol{x}}, \hat{\boldsymbol{y}}, \hat{\boldsymbol{z}})$ frame is related to the $(\hat{\boldsymbol{r}}, \hat{\boldsymbol{\theta}}, \hat{\boldsymbol{h}})$ frame by the three rotations Ω, i, and θ (Euler angles), where

$$\theta = \omega + \theta^* \tag{A11.30}$$

with ω being the argument of periapse and θ^* being the true anomaly of the spacecraft

$$\begin{pmatrix} \hat{\boldsymbol{x}} \\ \hat{\boldsymbol{y}} \\ \hat{\boldsymbol{z}} \end{pmatrix} = \begin{pmatrix} c_\Omega c_\theta - s_\Omega c_i s_\theta & -c_\Omega s_\theta - s_\Omega c_i c_\theta & s_\Omega s_i \\ s_\Omega c_\theta + c_\Omega c_i s_\theta & -s_\Omega s_\theta + c_\Omega c_i c_\theta & -c_\Omega s_i \\ s_i s_\theta & s_i c_\theta & c_i \end{pmatrix} \begin{pmatrix} \hat{\boldsymbol{r}} \\ \hat{\boldsymbol{\theta}} \\ \hat{\boldsymbol{h}} \end{pmatrix} \tag{A11.31}$$

Therefore

$$z = \boldsymbol{r} \cdot \hat{\boldsymbol{z}} = r s_i s_\theta \tag{A11.32}$$

so that the perturbation vector $\boldsymbol{f}$ in Eq. (A11.29) becomes

$$\begin{aligned} \boldsymbol{f} = & \frac{-3\mu J_2 R^2}{r^4}\left[\left(\frac{1}{2} - \frac{3}{2}s_i^2 s_\theta^2\right)\hat{\boldsymbol{r}} + s_i^2 s_\theta c_\theta \hat{\boldsymbol{\theta}} + s_i c_i s_\theta \hat{\boldsymbol{h}}\right] \\ & - \frac{\mu J_3 R^3}{r^5}\left[(6 s_i s_\theta - 10 s_i^3 s_\theta^3)\hat{\boldsymbol{r}} + \frac{1}{2}(15 s_i^3 s_\theta^2 c_\theta - 3 s_i c_\theta)\hat{\boldsymbol{\theta}} + \frac{1}{2}(15 s_i^2 c_i s_\theta^2 - 3c_i)\hat{\boldsymbol{h}}\right] \end{aligned} \tag{A11.33}$$

The oblateness perturbation components are now given by

$$\left.\begin{aligned} f_r &= -3\mu J_2 \frac{R^2}{r^4}\left[\left(\frac{1}{2} - \frac{3}{2}s_i^2 s_\theta^2\right)\right] \\ f_\theta &= -3\mu J_2 \frac{R^2}{r^4} s_i^2 s_\theta c_\theta \\ f_h &= -3\mu J_2 \frac{R^2}{r^4} s_i c_i s_\theta \end{aligned}\right\} \tag{A11.34}$$

and for the in-plane components, the J_3 zonal gives

$$\begin{aligned} f_r &= \mu J_3 \frac{R^3}{r^5} s_i \left[-6 s_\theta + \frac{5}{2} s_i^2 (3 s_\theta - s_{3\theta})\right] \\ f_\theta &= \mu J_3 \frac{R^3}{r^5} s_i \left[\frac{3}{2} c_\theta - \frac{15}{8} s_i^2 (c_\theta - c_{3\theta})\right] \end{aligned} \tag{A11.35}$$

Analysis of the Effects of the Solar Gravity Perturbation on a Spacecraft in Near-Circular Orbit about a Planet: Application to the VOIR Venus Orbiter Mission

NOMENCLATURE

F = disturbing function due to third body gravity
r = radius vector to spacecraft
$\mu_{\odot}$ = gravity constant of the Sun
$\mu_{♀}$ = gravity constant of the Venus
$\lambda_{\odot}$ = longitude of the Sun
$\hat{\boldsymbol{r}}, \hat{\boldsymbol{\theta}}, \hat{\boldsymbol{h}}$ = Euler-Hill frame
$f_{\odot}$ = perturbation vector due to the Sun's gravitational attraction
a = orbit semimajor axis
e = orbit eccentricity
i = orbit inclination
n = spacecraft mean motion, $\sqrt{\frac{\mu}{a^3}}$, rad/s
$n_{\odot}$ = mean motion of Venus-Sun line
$\mu_{\oplus}$ = gravity constant of Earth

12.1 INTRODUCTION

Concept studies for the Venus Orbiting Imaging Radar (VOIR) were carried out at NASA's Jet Propulsion Laboratory in the 1970s such as by Kerridge [1], who studied global mapping strategies. VOIR evolved later into the Venus Radar Mapper (VRM) with a reduced set of experiments, and in 1986, it was renamed Magellan and later launched successfully to orbit Venus. The analyses carried out here are applied to the original VOIR mission as an example.

This chapter evaluates the effects of additional perturbations experienced by a spacecraft's orbit, to determine their respective effects and to eventually ignore them in a first analysis.

The maneuver strategies will thus be designed accordingly. The net perturbation of the Sun's gravity is resolved into its out-of-plane and in-plane components, and the time history of the inclination i and eccentricity e are derived. It is shown that these elements have a twice-yearly variation. Furthermore, the semimajor axis remains constant (a well-known result for any orbit). The Sun's effect is thus shown to be negligible for low orbits.

The analysis presented in this chapter draws from the derivations provided by Breakwell [2]. Additional analyses were published by the author [3, 4], with Reference [4] applying a formulation based on nonsingular elements to carry out optimal low-thrust transfers.

12.2 ANALYSIS

The disturbing function due to a third body (solar gravity) is given by

$$\boldsymbol{F} = \nabla \frac{\mu_\odot}{|\boldsymbol{r}_\odot - \boldsymbol{r}|} \tag{12.1}$$

where $\boldsymbol{r}_\odot$ is the radius vector from the center of the planet to the Sun and $\boldsymbol{r}$ is the radius vector to the spacecraft from the same origin. The denominator appearing in Eq. (12.1) can be written as follows:

$$|\boldsymbol{r}_\odot - \boldsymbol{r}| = \left(r_\odot^2 - 2\boldsymbol{r}\boldsymbol{r}_\odot + r^2\right)^{1/2} = r_\odot\left(1 - 2\frac{\hat{\boldsymbol{r}}_\odot \cdot \boldsymbol{r}}{r_\odot} + \frac{r^2}{r_\odot^2}\right)^{1/2} \tag{12.2}$$

By definition, let

$$|\hat{\boldsymbol{r}} \cdot \hat{\boldsymbol{r}}_\odot| = \frac{\boldsymbol{r}}{r}\hat{\boldsymbol{r}}_\odot = \cos\phi \tag{12.3}$$

Now, Eq. (12.2) becomes

$$|\boldsymbol{r}_\odot - \boldsymbol{r}| = r_\odot\left(1 - 2\frac{r}{r_\odot}\cos\phi + \frac{r^2}{r_\odot^2}\right)^{1/2} \tag{12.4}$$

An expansion of Eq. (12.4) introduces the Legendre polynomials in $\cos\phi$ in the following form with $x = r/r_\odot$

$$\begin{aligned}(1 - 2x\cos\phi + x^2)^{-1/2} &= 1 + x\cos\phi + x^2\left(\frac{3}{2}\cos^2\phi - \frac{1}{2}\right) \\ &\quad + x^3\left(\frac{5}{2}\cos^3\phi - \frac{3}{2}\cos\phi\right) + \text{HOT}\end{aligned} \tag{12.5}$$

where HOT represents higher-order terms, so that finally Eq. (12.1) describing the disturbing potential is transformed into

$$\begin{aligned} \boldsymbol{F} &= \nabla\left\{\frac{\mu_\odot}{r_\odot} + \frac{\mu_\odot}{r_\odot}\frac{\boldsymbol{r}\cdot\hat{\boldsymbol{r}}_\odot}{r_\odot} + \frac{\mu_\odot}{r_\odot}\left[\frac{3}{2}\frac{(\boldsymbol{r}\cdot\hat{\boldsymbol{r}}_\odot)^2}{r_\odot^2} - \frac{1}{2}\frac{r^2}{r_\odot^2}\right] + \text{HOT}\right\} \\ &= 0 + \frac{\mu_\odot}{r_\odot^2}\hat{\boldsymbol{r}}_\odot + \frac{\mu_\odot}{r_\odot^3}[3(\boldsymbol{r}\cdot\hat{\boldsymbol{r}}_\odot)\hat{\boldsymbol{r}}_\odot - \boldsymbol{r}] + \text{HOT} \end{aligned} \tag{12.6}$$

The last term in Eq. (12.6) is the gravitational gradient term and is linear in $\boldsymbol{r}$; therefore, let the perturbation vector due to the Sun's gravitational attraction be $\boldsymbol{f}_\odot$, such that

$$\boldsymbol{f}_\odot = n_\odot^2[3(\hat{\boldsymbol{r}}_\odot\cdot\boldsymbol{r})\hat{\boldsymbol{r}}_\odot - \boldsymbol{r}] + \mathcal{O}(r^2) \tag{12.7}$$

where $\boldsymbol{n}_\odot$ is the angular rate of the planet–Sun line or the mean motion of the Sun. Returning to the Euler–Hill frame of Fig. 11.1, we have

$$\hat{\boldsymbol{x}} = (c_\Omega c_\theta - s_\Omega c_i s_\theta)\hat{\boldsymbol{r}} + (-c_\Omega s_\theta - s_\Omega c_i c_\theta)\hat{\boldsymbol{\theta}} + s_\Omega s_i \hat{\boldsymbol{h}} \tag{12.8}$$

$$\hat{\boldsymbol{y}} = (s_\Omega c_\theta + c_\Omega c_i s_\theta)\hat{\boldsymbol{r}} + (-s_\Omega s_\theta + c_\Omega c_i c_\theta)\hat{\boldsymbol{\theta}} - c_\Omega s_i \hat{\boldsymbol{h}} \tag{12.9}$$

Now, the unit vector in the Sun's direction can be evaluated as

$$\hat{\boldsymbol{r}}_\odot = c_{\lambda_\odot}\hat{\boldsymbol{x}} + s_{\lambda_\odot}\hat{\boldsymbol{y}} \tag{12.10}$$

where $\lambda_\odot$ is the longitude of the Sun and is given by the following linear function of time:

$$\lambda_\odot = n_\odot t \tag{12.11}$$

Using Eqs. (12.8) and (12.9), we can now write Eq. (12.10) in terms of the $\hat{\boldsymbol{r}}$, $\hat{\boldsymbol{\theta}}$, and $\hat{\boldsymbol{h}}$ components as

$$\hat{\boldsymbol{r}}_\odot = (c_{\lambda_\odot - \Omega} c_\theta + c_i s_\theta s_{\lambda_\odot - \Omega})\hat{\boldsymbol{r}} + (-s_\theta c_{\lambda_\odot - \Omega} + c_i c_\theta s_{\lambda_\odot - \Omega})\hat{\boldsymbol{\theta}} - s_i s_{\lambda_\odot - \Omega}\hat{\boldsymbol{h}} \tag{12.12}$$

The out-of-plane component f_h of $\boldsymbol{f}_\odot$ can be evaluated using Eq. (12.7) in the following way:

$$f_h = \boldsymbol{f}_\odot \cdot \hat{\boldsymbol{h}} = 3n_\odot^2(\hat{\boldsymbol{r}}_\odot\cdot\boldsymbol{r})(\hat{\boldsymbol{r}}_\odot\cdot\hat{\boldsymbol{h}}) \tag{12.13}$$

However, from Eq. (12.12)

$$\boldsymbol{r}\cdot\hat{\boldsymbol{r}}_\odot = r(\hat{\boldsymbol{r}}\cdot\hat{\boldsymbol{r}}_\odot) = r(c_{\lambda_\odot - \Omega} c_\theta + c_i s_\theta s_{\lambda_\odot - \Omega}) \tag{12.14}$$

$$\hat{\boldsymbol{r}}_\odot\cdot\hat{\boldsymbol{h}} = -s_i s_{\lambda_\odot - \Omega} \tag{12.15}$$

so that [5]

$$f_h = -3rn_{\odot}^2(c_{\lambda_{\odot}-\Omega}c_\theta s_i s_{\lambda_{\odot}-\Omega} + c_i s_i s_{\lambda_{\odot}-\Omega}^2 s_\theta) \tag{12.16}$$

For a circular polar orbit

$$\begin{aligned} i &= 90\ \text{deg} \\ r &= a \end{aligned} \tag{12.17}$$

$$f_h = -3an_{\odot}^2(c_\theta c_{\lambda_{\odot}-\Omega} s_{\lambda_{\odot}-\Omega}) \tag{12.18}$$

12.3 VARIATION OF THE INCLINATION DUE TO SOLAR GRAVITATIONAL PERTURBATIONS

The inclination i is affected only by the out-of-plane component f_h of $\boldsymbol{f}_{\odot}$ in the following way, where ♀ represents Venus

$$\frac{\mathrm{d}i}{\mathrm{d}\theta} = \frac{\dot{i}}{\dot{\theta}} = \frac{af_h c_\theta}{hn} = \frac{af_h c_\theta}{\sqrt{\mu_{♀} a}\sqrt{\mu_{♀}/a^3}} = \frac{a^2 f_h c_\theta}{\mu_{♀}} \tag{12.19}$$

Using Eq. (12.18) to replace f_h, we can convert Eq. (12.19) to the form

$$\frac{\mathrm{d}i}{\mathrm{d}\theta} = -3\left(\frac{a}{r_{\odot}}\right)^3 \frac{\mu_{\odot}}{\mu_{♀}} s_{\lambda_{\odot}-\Omega} c_{\lambda_{\odot}-\Omega} c_\theta^2 \tag{12.20}$$

so that averaging over the duration of one full orbit gives

$$\left.\frac{\mathrm{d}i}{\mathrm{d}\theta}\right|_{\text{av}} = \frac{\int_0^{2\pi} \frac{\mathrm{d}i}{\mathrm{d}\theta}\,\mathrm{d}\theta}{2\pi} = -\frac{3}{2}\left(\frac{a}{r_{\odot}}\right)^3 \frac{\mu_{\odot}}{\mu_{♀}} s_{\lambda_{\odot}-\Omega} c_{\lambda_{\odot}-\Omega} = -\frac{3}{4}\frac{n_{\odot}^2}{n^2} s_{2(n_{\odot}t-\Omega)} \tag{12.21}$$

indicating that $\mathrm{d}i/\mathrm{d}\theta$ achieves its maximum for

$$\sin 2(n_{\odot}t - \Omega) = 1 \tag{12.22}$$

It is now possible to evaluate the time history of i from

$$\frac{\mathrm{d}i}{\mathrm{d}t} = \dot{\theta}\frac{\mathrm{d}i}{\mathrm{d}\theta} = n\frac{\mathrm{d}i}{\mathrm{d}\theta} = -\frac{3}{4}\frac{n_{\odot}^2}{n}\sin 2(n_{\odot}t - \Omega) \tag{12.23}$$

The integration with respect to time yields

$$\Delta i|_0^t = +\frac{3}{8}\frac{n_{\odot}}{n}\left[c_{2(n_{\odot}t-\Omega)} - c_{2\Omega}\right] \tag{12.24}$$

The inclination i relative to the equator varies twice yearly from its nominal value of $\pi/2$ by the amount indicated in Eq. (12.24).

As an example, for the VOIR spacecraft, $n_{\odot}$ is the mean motion of the Venus–Sun line, and the relevant values are

$$n_{\odot} = 0.000000324 \text{ rad/s}$$
$$n = 0.001139277 \text{ rad/s}$$
$$\Omega = 193.0754 \text{ deg}$$

so that

$$(\Delta i)_{\max} = 0.00548 \text{ deg}$$

which is rather negligible.

12.4 VARIATION OF THE RIGHT ASCENSION OF THE NODE Ω

Following the same procedure as described in the previous section and using Eq. (12.18) again, we can write

$$\frac{\mathrm{d}\Omega}{\mathrm{d}\theta} = \frac{\dot{\Omega}}{\dot{\theta}} = \frac{r f_h s_\theta}{h s_i \sqrt{\mu_{♀}/a^3}} = 3\left(\frac{a}{r_{\odot}}\right)^3 s_{\lambda_{\odot}-\Omega} c_{\lambda_{\odot}-\Omega} \frac{\mu_{\odot}}{\mu_{♀}} s_\theta c_\theta \tag{12.25}$$

The average over one orbit is zero because the average of $s_\theta c_\theta$ between 0 and 2π is zero. Therefore

$$\left.\frac{\mathrm{d}\Omega}{\mathrm{d}\theta}\right|_{\mathrm{av}} = 0 \tag{12.26}$$

For a general inclination i, the expression for f_h in Eq. (12.16) must be used instead of that in Eq. (12.18) where the c_i term was ignored for polar orbits with $i = 90$ deg. Keeping $r = a$ for a circular orbit, the full expression for f_h in Eq. (12.16) will produce

$$\frac{\mathrm{d}\Omega}{\mathrm{d}\theta} = 3\left(\frac{a}{r_{\odot}}\right)^3 s_\theta s_{\lambda_{\odot}-\Omega}(c_{\lambda_{\odot}-\Omega} c_\theta + c_i s_{\lambda_{\odot}-\Omega} s_\theta) \frac{\mu_{\odot}}{\mu_{♀}}$$

such that

$$\left.\frac{\mathrm{d}\Omega}{\mathrm{d}\theta}\right|_{\mathrm{av}} = -\frac{3}{2}\left(\frac{a}{r_{\odot}}\right)^3 s^2_{\lambda_{\odot}-\Omega} c_i \frac{\mu_{\odot}}{\mu_{♀}}$$

Likewise, for

$$\frac{\mathrm{d}i}{\mathrm{d}\theta} = -3\left(\frac{a}{r_{\odot}}\right)^3 \frac{\mu_{\odot}}{\mu_{♀}} c_\theta s_i s_{\lambda_{\odot}-\Omega}(c_{\lambda_{\odot}-\Omega} c_\theta + c_i s_{\lambda_{\odot}-\Omega} s_\theta)$$

we have

$$\left.\frac{\mathrm{d}i}{\mathrm{d}\theta}\right|_{\mathrm{av}} = -\frac{3}{2}\left(\frac{a}{r_\odot}\right)^3 \frac{\mu_\odot}{\mu_♀} s_i s_{\lambda_\odot-\Omega} c_{\lambda_\odot-\Omega}$$

If the average value of $\mathrm{d}\Omega/\mathrm{d}\theta$ is applied to the case of the Moon orbiting Earth and perturbed by the Sun, then averaging over the yearly cycle of $\lambda_\odot$ and treating $\lambda_\odot$ on the right-hand side of the expression for $(\mathrm{d}\Omega/\mathrm{d}\theta)|_{\mathrm{av}}$ as constant, we can write

$$\left.\left.\frac{\mathrm{d}\Omega}{\mathrm{d}\theta}\right|_{\mathrm{av}}\right|_{\mathrm{av}} = \frac{3}{2}\frac{\mu_\odot}{\mu_\oplus}\left(\frac{r_D}{r_\odot}\right)^3 \frac{c_i}{2\pi}\int_0^{2\pi} s^2_{\Omega-\lambda_\odot}\,\mathrm{d}(\Omega-\lambda_\odot) = \frac{3}{4}\frac{\mu_\odot}{\mu_\oplus}\left(\frac{r_D}{r_\odot}\right)^3 c_i$$

which leads to

$$P_\Omega = P_D\left[\frac{4}{3}\frac{\mu_\oplus}{\mu_\odot}\left(\frac{r_\odot}{r_D}\right)^3\frac{1}{c_i}\right]$$

or 18.6 years for the precession period of the lunar orbit, where P_D is the period of the lunar orbit and a value of $i = 18.3$ deg was used.

12.5 IN-PLANE EFFECTS DUE TO SOLAR GRAVITATIONAL PERTURBATIONS

The component of the gravity term in the radial direction is obtained as follows:

$$f_r = \boldsymbol{f}_\odot\cdot\hat{\boldsymbol{r}} = n^2_\odot[3(\hat{\boldsymbol{r}}_\odot\cdot\boldsymbol{r})(\hat{\boldsymbol{r}}_\odot\cdot\hat{\boldsymbol{r}}) - (\boldsymbol{r}\cdot\hat{\boldsymbol{r}})] \tag{12.27}$$

However, from Eq. (12.12), we have

$$\boldsymbol{r} = r\hat{\boldsymbol{r}}$$
$$\hat{\boldsymbol{r}}_\odot\cdot\boldsymbol{r} = r(\hat{\boldsymbol{r}}_\odot\cdot\hat{\boldsymbol{r}})$$
$$\hat{\boldsymbol{r}}\cdot\hat{\boldsymbol{r}}_\odot = c_{\lambda_\odot-\Omega}c_\theta + c_i s_\theta s_{\lambda_\odot-\Omega}$$

so that

$$f_r = n^2_\odot\left[3r(c_{\lambda_\odot-\Omega}c_\theta + c_i s_\theta s_{\lambda_\odot-\Omega})^2 - r\right] \tag{12.28}$$

Once again, for the Venus VOIR spacecraft

$$i \simeq 90\ \mathrm{deg}$$
$$r = a$$
$$f_r = an^2_\odot\left(3c^2_\theta c^2_{\lambda_\odot-\Omega} - 1\right) \tag{12.29}$$

In a similar way, the third component ($\hat{\boldsymbol{\theta}}$ direction) is given by

$$f_\theta = \boldsymbol{f}_\odot\cdot\hat{\boldsymbol{\theta}} = n^2_\odot\left[3(\hat{\boldsymbol{r}}_\odot\cdot\boldsymbol{r})(\hat{\boldsymbol{r}}_\odot\cdot\hat{\boldsymbol{\theta}}) - \boldsymbol{r}\cdot\hat{\boldsymbol{\theta}}\right] \tag{12.30}$$

However, from Eq. (12.12)

$$\hat{\boldsymbol{r}}_{\odot} \cdot \hat{\boldsymbol{\theta}} = -s_{\theta} c_{\lambda_{\odot}-\Omega} + c_i c_{\theta} s_{\lambda_{\odot}-\Omega} \tag{12.31}$$

$$\hat{\boldsymbol{r}}_{\odot} \cdot \hat{\boldsymbol{r}} = c_{\lambda_{\odot}-\Omega} c_{\theta} + c_i s_{\theta} s_{\lambda_{\odot}-\Omega} \tag{12.32}$$

$$f_{\theta} = n_{\odot}^2 \left[3r(-s_{\theta} c_{\lambda_{\odot}-\Omega} + c_i c_{\theta} s_{\lambda_{\odot}-\Omega})(c_{\lambda_{\odot}-\Omega} c_{\theta} + c_i s_{\theta} s_{\lambda_{\odot}-\Omega}) \right] \tag{12.33}$$

For the VOIR spacecraft,

$$i \simeq 90 \text{ deg}$$

$$r = a$$

$$f_{\theta} = -3an_{\odot}^2 s_{\theta} c_{\theta} c_{\lambda_{\odot}-\Omega}^2 \tag{12.34}$$

In summary

$$\left.\begin{aligned} \boldsymbol{f}_{\odot} &= n_{\odot}^2 [3(\hat{\boldsymbol{r}}_{\odot} \cdot \boldsymbol{r})\hat{\boldsymbol{r}}_{\odot} - \boldsymbol{r}] \\ f_r &= an_{\odot}^2 \left(3c_{\theta}^2 c_{\lambda_{\odot}-\Omega}^2 - 1\right) \\ f_{\theta} &= -3an_{\odot}^2 s_{\theta} c_{\theta} c_{\lambda_{\odot}-\Omega}^2 \\ f_h &= -3an_{\odot}^2 \left[c_{\theta} c_{\lambda_{\odot}-\Omega} s_{\lambda_{\odot}-\Omega}\right] \end{aligned}\right\} \tag{12.35}$$

12.6 VARIATION OF THE ECCENTRICITY DUE TO SOLAR GRAVITATIONAL PERTURBATIONS

It is convenient to use the time rate of change of the eccentricity $\dot{e}$ given in terms of the perturbation vector components (Gaussian form) as

$$\dot{e} = \frac{r}{h} \left[s_{\theta^*}(1 + ec_{\theta^*})f_r + \left(e + 2c_{\theta^*} + ec_{\theta^*}^2\right) f_{\theta} \right] \tag{12.36}$$

where $\theta^* = \theta - \omega$ is the true anomaly and

$$\begin{aligned} r &= a \\ h &= \sqrt{\mu_{♀} a} \end{aligned} \tag{12.37}$$

$$\begin{aligned} \frac{\mathrm{d}e}{\mathrm{d}\theta} = \frac{n_{\odot}^2}{n^2} \Big\{ & [(s_{\theta} c_{\omega} - s_{\omega} c_{\theta}) + e(s_{\theta} c_{\omega} - s_{\omega} c_{\theta})(c_{\theta} c_{\omega} + s_{\theta} s_{\omega})] . \left(3c_{\theta}^2 c_{\lambda_{\odot}-\Omega}^2 - 1\right) \\ & + \left[e + 2(c_{\theta} c_{\omega} + s_{\theta} s_{\omega}) + e(c_{\theta} c_{\omega} + s_{\theta} s_{\omega})^2 \right] \left(-3s_{\theta} c_{\theta} c_{\lambda_{\odot}-\Omega}^2\right) \Big\} \end{aligned} \tag{12.38}$$

Here, $\mu_{♀}$ is, as before, the gravitational constant of Venus.

Terms not involving e do not contribute to the average value of $\mathrm{d}e/\mathrm{d}\theta$, so that the surviving terms are

$$\frac{\mathrm{d}e}{\mathrm{d}\theta} = e\frac{n_\odot^2}{n^2}\left[-3c^2_{\lambda_\odot-\Omega}s_\omega c_\omega\left(c_\theta^4 + s_\theta^2 c_\theta^2\right)\right] \tag{12.39}$$

$$\left.\frac{\mathrm{d}e}{\mathrm{d}\theta}\right|_{\mathrm{av}} = -\frac{3}{8}\frac{n_\odot^2}{n^2}es_{2\omega}c^2_{\lambda_\odot-\Omega} \tag{12.40}$$

Because $\lambda_\odot = n_\odot t$, we can now integrate with respect to time to obtain

$$\dot{e} = -\frac{3}{8}e\frac{n_\odot^2}{n^2}s_{2\omega}c^2_{\lambda_\odot-\Omega} \tag{12.41}$$

$$\Delta e|_0^t = -\frac{3}{8}\frac{n_\odot^2}{n}es_{2\omega}\int_0^t c^2_{n_\odot t-\Omega}\,\mathrm{d}t \tag{12.42}$$

$$\Delta e|_0^t = -\frac{3}{8}\frac{n_\odot}{n}es_{2\omega}\left\{\frac{n_\odot t}{2} + \frac{1}{4}\left[s_{2(n_\odot t-\Omega)} + s_{2\Omega}\right]\right\} \tag{12.43}$$

There is a twice-yearly variation contribution to the linear growth experienced by e. If e is perfectly circular, no variation will be experienced by e according to Eq. (12.43), so that a small departure from a circular orbit will provoke further variations in e. Finally, the total variation is proportional to $s_{2\omega}$, indicating that the position of periapse is a fundamental factor in the behavior of the orbit.

One can also use the Euler–Hill equations to determine the variation of the orbital parameters. This leads to the equations

$$\ddot{x} - 2n\dot{y} - 3n^2x = f_r = an_\odot^2\left(3c^2_{nt}c^2_{\lambda_\odot-\Omega} - 1\right) \tag{12.44}$$

$$\ddot{y} + 2n\dot{x} = f_\theta = -\frac{3}{2}an_\odot^2 s_{2nt}c^2_{\lambda_\odot-\Omega} \tag{12.45}$$

where θ has been replaced by nt, n being, of course, the mean orbital motion of the spacecraft around Venus.

Equation (12.45) can be integrated once to give

$$\dot{y} = -2nx + \frac{3}{4}\frac{an_\odot^2}{n}c^2_{\lambda_\odot-\Omega}c_{2nt} \tag{12.46}$$

which is now replaced in Eq. (12.44) to give

$$\ddot{x} + n^2x = \frac{3}{2}an_\odot^2c^2_{\lambda_\odot-\Omega}c_{2nt} + an_\odot^2\left(3c^2_{nt}c^2_{\lambda_\odot-\Omega} - 1\right) = f(t) \tag{12.47}$$

The general solution of Eq. (12.47) is

$$x = K_1\left(\frac{s_{nt}^2}{2n}+\frac{s_{nt}s_{3nt}}{6n}-\frac{c_{nt}^2}{2n}+\frac{c_{nt}c_{3nt}}{6n}+\frac{c_{nt}}{2n}-\frac{c_{nt}}{6n}\right)$$
$$+K_2\left(c^2_{\lambda_\odot-\Omega}\frac{s^2_{nt}c^2_{nt}}{n}+\frac{2c^2_{\lambda_\odot-\Omega}s^2_{nt}}{n}+\frac{c^2_{\lambda_\odot-\Omega}}{n}c^4_{nt}-\frac{c^2_{\lambda_\odot-\Omega}}{n}c_{nt}-\frac{1}{n}+\frac{c_{nt}}{n}\right) \tag{12.48}$$

where

$$K_1 = \frac{3}{2}\frac{an_\odot^2}{n}c^2_{\lambda_\odot-\Omega} \tag{12.49}$$

$$K_2 = \frac{an_\odot^2}{n} \tag{12.50}$$

If we take the derivative of Eq. (12.48) with respect to time, we obtain

$$\dot{x}=K_1\left(2s_{nt}c_{nt}+\frac{c_{nt}s_{3nt}+3s_{nt}c_{3nt}}{6}-\frac{1}{3}s_{nt}-\frac{s_{nt}c_{3nt}+3c_{nt}s_{3nt}}{6}\right)$$
$$+K_2\left[2c^2_{\lambda_\odot-\Omega}s_{nt}c_{nt}\left(c^2_{nt}-s^2_{nt}\right)+4c^2_{\lambda_\odot-\Omega}s_{nt}c_{nt}-4c^2_{\lambda_\odot-\Omega}c^3_{nt}s_{nt}+c^2_{\lambda_\odot-\Omega}s_{nt}-s_{nt}\right] \tag{12.51}$$

Compare these results to

$$\dot{x} = \dot{r} = \dot{a} - c_{nt}(\dot{a}e + a\dot{e}) + aens_{nt} \tag{12.52}$$

$$\dot{a} = 0 \tag{12.53}$$

$$\dot{e} = 0 \tag{12.54}$$

There is no variation in the semimajor axis or the eccentricity due to the Sun's gravity because it has been assumed that the initial orbit is perfectly circular. Equation (12.43) already showed the necessity of having a slight eccentricity to drive that term. However, if we consider the coefficient of s_{nt} appearing in Eqs. (12.51) and (12.52) for $\dot{x}$, then we have

$$\dot{x}_{s_{nt}} = -\frac{K_1}{3}s_{nt} + K_2\left(c^2_{\lambda_\odot-\Omega} - 1\right)s_{nt} = aens_{nt} \tag{12.55}$$

and using K_1 and K_2 in Eqs. (12.49) and (12.50), with $\lambda_\odot = n_\odot t$, gives

$$\Delta e = e(t) - e(0) = \frac{n_\odot^2}{2n^2}\left(c^2_{n_\odot t-\Omega} - c^2_{-\Omega}\right) \tag{12.56}$$

For example, for the VOIR spacecraft in a 250-km circular orbit

$$n = \sqrt{\frac{\mu_{♀}}{a^3}} = \sqrt{\frac{3.24858 \times 10^5}{(6302)^3}} = 0.001139277 \text{ rad/s}$$

$$n_{\odot} = \sqrt{\frac{\mu_{\odot}}{a_{♀}^3}} = \sqrt{\frac{1.3271544 \times 10^{11}}{(1.08210467 \times 10^8)^3}} = 0.000000324 \text{ rad/s}$$

The initial value of Ω (it remains constant, on average) is $\Omega = 193.0754$ deg. Initially, $\lambda_{\odot} = 195.18079$ deg, and after five days, $\Delta\lambda_{\odot} = n_{\odot}\Delta t = 8.01957$ deg, so that, at that time, $\lambda_{\odot} = 203.20036$ deg. Finally, $\Delta e = 0.000000002$, which agrees perfectly with the result obtained by numerically integrating this trajectory.

12.7 CONCLUSION

It is established that the Sun's gravitational perturbation can be safely neglected for a low circular orbit such as that of the VOIR spacecraft around Venus. As a consequence, maneuver strategies can be designed by considering only Venus's gravity field. However, solar gravitational perturbations on a large orbit can be highly significant and must be taken into account in the designs of both the trajectory and the maneuver.

REFERENCES

[1] Kerridge, S. J., "Global mapping strategies for a synthetic aperture radar system in orbit about Venus," *AIAA/AAS Astrodynamics Conference, AIAA Paper 80-1688*, Danvers, MA, Aug. 1980.

[2] Breakwell, J. V., "Lecture Notes: Space Mechanics," Course AA279A, Department of Aeronautics and Astronautics, Stanford University, Palo Alto, CA, 1974.

[3] Kéchichian, J. A., *Orbital Relative Motion and Terminal Rendezvous*, Springer Nature, Cham, Switzerland, 2021.

[4] Kéchichian, J. A., *Applied Nonsingular Astrodynamics: Optimal Low-Thrust Orbit Transfer*, Cambridge Aerospace Series, No. 45, Cambridge University Press, Cambridge, U.K., 2018.

[5] Kéchichian, J. A., "Combined effect of drag and zonal harmonics on a near-circular orbit," EM 314-216, Jet Propulsion Laboratory, Pasadena, CA, 1980.

CHAPTER 13

Theory of the Displacement of the Vacant Focus due to a Small Impulse and Relative Motion in Near-Circular Orbit

NOMENCLATURE

a	= orbit semimajor axis
μ	= gravity constant of Earth
ΔV_t, ΔV_n	= tangential and normal components of imparted ΔV.
θ^*	= true anomaly
e	= orbit ecccentricity
V	= orbit velocity, km/s
β	= flight path angle
n	= orbit mean motion, $\sqrt{\frac{\mu}{a^3}}$, rad/s
r_p, r_a	= perigee, apogee radii
x, y	= Euler - Hill frame

13.1 INTRODUCTION

This chapter also draws from Breakwell's class notes [1] and provides useful analytic expressions for the changes experienced by the various orbit elements due to the application of a small impulse at a given location on the initial elliptical orbit. Other useful expressions are provided for the relative motion of two close satellites in near-circular orbits using the Euler–Hill rotating frame [1]. Finally, a discussion of the sensitivity of the velocity changes for the classic Hohmann transfer is provided at the end of the chapter.

13.2 DISPLACEMENT OF VACANT FOCUS DUE TO A SMALL IMPULSE

A small impulsive change in velocity ΔV applied at position P, as shown in Fig. 13.1, can be resolved into normal and tangential components ΔV_n and ΔV_t, respectively, where $\Delta V_n = f_n \Delta t$ and $\Delta V_t = f_t \Delta t$ with f_n and f_t representing the corresponding normal and tangential components of the acceleration imparted to the spacecraft during time Δt. From the energy equation

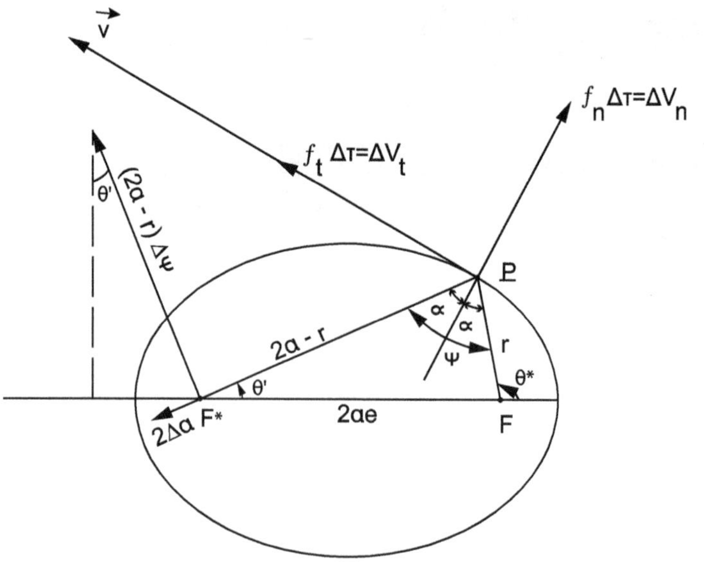

Fig. 13.1 Displacement of vacant focus due to small impulse.

$E = -\mu/(2a)$, where a is the semimajor axis and μ is the gravitational constant of Earth, we have

$$dE = \frac{\mu}{2a^2}\Delta a = \boldsymbol{V} \cdot \boldsymbol{f}\Delta t = Vf_t\Delta t$$

$$\Delta a = \frac{2a^2}{\mu} Vf_t\Delta t \tag{13.1}$$

which shows that a is affected only by f_t. Moreover, because P remains fixed

$$FP + F^*P = 2a$$

$$d(FP + F^*P) = d(F^*P) = 2\Delta a \tag{13.2}$$

indicating that F^*P, which was equal in length to $2a - r$ before the application of ΔV, is now equal to $2a - r + 2\Delta a$, with Δa given by Eq. (13.1).

On the other hand, from the triangle F^*PF with fixed angle θ^*, $\theta^* = \theta' + 2\alpha$, and $d\theta' = -2\,d\alpha$. The change in the angle α is due to ΔV_n because

$$V\Delta\alpha = \Delta V_n = f_n\Delta t \tag{13.3}$$

$$\Delta\alpha = \frac{f_n\Delta t}{V} \tag{13.4}$$

The displacement of F^* due to ΔV_n is orthogonal to F^*P and is given by $(2a - r)$ $\Delta\psi$, with $\psi = 2\alpha$

$$\Delta\psi = 2\Delta\alpha = \frac{2f_n\Delta t}{V} = \Delta\theta' \tag{13.5}$$

Adding the effects of ΔV_n and ΔV_t and observing that $FF^* = 2ae$, the displacement of F^* is formulated by the following two equations:

$$\left.\begin{aligned} \mathrm{d}(2ae) &= 2c_{\theta'}\,\mathrm{d}a - (2a - r)s_{\theta'}\,\mathrm{d}\theta' \\ 2ae\,\mathrm{d}\omega &= 2s_{\theta'}\,\mathrm{d}a + (2a - r)c_{\theta'}\,\mathrm{d}\theta' \end{aligned}\right\} \tag{13.6}$$

However

$$\begin{aligned} (2a - r)c_{\theta'} &= 2ae + rc_{\theta^*} \\ (2a - r)s_{\theta'} &= rs_{\theta^*} \\ V^2 &= \frac{\mu}{a}\left(\frac{2a - r}{r}\right) \end{aligned}$$

These equations combined with Eqs. (13.1) and (13.5) transform the set in Eq. (13.6) into

$$\left.\begin{aligned} \dot{e} &= 2(e + c_{\theta^*})\frac{f_t}{V} - \frac{r}{a}s_{\theta^*}\frac{f_n}{V} \\ e\dot{\omega} &= 2s_{\theta^*}\frac{f_t}{V} + \left(2e + \frac{r}{a}c_{\theta^*}\right)\frac{f_n}{V} \end{aligned}\right\} \tag{13.7}$$

The variation in the heights of periapse and apoapse can be obtained from Eq. (13.7) as follows

$$\begin{aligned} r_p &= a(1 - e) \\ \dot{r}_p &= (1 - e)\dot{a} - a\dot{e} \end{aligned}$$

$$\dot{r}_p = \frac{f_t}{V}\left[(1 - e)\frac{2a^2}{\mu}V^2 - 2a(e + c_{\theta^*})\right] + rs_{\theta^*}\frac{f_n}{V} \tag{13.8}$$

and

$$\begin{aligned} r_a &= a(1 + e) \\ \dot{r}_a &= \dot{a}(1 + e) + a\dot{e} \end{aligned}$$

$$\dot{r}_a = \frac{f_t}{V}\left[(1 + e)\frac{2a^2}{\mu}V^2 + 2a(e + c_{\theta^*})\right] - rs_{\theta^*}\frac{f_n}{V} \tag{13.9}$$

The change in the rotation or inclination of the orbit plane is affected only by the out-of-plane component of the impulsive velocity change. Let f_h be the out-of-plane component of the acceleration vector. Then

$$\dot{i} = \frac{f_h}{V_\theta} c_\theta \tag{13.10}$$

where $\theta = \theta^* + \omega$, θ^* is the true anomaly, ω is the argument of periapse, and

$$V_\theta = V c_\beta = \frac{V(1 + e c_{\theta^*})}{(1 + e^2 + 2e c_{\theta^*})^{1/2}} \tag{13.11}$$

where β is the flight path angle.

Combining the energy equation

$$\frac{V^2}{2} - \frac{\mu}{r} = -\frac{\mu}{2a} \tag{13.12}$$

with Eq. (13.11) to eliminate the velocity V_θ, we can reduce Eq. (13.10) to the form

$$\Delta i = \frac{f_h \Delta t (1 + e^2 + 2e c_{\theta^*})^{1/2} c_{\theta^* + \omega}}{V(1 + e c_{\theta^*})} \tag{13.13}$$

13.2.1 CHANGE IN SEMIMAJOR AXIS DUE TO AN IMPULSIVE CHANGE IN VELOCITY

Equation (13.1) describes how an impulsive change in velocity $\Delta V = f_t \Delta t$, where f_t is the tangential component of the acceleration applied to the spacecraft and Δt the duration of the firing, perturbs the semimajor axis

$$\Delta a = \frac{2a^2}{\mu} V f_t \Delta t = \frac{2a^2}{\mu} V \Delta V \tag{13.14}$$

This equation is a good approximation for small Δt and small ΔV, because a is constant on the right-hand side of Eq. (13.14) and thus represents a linearized perturbation of the semimajor axis.

An exact formula can be obtained from the energy equation as follows:

$$\frac{V^2}{2} - \frac{\mu}{r} = -\frac{\mu}{2a} \tag{13.15}$$

$$V \, \mathrm{d}V = \frac{\mu}{2a^2} \, \mathrm{d}a \tag{13.16}$$

$$\frac{\mathrm{d}a}{a^2} = \frac{2}{\mu} V \, \mathrm{d}V \tag{13.17}$$

Integrating Eq. (13.17) between a_0, the initial value, and a, the current semimajor axis gives

$$a = \frac{a_0}{1 - \dfrac{a_0}{\mu}(V^2 - V_0^2)} \tag{13.18}$$

where V_0 is the initial velocity at the point of application of the impulsive change and V is the value reached at the end of the burn. Therefore, $V = V_0 + \Delta V$, and $V^2 - V_0^2 = (V - V_0)(V + V_0) = \Delta V(2V_0 + \Delta V)$such that Eq. (13.18) is converted to

$$\Delta a = a - a_0 = \frac{\frac{a_0^2}{\mu}(V^2 - V_0^2)}{1 - \frac{a_0}{\mu}(V^2 - V_0^2)} = \frac{\frac{a_0^2}{\mu}\Delta V(2V_0 + \Delta V)}{\left[1 - \frac{a_0}{\mu}\Delta V(2V_0 + \Delta V)\right]} \tag{13.19}$$

For example, a 100-s burn corresponding to $\Delta V_1 = f_t \Delta t = 0.3238 \times 10^{-3} \times 10^2 = 0.03238$ km/s applied to an initial circular orbit with $a = 6563$ km yields $\Delta a_1 = (2a_0^2/\mu)V_0 \Delta V = 54.5369$ km using Eq. (13.14) and

$$\Delta a_2 = \frac{\frac{a_0^2}{\mu}\Delta V(2V_0 + \Delta V)}{\left[1 - \frac{a_0}{\mu}\Delta V(2V_0 + \Delta V)\right]} = 55.1091 \text{ km}$$

using Eq. (13.19) . This shows that the linearized form in Eq. (13.14) is very accurate for small burns.

13.2.2 CHANGE IN ANGULAR MOMENTUM DUE TO AN IMPULSIVE CHANGE IN VELOCITY

When only tangential accelerations are considered, $f_n = 0$ reduces Eqs. (13.7–13.9) to

$$\dot{e} = 2(e + c_{\theta^*})\frac{f_t}{V} \tag{13.20}$$

$$e\dot{\omega} = s_{\theta^*}\frac{2}{V}f_t \tag{13.21}$$

$$\dot{r}_p = \frac{f_t}{V}\left[(1 - e)\frac{2a^2}{\mu}V^2 - 2a(e + c_{\theta^*})\right] \tag{13.22}$$

$$\dot{r}_a = \frac{f_t}{V}\left[(1 + e)\frac{2a^2}{\mu}V^2 + 2a(e + c_{\theta^*})\right] \tag{13.23}$$

with $\Delta V_t = f_t \Delta t$.

Introducing the flight path angle β to relate the tangential and normal components of the acceleration to the radial and 90 deg ahead directions in the orbital plane, it is possible to write

$$f_r = f_t s_\beta - f_n c_\beta \tag{13.24}$$

$$f_\theta = f_t c_\beta + f_n s_\beta \tag{13.25}$$

The rate of change of the angular momentum is given by

$$\dot{h} = r f_\theta = r(f_t c_\beta + f_n s_\beta) \tag{13.26}$$

but with $f_n = 0$

$$\dot{h} = r f_t c_\beta \tag{13.27}$$

The complement of β, denoted as ϕ, is equal to $\beta + 90$ deg, so that

$$s_\phi = \left(\frac{1 - e^2}{1 - e^2 c_E^2}\right)^{1/2} = \frac{1 + e c_{\theta^*}}{(1 + e^2 + 2e c_{\theta^*})^{1/2}} \tag{13.28}$$

$$c_\phi = \frac{-e s_E}{(1 - e^2 c_E^2)^{1/2}} = \frac{-e s_{\theta^*}}{(1 + e^2 + 2e c_{\theta^*})^{1/2}} \tag{13.29}$$

where E is the eccentric anomaly and θ^* is the true anomaly. This implies that

$$c_\beta = \frac{(1 + e c_{\theta^*})}{(1 + e^2 + 2e c_{\theta^*})^{1/2}} \tag{13.30}$$

such that Eq. (13.27) becomes

$$\dot{h} = \frac{r f_t (1 + e c_{\theta^*})}{(1 + e^2 + 2e c_{\theta^*})^{1/2}}$$

$$\Delta h = \frac{r \Delta V_t (1 + e c_{\theta^*})}{(1 + e^2 + 2e c_{\theta^*})^{1/2}} \tag{13.31}$$

where position r is given by

$$r = \frac{h^2/\mu}{(1 + e c_{\theta^*})} \tag{13.32}$$

13.2.3 SENSITIVITY ANALYSIS OF VACANT FOCUS THEORY

Considering only tangential components of the impulsive changes in velocity, the in-plane perturbations from Section 13.2 reduce to

$$\Delta a = \frac{2a^2}{\mu} V \Delta V \tag{13.33}$$

$$\Delta e = \frac{2(e + c_{\theta^*})\Delta V}{V} \tag{13.34}$$

Δa is affected only by the magnitude of the tangential change in velocity ΔV_t, whereas Δe is also dependent on the location of that same velocity change. Analytic partial derivatives are obtained as

$$\frac{\partial(\Delta a)}{\partial(\Delta V)} = \frac{2a^2}{\mu} V \tag{13.35}$$

$$\frac{\partial(\Delta e)}{\partial(\Delta V)} = \frac{2(e + c_{\theta^*})}{V} \tag{13.36}$$

$$\frac{\partial(\Delta e)}{\partial\theta^*} = -\frac{2s_{\theta^*}}{V}\Delta V \tag{13.37}$$

such that perturbations in the magnitude of the velocity change and in the firing location yield

$$\mathrm{d}(\Delta a) = \frac{2a^2}{\mu} V\,\mathrm{d}(\Delta V) \tag{13.38}$$

$$\mathrm{d}(\Delta e) = \frac{2(e + c_{\theta^*})\,\mathrm{d}(\Delta V)}{V} - 2\frac{\Delta V}{V} s_{\theta^*}\,\mathrm{d}\theta^* \tag{13.39}$$

where use is made of the relation

$$\mathrm{d}(\Delta e) = \frac{\partial(\Delta e)}{\partial(\Delta V)}\,\mathrm{d}(\Delta V) + \frac{\partial(\Delta e)}{\partial\theta^*}\,\mathrm{d}\theta^* \tag{13.40}$$

Equation (13.13) of Section 13.2 allows us to express the perturbation in Δi as

$$\mathrm{d}(\Delta i) = \frac{\partial(\Delta i)}{\partial(\Delta V_h)}\,\mathrm{d}(\Delta V_h) + \frac{\partial(\Delta i)}{\partial\theta^*}\,\mathrm{d}\theta^* \tag{13.41}$$

with

$$\frac{\partial(\Delta i)}{\partial(\Delta V_h)} = \frac{(1 + e^2 + 2ec_{\theta^*})^{1/2} c_{\theta^*+\omega}}{V(1 + ec_{\theta^*})} \tag{13.42}$$

$$\frac{\partial(\Delta i)}{\partial\theta^*} = \frac{\Delta V_h}{V}\left[\frac{e^2 s_{\theta^*} c_{\theta^*+\omega}}{(1 + ec_{\theta^*})^2}\frac{(e + c_{\theta^*})}{(1 + e^2 + 2ec_{\theta^*})^{1/2}} - \frac{s_{\theta^*+\omega}(1 + e^2 + 2ec_{\theta^*})^{1/2}}{(1 + ec_{\theta^*})}\right] \tag{13.43}$$

In summary, then, the perturbation equations are

$$
\begin{aligned}
d(\Delta a) &= \frac{2a^2}{\mu} V\, d(\Delta V) \\
d(\Delta e) &= \frac{2(e + c_{\theta^*})\, d(\Delta V)}{V} - 2\frac{\Delta V}{V} s_{\theta^*}\, d\theta^* \\
d(\Delta i) &= \frac{(1 + e^2 + 2ec_{\theta^*})^{1/2} c_{\theta^*+\omega}\, d(\Delta V_h)}{V(1 + ec_{\theta^*})} \\
&\quad + \frac{\Delta V_h}{V}\left[\frac{e^2 s_{\theta^*} c_{\theta^*+\omega}(e + c_{\theta^*})}{(1 + ec_{\theta^*})^2(1 + e^2 + 2ec_{\theta^*})^{1/2}} - \frac{s_{\theta^*+\omega}(1 + e^2 + 2ec_{\theta^*})^{1/2}}{(1 + ec_{\theta^*})}\right] d\theta^*
\end{aligned} \tag{13.44}
$$

13.3 RELATIVE MOTION OF TWO CLOSE SATELLITES IN NEAR-CIRCULAR ORBITS

Suppose that one satellite is in a perfectly circular orbit with semimajor axis a.

Let (x, y) be a frame attached to the spacecraft with x pointing in the radial direction and y along the velocity vector (Fig. 13.2). This is called the Euler–Hill frame. A second satellite is assumed to be close to the first one and in a near-circular orbit with instantaneous position vector $\boldsymbol{r}(t)$. Let $\boldsymbol{\rho}$ be the position of this satellite in the (x, y) frame. Let x_0 and y_0 be the components of $\boldsymbol{\rho}$ at time $t = 0$, and let $\dot{x}_0$ and $\dot{y}_0$ be the corresponding components of the relative initial velocity vector.

The motion of the spacecraft in the (x, y) frame is described by the Euler–Hill equations

$$
\left.\begin{aligned}
\ddot{x} - 2n\dot{y} - 3n^2 x = 0 \\
\ddot{y} + 2n\dot{x} = 0
\end{aligned}\right\} \tag{13.45}
$$

where $n = \sqrt{\mu/a^3}$ is the mean motion of the reference spacecraft.

The solution of Eq. (13.45) is given by

$$
\begin{aligned}
x &= \frac{\dot{x}_0}{n} s_{nt} - \left(\frac{2\dot{y}_0}{n} + 3x_0\right) c_{nt} + \left(\frac{2\dot{y}_0}{n} + 4x_0\right) \\
y &= \frac{2\dot{x}_0}{n} c_{nt} + \left(\frac{4\dot{y}_0}{n} + 6x_0\right) s_{nt} + \left(y_0 - \frac{2\dot{x}_0}{n}\right) - (3\dot{y}_0 + 6nx_0)t
\end{aligned} \tag{13.46}
$$

The eccentric anomaly is related to the mean anomaly by

$$
E = M + e \sin M + O(e^2) \tag{13.47}
$$

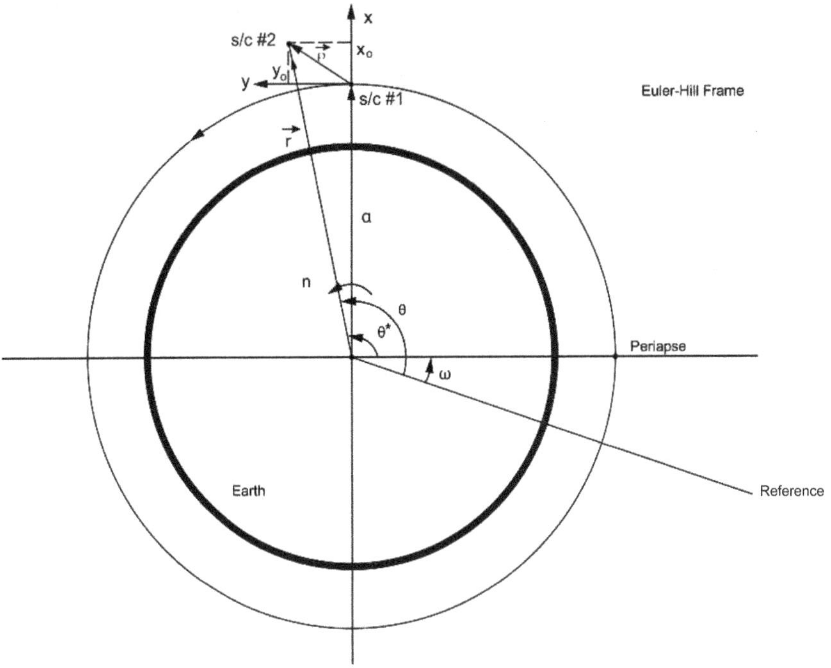

Fig. 13.2 Relative motion geometry.

where

$$M = n(t - t_p) \tag{13.48}$$

with t_p representing the time at periapse.

The true anomaly θ^* is also given by

$$\theta^* = \theta - \omega = M + 2e \sin M + O(e^2) \tag{13.49}$$

where ω is the argument of periapse and θ is the angular position measured from a suitable reference direction. As ω is poorly defined for a near-circular orbit, let

$$\xi = ec_\omega = ec_{nt_p} \tag{13.50}$$

$$\eta = es_\omega = es_{nt_p} \tag{13.51}$$

From Eq. (13.49) , one obtains

$$\begin{aligned} \theta = \omega + M + 2e \sin M + O(e^2) &\simeq \omega - nt_p + nt + 2e \sin n(t - t_p) \\ &\simeq \theta_0 + nt + 2e(s_{nt}c_{nt_p} - s_{nt_p}c_{nt}) \simeq \theta_0 + nt + 2(\xi s_{nt} - \eta c_{nt}) \end{aligned} \tag{13.52}$$

On the other hand, with the use of Eqs. (13.47) and (13.49) , the orbit equation gives

$$\begin{aligned} r &= a(1 - e\cos E) \simeq a\left[1 - e\cos M + O(e^2)\right] \\ &\simeq a\left[1 - ec_{\theta^*} + O(e^2)\right] \simeq a[1 - e\cos(M + 2e\sin M)] \\ &\simeq a(1 - e\cos M) \simeq a\left[1 - e\cos n(t - t_p)\right] \\ &\simeq a\left[1 - e\left(c_{nt}c_{nt_p} + s_{nt}s_{nt_p}\right)\right] \simeq a(1 - \xi c_{nt} - \eta s_{nt}) \end{aligned} \tag{13.53}$$

From Eq. (13.52)

$$\delta\theta = \delta\theta_0 + (\delta n)t + 2(\delta\xi)s_{nt} - 2(\delta\eta)c_{nt}$$

However, from $n = \sqrt{\mu/a^3}$, we have $\delta n = -(3/2)(n/a)\,\delta a$ such that

$$\delta\theta = \delta\theta_0 - \frac{3}{2}\frac{\delta a}{a}nt + 2(\delta\xi)s_{nt} - 2(\delta\eta)c_{nt} \tag{13.54}$$

It is now possible to obtain an expression for y as follows:

$$y = a\,\delta\theta = a\,\delta\theta_0 - \frac{3}{2}(\delta a)nt + 2a(\delta\xi)s_{nt} - 2a(\delta\eta)c_{nt} \tag{13.55}$$

Comparing this expression to the general solution given in Eq. (13.46), we obtain expressions for $\delta\xi$ and $\delta\eta$ as functions of the initial position and velocity as

$$a\,\delta\xi = \frac{2\dot{y}_0}{n} + 3x_0 \tag{13.56}$$

$$a\,\delta\eta = -\frac{\dot{x}_0}{n} \tag{13.57}$$

From Eq. (13.53) , we have

$$\frac{\delta r}{a} = \frac{\delta a}{a} - c_{nt}\,\delta\xi - s_{nt}\,\delta\eta \tag{13.58}$$

Let $x = \delta r$, then Eq. (13.58) yields

$$\frac{\delta a}{a} = \frac{x}{a} + c_{nt}\,\delta\xi + s_{nt}\,\delta\eta \tag{13.59}$$

However, Eq. (13.54) provides

$$\delta\dot{\theta} = -\frac{3}{2}n\frac{\delta a}{a} + 2n(c_{nt}\,\delta\xi + s_{nt}\,\delta\eta)$$

and using Eq. (13.59) to eliminate $\delta\xi$ and $\delta\eta$,

$$\delta\dot{\theta} = -\frac{3}{2}n\frac{\delta a}{a} + 2n\frac{\delta a}{a} - 2n\frac{x}{a} \tag{13.60}$$

such that the forward velocity measured in a rotating frame is given by

$$\dot{y} = a\,\delta\dot{\theta} = \frac{1}{2}n\,\delta a - 2nx \tag{13.61}$$

$$\frac{\delta a}{a} = \frac{2\dot{y}}{na} + \frac{4x}{a} = \frac{2\dot{y}_0}{na} + \frac{4x_0}{a} \tag{13.62}$$

The initial velocity in the θ direction is then

$$\delta V_{\theta_0} = \delta(r_0\dot{\theta}_0) = r_0\,\delta\dot{\theta}_0 + \dot{\theta}_0\,\delta r_0 = \dot{y}_0 + nx_0 \tag{13.63}$$

Finally, with Eq. (13.62), the relative change in the semimajor axis is obtained as

$$\frac{\delta a}{a} = \frac{2\delta V_{\theta_0}}{na} + \frac{2x_0}{a} \tag{13.64}$$

13.4 SENSITIVITY ANALYSIS OF HOHMANN TRANSFER VELOCITY CHANGES

Let us now consider a Hohmann transfer from a low circular orbit of radius r_1 to a higher circular orbit of radius r_2. The first velocity change ΔV_1 is such that the transfer ellipse reaches an apogee of radial distance r_2, where the second velocity change ΔV_2 is applied tangentially to carry out the transfer. Thus

$$\Delta V_1 = \sqrt{2\mu\left[\frac{r_2}{r_1(r_1+r_2)}\right]} - \sqrt{\left(\frac{\mu}{r_1}\right)} \tag{13.65}$$

$$\Delta V_2 = \sqrt{\left(\frac{\mu}{r_2}\right)} - \sqrt{2\mu\left[\frac{1}{r_2} - \frac{1}{(r_1+r_2)}\right]} \tag{13.66}$$

Following Battin [2], the equations that relate position and velocity to initial position and velocity as functions of the true anomaly for elliptical orbits are as follows

$$\left.\begin{aligned} \boldsymbol{r} &= \left\{1 - \frac{r}{p}\left[1 - \cos(\theta^* - \theta_0^*)\right]\right\}\boldsymbol{r}_0 + \frac{rr_0}{\sqrt{\mu p}}\sin(\theta^* - \theta_0^*)\boldsymbol{V}_0 \\ \boldsymbol{V} &= \left\{\frac{\boldsymbol{r}_0 \cdot \boldsymbol{V}_0}{pr_0}\left[1 - \cos(\theta^* - \theta_0^*)\right] - \frac{1}{r_0}\sqrt{\frac{\mu}{p}}\sin(\theta^* - \theta_0^*)\right\}\boldsymbol{r}_0 \\ &\quad + \left\{1 - \frac{r_0}{p}\left[1 - \cos(\theta^* - \theta_0^*)\right]\right\}\boldsymbol{V}_0 \end{aligned}\right\} \tag{13.67}$$

where θ^* is the true anomaly and $p = h^2/\mu$, with h being the angular momentum. In a more compact form, Eq. (13.67) can be written using the state transition matrix representation as in the equation

$$\begin{aligned} \begin{pmatrix} \boldsymbol{r} \\ \boldsymbol{V} \end{pmatrix} &= \Phi(t,\ t_0)\begin{pmatrix} \boldsymbol{r}_0 \\ \boldsymbol{V}_0 \end{pmatrix} = \frac{\partial\begin{pmatrix} \boldsymbol{r} \\ \boldsymbol{V} \end{pmatrix}}{\partial(\boldsymbol{r}_0,\ \boldsymbol{V}_0)}\begin{pmatrix} \boldsymbol{r}_0 \\ \boldsymbol{V}_0 \end{pmatrix} \\ &= \begin{pmatrix} \Phi_{rr_0} & \Phi_{rV_0} \\ \Phi_{Vr_0} & \Phi_{VV_0} \end{pmatrix}\begin{pmatrix} \boldsymbol{r}_0 \\ \boldsymbol{V}_0 \end{pmatrix} \end{aligned} \tag{13.68}$$

For the apoapse, Eq. (13.67) becomes

$$\boldsymbol{r}_a = \left(1 - 2\frac{r_0}{p}\right)\boldsymbol{r}_0 \tag{13.69}$$

$$\boldsymbol{V}_a = \left(2\frac{\boldsymbol{r}_0 \cdot \boldsymbol{V}_0}{pr_0}\right)\boldsymbol{r}_0 + \left(1 - 2\frac{r_0}{p}\right)\boldsymbol{V}_0 \tag{13.70}$$

Here, $\boldsymbol{V}_0 = \boldsymbol{V}_p^+ = \boldsymbol{V}_p^- + \Delta\boldsymbol{V}_1$ is the velocity at periapse just after the application of the first velocity change $\Delta\boldsymbol{V}_1$, and $\boldsymbol{V}_p^-$ is the orbital velocity before the change. From Eq. (13.69)

$$\begin{aligned} \boldsymbol{r}_a \cdot \boldsymbol{r}_a &= \left(1 - 2\frac{r_a}{p}\right)^2 \boldsymbol{r}_0 \cdot \boldsymbol{r}_a \\ r_a &= \frac{-r_0}{\left(1 - 2\dfrac{r_0}{p}\right)} \end{aligned} \tag{13.71}$$

where the minus sign was chosen because r_a is 180 deg away from the periapse of the transfer ellipse.

The angular momenta at point A, the point at which ΔV_1 is applied, before and after the velocity change, are given by

$$h_A^- = r_0 V_p^- = r_0\sqrt{\frac{\mu}{r_0}} = \sqrt{\mu r_0} \tag{13.72}$$

$$h_A^+ = r_0 V_p^+ = r_0\left(\sqrt{\frac{\mu}{r_0}} + \Delta V_1\right) \tag{13.73}$$

respectively, such that, from $p^+ = (h_A^+)^2/\mu$, we can write

$$p^+ = r_0 + \frac{r_0^2 \Delta V_1^2}{\mu} + 2\frac{r_0^{3/2}}{\sqrt{\mu}}\Delta V_1 \tag{13.74}$$

Finally, using Eqs. (13.71) and (13.74), the achieved apoapse of the transfer ellipse is obtained as

$$r_a = \frac{-r_0}{1 - \dfrac{2}{\left(1 + \dfrac{r_0 \Delta V_1^2}{\mu} + 2\sqrt{\dfrac{r_0}{\mu}}\Delta V_1\right)}} \tag{13.75}$$

and using p for p^+, we have

$$\left(\frac{\partial r_a}{\partial p}\right)_p = \frac{2r_0^2}{p^2\left(1 - \dfrac{2r_0}{p}\right)^2} \tag{13.76}$$

However

$$p^+ = \frac{h^{+2}}{\mu} = \frac{r_0^2(V_p^- + \Delta V_1)^2}{\mu}$$

$$\frac{\partial h}{\Delta V_1} = r_0$$

$$\frac{\partial p}{\partial h} = \frac{2h}{\mu}$$

such that

$$\left[\frac{\partial r_a}{\partial(\Delta V_1)}\right]_p = \frac{\partial r_a}{\partial h}\frac{\partial h}{\partial(\Delta V_1)} = \frac{\partial r_a}{\partial p}\frac{\partial p}{\partial h}\frac{\partial h}{\partial(\Delta V_1)} = \frac{4\mu r_0^3}{h^3\left(1 - 2\dfrac{r_0}{p}\right)^2} \tag{13.77}$$

where h has been used for h^+ and p for p^+.

The sensitivity of the second burn (ΔV_2) is analyzed in a similar manner using the same equation [i.e., Eq. (13.77)] with the following initial conditions at the apoapse of the transfer ellipse, which is denoted by point B

$$r_0 = r_B$$

$$V_0 = V_{B_+} = V_{B_-} + \Delta V_2$$

Before the application of ΔV_2, we have

$$h_B^- = h_A^+ = r_A V_A^+ = r_0 \left(\sqrt{\frac{\mu}{r_0}} + \Delta V_1 \right) \tag{13.78}$$

$$p_B^- = \frac{h_A^{+2}}{\mu} = \frac{r_0^2}{\mu} \left(\frac{\mu}{r_0} + \Delta V_1^2 + 2\sqrt{\frac{\mu}{r_0}} + \Delta V_1 \right) \tag{13.79}$$

After ΔV_2, the angular momentum becomes

$$h_B^+ = r_B V_B^+ = r_B (V_B^- + \Delta V_2) \tag{13.80}$$

V_B^- is calculated from the energy equation (transfer ellipse) as

$$V_B^- = \sqrt{2\left(\frac{\mu}{r_B} - \frac{\mu}{2a} \right)} \tag{13.81}$$

such that, using r to designate the apogee of the new transfer ellipse, meaning that reached after the application of ΔV_2, we obtain

$$\left[\frac{\partial r}{\partial (\Delta V_2)} \right]_{p^+} = \frac{4\mu r_B^3}{h_B^{+3} \left(1 - 2\dfrac{r_B}{p_B^+} \right)^2} \tag{13.82}$$

REFERENCES

[1] Breakwell, J. V., "Lecture Notes: Space Mechanics," *Course AA279A, Department of Aeronautics and Astronautics*, Stanford University, Palo Alto, CA, 1974.

[2] Battin, R. H., *An Introduction to the Mathematics and Methods of Astrodynamics, AIAA Education Series*, AIAA, New York, 1987, p. 452.

INDEX

Page numbers like **123** refer to figures; page numbers like ***123*** refer to tables

SUPPORTING MATERIALS

A complete listing of titles in the Progress in Astronautics and Aeronautics series is available from AIAA's electronic library, Aerospace Research Central (ARC) at arc.aiaa.org. Visit ARC frequently to stay abreast of product changes, corrections, special offers, and new publications.

AIAA is committed to devoting resources to the education of both practicing and future aerospace professionals. In 1996, the AIAA Foundation was founded. Its programs enhance scientific literacy and advance the arts and sciences of aerospace. For more information, please visit www.aiaafoundation.org.

CPSIA information can be obtained
at www.ICGtesting.com
Printed in the USA
BVHW052022020423
661528BV00003B/7

9 781624 106972